Biosecurity

Biosecurity: A Systems Perspective provides an overview of biosecurity as a system of related components, actors, and risks. This book—directed to the biosecurity practitioner, generalist scientist, and student—introduces overall features of the biosecurity system while walking the reader through the most up-to-date research on each step of the continuum (i.e. pre-border, border, and post-border activities). This book, which explicitly incorporates economic and social dimensions as well as varied decision-making contexts, paves the way for a more systemic approach to biosecurity risk management. The work spans statistics, ecology, mathematics, economics, veterinary science, human medicine, and sociology, involving collaborators across government, academia, and the private sector.

- This book uses a broad definition of biosecurity, rather than solely focusing on plant health, animal health, security, or one step of the biosecurity system (e.g. surveillance). As such, this book is a one-stop shop for readers interested in all aspects of biosecurity.
- The content and language are accessible to a wide range of audiences, including generalist scientists, biosecurity practitioners, and graduate students. More complex content is introduced in standalone boxes or chapters.
- The book follows a clear, simple structure within and among chapters (i.e. following the structure of the biosecurity system), making it a preferred option for graduate students.
- This book introduces novel cross-cutting themes, such as the importance of economic efficiency and institutional and social factors, going beyond the traditional science-based approach to biosecurity.
- Written in partnership with government agency practitioners, this book uses on-the-ground case studies to demonstrate how biosecurity principles are applied in practice.
- The book addresses challenges common to regulators in general, namely efficient regulation in uncertain and rapidly changing environments with multiple stakeholders who have—at times—conflicting priorities.

Biosecurity
A Systems Perspective

Edited by
Susan M. Hester, Lucie M. Bland, Edith Arndt, Sana Bau,
James S. Camac, Evelyn Mannix, Raphaël Trouvé
and Andrew P. Robinson

CRC Press
Taylor & Francis Group
Boca Raton London New York

CRC Press is an imprint of the
Taylor & Francis Group, an **informa** business

First edition published 2024
by CRC Press
2385 Executive Center Drive, Suite 320, Boca Raton, FL 33431

and by CRC Press
4 Park Square, Milton Park, Abingdon, Oxon, OX14 4RN

ISBN: 9781032181684 (hbk)
ISBN: 9781032181691 (pbk)
ISBN: 9781003253204 (ebk)

DOI: 10.1201/9781003253204

Typeset in Times
by KnowledgeWorks Global Ltd.

Contents

SECTION I   Pre-Border

SECTION II   Border

SECTION III   Post-Border

SECTION IV   Whole of System

SECTION V   Innovative Methods

SECTION VI   Conclusion

Foreword

In a rapidly changing and increasingly interconnected world, growing trade and travel continue to put enormous strain on the capacity of biosecurity agencies to assess and manage risks, and to do so efficiently from the viewpoint of their funders and those most impacted by biosecurity measures.

Biosecurity agencies are expected to protect domestic economies from hundreds of exotic biological threats, many of which will arrive in novel and unexpected ways. Many will also have the potential to cause disastrous impacts should they establish in natural, agricultural, marine, forestry, and built infrastructure environments.

The complexity and dimensions of the problems facing biosecurity agencies cannot be overstated:

- The global trading regime established under the World Trade Organisation will continue to be impacted by major, globally significant events such as pandemics and wars.
- Predicting the impacts of biosecurity threats in countries and regions where they have not been seen before is difficult and often highly speculative.
- It is unclear how much preventative biosecurity measures typically applied on importers will actually reduce risk.
- Biosecurity agencies have historically been dominated by science-trained staff, yet a diversity of professional and technical disciplines is now required more than ever.
- Useful methodologies from other disciplines often go unnoticed or undervalued.
- Time is often not available for biosecurity agency staff to interact with cutting-edge researchers and research.
- New technologies that provide potential improvements in efficiency are often not immediately available nor readily implementable.
- Retrieval and analysis of data that would allow proactive decision-making are often plagued by outmoded systems, slow development of new software tools, and poor interface between commercial and government systems.
- Available budgets need to be used as efficiently as possible across the biosecurity continuum, across threats, and across activities within management programmes.

Successfully addressing ever-evolving biosecurity challenges will require a shift away from the incremental approaches commonly used by biosecurity agencies, towards approaches that are more agile, evidence-based, multidisciplinary, and both logistically and commercially practical.

Biosecurity agencies are unlikely to have access to the required expertise in-house or the ability to effect whole-scale changes without drawing on the best knowledge and technology available nationally, if not globally. Biosecurity, academic, and political leaders recognised almost two decades ago that a fundamental change was necessary. Such a change was only practical via long-term collaborations with external organisations whose combined disciplines, talents, networks, and processing capacity could be focused on finding solutions to complicated biosecurity problems.

A powerful and globally relevant example of this approach was put in place in 2006 as the Australian Centre of Excellence for Risk Analysis, now the Centre of Excellence for Biosecurity Risk Analysis (CEBRA). Since then, CEBRA has evolved into a well-established and ongoing collaboration supported by the Australian and New Zealand Governments.

CEBRA hosts a body of academics, based at several universities and centred at the University of Melbourne, from various disciplines (including mathematics, economics, statistics, ecology, and epidemiology) who apply their diverse expertise to solve problems outlined in project proposals that are developed, carried out, and implemented in full collaboration with biosecurity agency staff.

Projects typically address complicated and pressing biosecurity problems for which in-house agency expertise is not available, and pragmatic policy outcomes are generated. This enduring collaboration has produced cutting-edge work on intelligence gathering, expert judgement, data management, novel inspection regimes, designing efficient and strategy-proof biosecurity rules and regulation, risk mapping, and spatial spread models for animal and plant diseases.

CEBRA's abilities to draw from a wide academic base and to partner more flexibly than a government department are critical to understanding and changing the ways that biosecurity risk is mitigated. CEBRA is a standout global capability, and while directly benefiting Australian and New Zealand biosecurity agencies, there are flow-on benefits to biosecurity agencies worldwide. Imagine what a network of CEBRA-like organisations across the globe could achieve!

To assist biosecurity practitioners across the globe to address the biosecurity challenges they face, CEBRA has produced *Biosecurity: A Systems Perspective*. This publication is the result of more than 15 years of excellent work by a diversity of specialists. It is a must-read for emerging biosecurity specialists worldwide, for managers responsible for delivering biosecurity innovations, and for senior leaders in biosecurity agencies. It is also a valuable reference book for academics and teachers of biosecurity around the world.

I highly commend the contributors of *Biosecurity: A Systems Perspective* and recommend it to all involved in or planning a career in biosecurity risk mitigation.

Rob Delane
Australia's Inspector-General of Biosecurity
2019–2022

Acknowledgements

This book was inspired by nearly two decades of research undertaken by the Centre of Excellence for Biosecurity Risk Analysis (CEBRA) and its predecessor organisation, the Australian Centre of Excellence for Risk Analysis. The research has significantly extended global knowledge about biosecurity problems and practice across the biosecurity continuum. The "CEBRA model" involves a shared responsibility for project design, development, and implementation between biosecurity agency staff and researchers. This close science-government interaction has ensured that research outputs will generate pragmatic policy outcomes. The co-editors acknowledge the financial support of CEBRA in producing this book.

Individual book chapters benefited greatly from input provided by many external biosecurity experts, namely Chris Baker, Oscar Cacho, Susie Collins, Aaron Dodd, Lee Failing, Robin Gregory, Cindy Hauser, Monique Ladds, Graham Long, Ajay Niranjane, David Orme, Mark Stanaway, and Ranjith Subasinghe.

As editors and authors of books are acutely aware, the support of friends and families is essential if the idea of a book is to ever become a reality. We do not understate this support or the many sacrifices that were made in completing this book.

Author Biographies

Susan M. Hester
Susan M. Hester is Deputy CEO of the Centre of Excellence for Biosecurity Risk Analysis (CEBRA) at the University of Melbourne, and Associate Professor at the UNE Business School at the University of New England. She has a PhD in Agricultural Economics from the University of New England and an Honours degree in Economics from the University of Western Australia. Susan is an applied economist and has worked almost exclusively on issues related to invasive-species management since 2002. This has largely been via her role as a Chief Investigator with CEBRA, commencing in 2009. Recent projects have involved: re-evaluating management of European wasps using biocontrol agents; understanding the value of passive surveillance; and introducing incentive-compatible policies into the Australian biosecurity system in order to maximise value for money from biosecurity budgets.

Lucie M. Bland
Lucie is an accomplished Animal Biologist. She completed a BA (Hons) in Biological Sciences from the University of Oxford, and a PhD in Ecology & Evolution from Imperial College London and London Zoo. Lucie's research involves increasing knowledge about zoo animal biology, management, and conservation, and she also investigates the role of the wildlife trade in spreading zoonotic diseases. Lucie is also a natural health practitioner and animal lover.

Edith Arndt
Edith Arndt is a Research Fellow specialising in biosecurity at the Centre of Excellence for Biosecurity Risk Analysis (CEBRA) at the University of Melbourne. She earned her Master's degree from the University of Vienna, and later obtained her PhD from the University of Melbourne. Within CEBRA, she engages in various biosecurity-related research projects through a collaborative co-production model alongside research partners at the Australian Government Department of Agriculture, Fisheries and Forestry. Her prior projects have encompassed a wide array of topics, including the development of an evaluation framework for Australia's national biosecurity system, the examination of factors influencing marine vessel biofouling and its prevention and management, as well as the optimisation of resource allocation for border biosecurity risk controls. Before joining CEBRA, Edith worked within the public service sector at the state government level for nine years. Her responsibilities included monitoring, evaluation, and reporting; fire severity mapping; and database development. She has an interest in the interface between research and policy, and she actively engages in outreach activities aimed at inspiring high school students to pursue careers in the field of science.

Sana Bau
Sana is a Researcher in decision science with a special interest in the use of science in environmental policy and practice. Sana has a BSc degree from the University of Otago and a Master of Environment degree from the University of Melbourne and is completing a PhD at the University of Melbourne. Sana's other pursuits include web content creation and communications, lecturing in environmental risk assessment, and fashion and textiles.

James S. Camac
James is a Senior Research Fellow and Chief Investigator within the Centre of Excellence for Biosecurity Risk Analysis (CEBRA). As a trained quantitative and applied ecologist, he has extensive experience in field ecology, long-term ecological monitoring, experimental design, ecological

statistics, reproducible data science, biosecurity risk analysis, and project management. His research tackles problems in two broad areas: understanding and forecasting changes in biodiversity, especially under climate change and changing fire regimes; and enhancing post-border surveillance networks to protect Australia's environmental, social, and economic values from exotic pests and diseases.

Evelyn Mannix

Evelyn Mannix is currently a Research Fellow at the Centre of Excellence for Biosecurity Risk Analysis, working on solving biosecurity challenges using the latest deep learning and AI approaches. They received their honours degree from the Australian National University with a university medal, and worked for several years at the Australian Bureau of Agricultural and Resource Economics and Sciences as a research officer, applying mathematical models to address biosecurity risks. They are currently undertaking a PhD at the University of Melbourne, with the Melbourne Centre for Data Science.

Raphaël Trouvé

Raphaël is a Research Fellow in forest ecology and statistical modelling at the University of Melbourne. He received a Master and PhD in Forest Sciences from AgroParisTech. Raphaël develops and applies quantitative methods to solve environmental problems, with applications to forest conservation and management and biosecurity.

Andrew P. Robinson

Andrew P. Robinson is CEO of the Centre of Excellence for Biosecurity Risk Analysis (CEBRA), and Professor in Applied Statistics at the University of Melbourne. He has a PhD in Forestry and a Master's in Statistics from the University of Minnesota, and has published four books, 90 research articles, and 50 ACERA/CEBRA technical reports on various aspects of risk analysis and biosecurity. He is an elected member of the International Statistical Institute. He joined the University of Melbourne in 2005 from the University of Idaho, where he was an Associate Professor in Forest Inventory and Forest Biometrics. Andrew spends much of his time thinking about biosecurity at national borders, including analysing inspection and interception data using statistical tools, designing and trialling inspection surveillance systems, developing metrics by which regulatory inspectorates can assess their performance, and discussing all of the above with, and indeed at, interested parties.

List of Contributors

Edith Arndt
CEBRA
The University of Melbourne
Melbourne, Australia

Sana Bau
CEBRA
The University of Melbourne
Melbourne, Australia

John B. Baumgartner
CEBRA
The University of Melbourne
Melbourne, Australia

Lucie M. Bland
CEBRA
The University of Melbourne
Melbourne, Australia

Mark Burgman
School of Life Sciences
University of Hawai'i at Mānoa
Honolulu, Hawai'i

James S. Camac
CEBRA
The University of Melbourne
Melbourne, Australia

Arthur Campbell
Monash University
Clayton, Australia

Long Chu
Crawford School of Public Policy
Australian National University
Canberra, Australia

Jocelyn Cranefield
Wellington School of Business and
    Government Ōrauariki
Victoria University of Wellington
Wellington, New Zealand

Mark J. Ducey
University of New Hampshire
Durham, New Hampshire, USA

Anca Hanea
CEBRA
The University of Melbourne
Melbourne, Australia

Susan M. Hester
CEBRA
The University of Melbourne
Melbourne, Australia
and
University of New England
    Business School
University of New England
Armidale, Australia

Les Kneebone
CEBRA
The University of Melbourne
Melbourne, Australia

Tom Kompas
CEBRA
The University of Melbourne
Melbourne, Australia

Christine Li
CEBRA
The University of Melbourne
Melbourne, Australia

Evelyn Mannix
CEBRA
The University of Melbourne
Melbourne, Australia

Allan Mooney
Independent
Kudardup, Australia

Hoa-Thi-Minh Nguyen
Crawford School of Public Policy
Australian National University
Canberra, Australia

Natasha Page
People, Movement and Place
Aurecon
Docklands, Australia

Estibaliz Palma
CEBRA
The University of Melbourne
Melbourne, Australia

Andrew P. Robinson
CEBRA
The University of Melbourne
Melbourne, Australia

Anthony Rossiter
CEBRA
The University of Melbourne
Melbourne, Australia

Karen Schneider
CEBRA
The University of Melbourne
Melbourne, Australia

Gary Stoneham
The Centre for Market Design
The University of Melbourne
Melbourne, Australia

Raphaël Trouvé
School of Ecosystem and Forest Sciences
The University of Melbourne
Richmond, Australia

1 Introduction

Edith Arndt, Evelyn Mannix, Andrew P. Robinson,
Lucie M. Bland, Susan M. Hester, and Raphaël Trouvé

The global COVID-19 pandemic of 2020–2021 brought with it a new level of awareness about biosecurity—the protection of the economy, environment, and human health from the negative impacts associated with the entry, establishment, or spread of exotic pests and diseases (as defined by Beale et al. 2008). The pandemic also brought into sharp focus the complex issues tackled by those responsible for managing biosecurity risks. Biosecurity managers have a range of tools at their disposal (e.g. movement restrictions, post-border quarantine, contact tracing, pre- and post-border testing regimes, lockdowns, and vaccine development), many of which were used to manage COVID-19. Managers often must deploy these tools rapidly and in an environment of great uncertainty, with limited knowledge about threats. The pandemic was also a reminder of the interconnectedness of people via trade and travel and of the enormous pressure on biosecurity systems to pivot quickly when new challenges arise. With biosecurity threats only increasing in the coming decades due to increased trade and travel, population growth, urbanisation, climate change, and antimicrobial resistance (CSIRO 2020), creating and maintaining effective and efficient biosecurity systems is more important than ever.

Biosecurity failures can have dire consequences, because the economic, social, and environmental costs of pest and disease incursions can be extremely high. For example, the 2001 outbreak of foot-and-mouth disease (which affects cattle, sheep, goats, and pigs) in the United Kingdom had a profound national impact. Over 6.5 million animals were culled to control the epidemic and the economic toll was estimated to be at least GBP 7.7 billion, including the direct and indirect impacts on agricultural, food, and tourism industries (Haydon, Kao, and Kitching 2004, Thompson et al. 2002).

Similarly, the red imported fire ant (RIFA; *Solenopsis invicta*), native to South America, has invaded many parts of the world, including the southern United States, the Caribbean and parts of Asia and Australia. Not only do the ants attack and sting livestock (Jetter, Hamilton, and Klotz 2002), but their voracious feeding and mound-building behaviours affect native animals and habitats, crops, irrigation systems, machinery, electrical equipment, and urban environments (Wylie and Janssen-May 2017). RIFA stings are painful for humans and can lead to infections and, in severe cases, to death (Jemal and Hugh-Jones 1993). A recent comprehensive analysis of reported economic costs incurred by invasive ants worldwide since 1930 estimated costs of *Solenopsis* spp. (primarily *S. invicta*) amounting to USD 31.89 billion, which is most likely a gross underestimate due to widespread underreporting (Angulo et al. 2022). Other analyses estimate these costs to be much greater. In the United States, RIFA was accidentally introduced in the 1930s and has since spread across 13 states in the southern United States (Allen, Epperson, and Garmestani 2004), costing in excess of USD 6 billion per year in agricultural, infrastructure, household, and a range of other damages (in 2006 dollars; Lard et al. 2006). In Australia, an eradication programme—the largest ant eradication ever attempted globally, commenced upon first detection of this pest in 2001 at a cost of approximately AUD 800 million, aiming to avert annual impact and control costs of AUD 2 billion should the ant spread across the country (Scott-Orr, Gruber, and Zacharin 2021).

Biosecurity presents a very difficult problem to decision makers when there are no clear-cut solutions for avoiding or minimising the impacts of pests and diseases, or when potential solutions are infeasible or prohibitively expensive (DeFries and Nagendra 2017). Biosecurity can be considered a *wicked problem*, in that many risks are involved, often with uncertain or unintended consequences (DeFries and Nagendra 2017, Head 2008, Rittel and Webber 1973). Here, a biosecurity risk is defined as the combined likelihood of occurrence and the consequence of a pest or disease entering, establishing, and spreading in a region. Biosecurity funding is used to create rules and regulations and to undertake targeted programmes and activities, all of which are aimed at reducing biosecurity risk to a level that is acceptable to a particular country or region. Biosecurity systems manage risks across a continuum encompassing pre-border, border, and post-border components (Beale et al. 2008), and multiple measures are implemented across this continuum to manage a particular risk.

For many developed countries, throwing more money at the problem and scaling existing biosecurity activities will not be sufficient or efficient to manage increasing risks. Predictions from the Australian Government Department of Agriculture, Fisheries and Forestry showed that even increasing investment into biosecurity threefold by 2025 would not suffice to keep residual risk at 2014–2015 levels (Craik, Palmer, and Sheldrake 2017). As global pathway volumes and supply chain complexities increase, the resources invested in biosecurity activities will need to increase at an even faster rate to ensure absolute risk remains identical (Hulme 2009, Dodd et al. 2015, IGB 2019). Changes in average and extreme air temperatures are also likely to offer more opportunities for exotic pests and diseases to establish and spread outside of their native ranges (Bergot et al. 2004, Hellmann et al. 2008).

Staring down the barrel of increased threat from pests and diseases, it becomes clear that a more fundamental change to biosecurity needs to happen—one that incorporates a systemic perspective on risk. Recently, the Centre of Excellence for Biosecurity Risk Analysis (CEBRA) developed a framework to evaluate the performance of the Australian biosecurity system (Schneider and Arndt 2020). During this process, it became evident that coordinated and effective action relies on a comprehensive understanding of how biosecurity systems work.

Although previous biosecurity books have looked at specific taxa (see Wilson, Panetta, and Lindgren 2016 for an in-depth investigation of plant incursions), specific regions (see Poland et al. 2021 for a look at invasive species in the United States), specific activities (see Jarrad, Low-Choy, and Mengersen 2015) or outlined scientific solutions to the risk assessment and management of invasive species (see Robinson et al. 2017), this book breaks new ground by explicitly considering biosecurity as a complex system of multiple interconnected components, actors, incentives, feedback loops, and regulatory pathways.

Specifically, this book uncovers the overarching roles of resource allocation and prioritisation; incorporating social considerations (including policy makers and stakeholder responses); and ensuring biosecurity objectives are met and actions are improved through monitoring, evaluation, and the uptake of research findings. By explicitly incorporating human and economic dimensions as well as varied decision-making contexts, we provide clear paths forward for the development of robust, comprehensive, and effective biosecurity systems.

The book starts with a synthetic overview of biosecurity systems, ranging from the influence of international regulations to the setup of biosecurity activities along the pre-border, border, and post-border continuum. We then take a sequential approach to describing and addressing the problems biosecurity agencies face, walking the reader through pre-border, border, and post-border activities and challenges. We do this for a number of reasons: first, so that the novice reader can easily grapple with the complexities of biosecurity systems, and, second, because this sequential view has already been adopted by many countries (Beale et al. 2008).

We then explore cross-cutting themes across biosecurity systems, including the role of incentive-compatible regulations in shaping stakeholder compliance; the use of economic prioritisation to allocate biosecurity funding; refining biosecurity activities through monitoring and evaluation; and current barriers to research uptake by practitioners. We dedicate a final book section to outlining innovative methods in biosecurity research, including the use of expert elicitation techniques, automation with artificial intelligence, and risk mapping. We complete the journey by providing practical recommendations to improve decision making across the biosecurity continuum.

We believe this book is much needed, both as a systems perspective to biosecurity and as enlightening material for researchers, practitioners, and policy makers tasked with responding to biosecurity threats. Our book is also tailored to graduate students enrolled in biosecurity courses, which are becoming more popular worldwide as countries realise the importance of protecting their people, economy, and environment from biosecurity threats.

This book represents a culmination and waypoint of more than a decade of research by CEBRA, a leader in the analysis of biosecurity risks in Australia and New Zealand. For many years, CEBRA has been championing an interdisciplinary approach to biosecurity risk analysis in collaboration with its government and industry partners (Arndt et al. 2020). This book, co-authored by more than 20 CEBRA academics and industry partners, showcases this collaborative approach to research and implementation. While our book focuses on Australian and New Zealand perspectives, we also draw extensively on experiences and policies from other countries to provide a comprehensive overview of biosecurity systems.

We hope that this book will help create a common language between biosecurity researchers and practitioners to enhance collaboration, innovation, and effective decision making. And in turn, we hope to prepare biosecurity students for an exciting career ahead with a text that is accessible and thought provoking.

REFERENCES

Allen, C. R., D. M. Epperson, and A. S. Garmestani. 2004. "Red imported fire ant impacts on wildlife: A decade of research." *The American Midland Naturalist* 152 (1):88–103.

Angulo, E., B. D. Hoffmann, L. Ballesteros-Mejia, A. Taheri, P. Balzani, A. Bang, D. Renault, M. Cordonnier, C. Bellard, C. Diagne, D. A. Ahmed, Y. Watari, and F. Courchamp. 2022. "Economic costs of invasive alien ants worldwide." *Biological Invasions* 24 (7):2041–2060. https://doi.org/10.1007/s10530-022-02791-w.

Arndt, E., M. Burgman, K. Schneider, and A. Robinson. 2020. "Working with government — innovative approaches to evidence-based policy-making." In *Conservation Research, Policy and Practice*, edited by William J Sutherland, Peter N M Brotherton, Zoe G Davies, Nancy Ockendon, Nathalie Pettorelli and Juliet A Vickery, 216–229. Cambridge, UK: Cambridge University Press.

Beale, R., J. Fairbrother, A. Inglis, and D. Trebeck. 2008. One Biosecurity. A working partnership. The independent review of Australia's quarantine and biosecurity arrangements. Report to the Australian Government: Quarantine and Biosecurity Review Panel, Department of Agriculture, Fisheries and Forestry (Australia).

Bergot, M., E. Cloppet, V. Pérarnaud, M. Déqué, B. Marçais, and M.-L. Desprez-Loustau. 2004. "Simulation of potential range expansion of oak disease caused by *Phytophthora cinnamomi* under climate change." *Global Change Biology* 10 (9):1539–1552. https://doi.org/10.1111/j.1365-2486.2004.00824.x

Craik, W., D. Palmer, and R. Sheldrake. 2017. *Priorities for Australia's biosecurity system. An independent review of the capacity of the national biosecurity system and its underpinning intergovernmental agreement.* Canberra: Department of Agriculture and Water Resources.

CSIRO. 2020. "Australia's biosecurity future. Unlocking the next decade of resilience." Commonwealth Scientific and Industrial Research Organisation, accessed 19 February 2020. https://www.csiro.au/en/Do-business/Futures/Reports/Health/Biosecurity-Futures.

DeFries, R., and H. Nagendra. 2017. "Ecosystem management as a wicked problem." *Science* 356:265–270.

Dodd, A. J., M. A. Burgman, M. A. McCarthy, and N. Ainsworth. 2015. "The changing patterns of plant naturalization in Australia." *Diversity and Distributions* 21 (9):1038–1050. https://doi.org/10.1111/ddi.12351.

Haydon, D. T., R. R. Kao, and R. P. Kitching. 2004. "The UK foot-and-mouth disease outbreak — the aftermath." *Nature Reviews Microbiology* 2 (8):675–681. http://doi.org/10.1038/nrmicro960.

Head, B. W. 2008. "Wicked problems in public policy." *Public Policy* 3 (2):101–118.

Hellmann, J. J., J. E. Byers, B. G. Bierwagen, and J. S. Dukes. 2008. "Five potential consequences of climate change for invasive species." *Conservation Biology* 22 (3):534–543. https://doi.org/10.1111/j.1523-1739.2008.00951.x.

Hulme, P. E. 2009. "Trade, transport and trouble: Managing invasive species pathways in an era of globalization." *Journal of Applied Ecology* 46 (1):10–18.

IGB. 2019. *Pest and disease interceptions and incursions in Australia.* Canberra: Inspector-General of Biosecurity. Department of Agriculture and Water Resources.

Jarrad, F., S. Low-Choy, and K. Mengersen, eds. 2015. *Biosecurity Surveillance: Quantitative Approaches. CABI invasive species series.* Wallingford, Oxfordshire: CABI. http://doi.org/10.1079/9781780643595.0000.

Jemal, A., and M. Hugh-Jones. 1993. "A review of the red imported fire ant (*Solenopsis invicta* Buren) and its impacts on plant, animal, and human health." *Preventive Veterinary Medicine* 17:19–32.

Jetter, K. M., J. Hamilton, and J. H. Klotz. 2002. "Eradication costs calculated: Red imported fire ants threaten agriculture, wildlife and homes." *California Agriculture* 56 (1):26–34. http://doi.org/10.3733/ca.v056n01p26.

Lard, C. F., J. Schmidt, B. Morris, L. Estes, C. Ryan, and D. Bergquist. 2006. *An economic impact of imported fire ants in the United States of America.* College Station, TX: Department of Agricultural Economics, Texas A&M University.

Poland, T. M., T. Patel-Weynand, D. M. Finch, C. F. Miniat, D. C. Hayes, and V. M. Lopez, eds. 2021. *Invasive Species in Forests and Rangelands of the United States: A Comprehensive Science Synthesis for the United States Forest Sector.* Cham: Springer Nature.

Rittel, H. W. J., and M. M. Webber. 1973. "Dilemmas in a general theory of planning." *Policy Sciences* 4 (2):155–169. http://doi.org/10.1007/BF01405730.

Robinson, A. P., T. Walshe, M. A. Burgman, and M. Nunn, eds. 2017. *Invasive Species: Risk Assessment and Management.* Cambridge: Cambridge University Press.

Schneider, K., and E. Arndt. 2020. Evaluating the health of Australia's biosecurity system. Final report for CEBRA project 170714. Centre of Excellence for Biosecurity Risk Analysis. The University of Melbourne.

Scott-Orr, H., M. Gruber, and W. Zacharin. 2021. National Red Imported Fire Ant Eradication Program Strategic Review August 2021. Independent Review Panel for the National Red Imported Fire Ant Eradication Program.

Thompson, D., P. Muriel, D. Russell, P. Osborne, A. Bromley, M. Rowland, S. Creigh-Tyte, and C. Brown. 2002. "Economic costs of the foot and mouth disease outbreak in the United Kingdom in 2001." *Revue Scientifique et Technique (International Office of Epizootics)* 21 (3):675–687. https://doi.org/10.20506/rst.21.3.1353.

Wilson, J., F. Panetta, and C. Lindgren. 2016. "Evaluation of management performance." In *Detecting and responding to alien plant incursions (Ecology, biodiversity and conservation)*, 111–138. Cambridge, UK: Cambridge University Press.

Wylie, F. R., and S. Janssen-May. 2017. "Red imported fire ant in Australia: What if we lose the war?" *Ecological Management & Restoration* 18 (1):32–44. https://doi.org/10.1111/emr.12238.

2 Biosecurity Systems and International Regulations

Lucie M. Bland, Edith Arndt, Karen Schneider,
Allan Mooney, and Susan M. Hester

ABSTRACT

This chapter provides a summary of the structure of national biosecurity systems, using Australian and international examples. We define a biosecurity system as the suite of management activities implemented by a jurisdiction to protect its economy, environment, and human health from the damaging impacts of pests and diseases. A biosecurity system includes key activities conducted in pre-border, border, and post-border settings, as well as other components such as human, financial, and physical resources and broader influencing factors. Pre-border activities are primarily influenced by international regulations, including the World Trade Organisation's Agreement on the Application of Sanitary and Phytosanitary Measures. Border and post-border biosecurity measures are typically the domains of national and local participants. Biosecurity systems are organised hierarchically (incorporating participants at various levels of authority), with interconnected activities that contribute to the overall performance and efficiency of the system. An advanced way of dealing with this dense web of interactions is to implement a fully integrated biosecurity framework, featuring legislative authority, a budget, and integrated institutional arrangements. In mature biosecurity systems, all phases of the biosecurity continuum work in a coordinated fashion to ensure high performance and efficient resource allocation.

GLOSSARY

Biosecurity system The suite of management activities implemented by a jurisdiction to protect its economy, environment, and human health from the damaging impacts of pests and diseases. As such, a biosecurity system consists of interacting components, including participants, resources, and influencing factors.

SPS Agreement The Agreement on the Application of Sanitary and Phytosanitary Measures entered into force with the establishment of the World Trade Organization on 1 January 1995, and concerns the application of food safety and animal and plant health regulations.

Appropriate level of protection The SPS Agreement defines the appropriate level of protection (ALOP) as "the level of protection deemed appropriate by the Member establishing a sanitary or phytosanitary measure to protect human, animal or plant life or health within its territory" (SPS Agreement, Annex A, paragraph 5).

Biosecurity risk material (BRM) Living organisms exotic to a jurisdiction, or biological material carrying an exotic pest or disease. Biosecurity risk material includes live animals and plants, animal and plant material (e.g. fur, feathers, pollen, and flowers), food scraps, and soil.

Market access A jurisdiction's ability to enter a foreign market and sell its goods and services.

Least-restrictive trade Under the SPS Agreement, it is a requirement that sanitary and phytosanitary measures implemented to achieve ALOP are no more trade restrictive than required.

DOI: 10.1201/9781003253204-2

Risk assessment The SPS Agreement defines risk assessment as "the evaluation of the likelihood of entry, establishment or spread of a pest or disease within the territory of an importing Member according to the sanitary or phytosanitary measures which might be applied, and of the associated potential biological and economic consequences" (Annex A, paragraph 4). Pest risk analysis is prescribed by the International Plant Protection Convention (IPPC 2019). Import risk analysis is prescribed by the World Organisation for Animal Health (WOAH 2018).

Precautionary principle Under the precautionary principle, a conservative approach or decision is adopted when scientific information is unavailable or uncertain and stakes are high. The greater the uncertainty, the more conservative the decision should be under this principle.

Risk profile A jurisdiction's risk profile is the collective of pests and diseases that threaten the biosecurity status of the jurisdiction, including entry pathways, exposure scenarios, and mitigation measures.

Non-compliance The failure of participants in the biosecurity system to act in accordance with biosecurity requirements.

Competent Authority An organisation recognised by national governments as responsible for developing and administering sanitary and phytosanitary measures within a country. For animals, this is usually a national Veterinary Authority. For plants, it is a National Plant Protection Organisation. For food, it is typically a Ministry or Department of Health.

Notifiable pests and diseases have the potential for rapid spread and serious socio-economic or public health consequences. They are of major importance in the international trade of plants, plant products, animals, and animal products (WOAH 2023).

INTRODUCTION

Across the globe, governments manage the risks associated with pests and diseases by adopting international regulations, creating national regulations, and implementing biosecurity management activities. Under the World Trade Organisation (WTO) Agreement on the Application of Sanitary and Phytosanitary Measures (SPS Agreement), signatory countries are entitled to set an appropriate level of protection (ALOP) to protect their economy, society, and environment from risks from pests and diseases. To meet their ALOP, national regulators undertake a variety of pre-border, border, and post-border activities, which together form a "biosecurity system". National biosecurity systems aim to achieve four key outcomes:

- Reduce the likelihood of exotic pests and diseases entering the jurisdiction.
- Prepare stakeholders to detect, respond to, and manage exotic pests and diseases that might enter.
- Reduce the impact of already established priority pests and diseases.
- Minimise disruption to market access.

The main activities of biosecurity systems are sequential and cumulative (see Box 2.1), which means that activities in one part of the system (e.g. pre-border and border) can influence the outcomes of subsequent activities (e.g. post-border). Unfortunately, pre-border, border, and post-border activities are often treated separately in the academic literature (Anderson et al. 2017), as are animal and plant biosecurity (Gordh and McKirdy 2014). Recently, the Centre of Excellence for Biosecurity Risk Analysis (CEBRA) developed a framework to evaluate the performance of the Australian biosecurity system as a whole (Schneider and Arndt 2020). During this process, it became evident

BOX 2.1. THE SEVEN MAIN ACTIVITIES OF BIOSECURITY SYSTEMS

Based on Schneider and Arndt (2020), the seven main activities of a biosecurity system are:

1. *Anticipating biosecurity risk*: Anticipating biosecurity risk relies on understanding the context in which a biosecurity system operates and using this information to guide pest risk analyses and import risk analyses (see Chapter 3. Anticipate).
2. *Preventing biosecurity risk material (BRM) from reaching the border*: Biosecurity managers aim to handle biosecurity risk offshore and to prevent threats from reaching the border by implementing pre-border interventions (e.g. certification and treatment; see Chapter 4. Prevent).
3. *Screening entry pathways*: Pre-border activities are not designed to be completely effective in preventing BRM from reaching the border, so border screening is implemented to detect non-compliance with regulations and to prevent pest and diseases from passing through the border (see Chapter 5. Screen and Chapter 13. Profiling and Automation).
4. *Preparing for an incursion or outbreak*: To effectively respond to incursions or outbreaks of pests and diseases, biosecurity agencies invest in activities that prepare system participants for incursions (see Chapter 7. Prepare, Respond, and Recover).
5. *Detecting incursions or outbreaks*: Detecting an incursion or outbreak early (i.e. before a pest or disease has spread) can improve biosecurity response outcomes. National and regional governments usually invest in a mix of surveillance methods to detect incursions (see Chapter 6. Detect).
6. *Responding to an incursion or outbreak*: When an invasion of a pest or disease is first discovered, the immediate response involves determining the extent of the incursion and controlling the outbreak. Ideally, this initial response preserves all longer-term management options (see Chapter 7. Prepare, Respond, and Recover).
7. *Recovering from and adapting to an incursion or outbreak*: Pests and diseases that circumvent all measures to prevent their entry and efforts to eradicate them may require ongoing management to reduce impacts (see Chapter 7. Prepare, Respond, and Recover).

that coordinated and effective action relies on a comprehensive understanding of how a biosecurity system works.

This chapter provides a summary of the structure of national biosecurity systems, using Australian and international examples. We first summarise the importance of international regulations (including the SPS Agreement) in shaping national standards for animal, plant, and environmental biosecurity. We then describe the main components of a typical biosecurity system, including its main activities, inputs, outcomes, and influencers. Finally, we outline how considering biosecurity as a complex system and a wicked problem (Head 2008) can improve coordination across all phases of the biosecurity continuum, with the aim of ensuring high management performance and efficient resource allocation.

INTERNATIONAL REGULATIONS FOR BIOSECURITY

International regulations are the first line of defence against the impacts of pests and diseases, and they represent the foundation of global biosecurity risk management. World trade is governed by a set of rules designed by member countries of the WTO, a body whose 164 signatory countries represent around 98% of global trade (WTO 2021). Its predecessor, the General Agreement on Tariffs and Trade (GATT), was formed by 23 countries in 1948 following the Second World War to create an economic and trade cohesion that hadn't existed in the pre-war period. The objective of the WTO (as was the case with the GATT) is to promote international trade by reducing or eliminating trade barriers. This is achieved by free trade or least-restrictive trade (Fraser, Cook, and Haddock-Fraser 2019).

The transformation of the GATT into the WTO in 1995 saw the introduction of the Agreement on the Application of Sanitary and Phytosanitary Measures (i.e. the SPS Agreement; Box 2.2). The SPS Agreement sets out the measures that may be applied by signatory countries to protect themselves from risks to animal, plant, or human health arising from the entry, establishment, or spread of pests and diseases (Box 2.2; WTO 1995). The establishment of the SPS Agreement was a game changer for quarantine policy making by national governments (Anderson, McRae, and Wilson 2012).

BOX 2.2. PRINCIPLES OF THE AGREEMENT ON THE APPLICATION OF SANITARY AND PHYTOSANITARY MEASURES (SPS AGREEMENT)

The SPS Agreement consists of 14 principles that guide WTO member countries in their application of sanitary and phytosanitary measures. Of those, nine central principles can affect international trade (Articles 2–10; Nunn 2012; WTO 1995).

Article 2: Basic rights and obligations. All members have the right to determine sanitary and phytosanitary measures, but they are also obligated not to discriminate among members when trade conditions are similar.

Article 3: Harmonisation. Sanitary or phytosanitary measures should be based on international standards, guidelines, or recommendations, but members may introduce measures to achieve higher levels of protection.

Article 4: Equivalence. Members are expected to accept other countries' measures (which may be different to their own) as equivalent if these measures achieve the ALOP.

Article 5: Risk assessment and determination of ALOP. Sanitary and phytosanitary measures should be based on risk assessments that draw on scientific evidence and information on production, screening activities, prevalence of pests and diseases, ecology, and treatments. When determining measures, members are to consider economic factors in the event of entry, establishment, and spread of pests and diseases (e.g. potential damages, control, and eradication costs). Members also need to ensure that determining their ALOP does not result in discrimination or restrictions to international trade.

Article 6: Regional conditions. Members should adapt their measures to the sanitary or phytosanitary characteristics of both exporting and destination areas, such as by recognising areas of pest and disease absence or low prevalence.

Article 7: Transparency. Sanitary or phytosanitary regulations and changes to measures need to be published by members. Each member should establish a single enquiry point, provide relevant documents, respond to questions, and adhere to notification procedures.

Article 8: Control, inspection, and approval procedures. Members need to undertake and complete these procedures as prescribed in Annex C of the Agreement, for example in a timely manner and by disclosing processing periods, respecting confidentiality, and charging appropriate fees.

Article 9: Technical assistance. Developing countries can receive technical assistance (e.g. advice, credits, training, and equipment) from other members to achieve their ALOP.

Article 10: Special and differential treatment. Developing countries can be granted leniency towards their efforts to comply with the provisions of the Agreement, for example through a phased introduction of new measures or longer timeframes for compliance.

The remainder of articles in the Agreement prescribe a process for consultations and dispute settlement, encourage observation of the provisions of the Agreement during implementation, and specify an overarching administrative body under the Agreement—the Committee on Sanitary and Phytosanitary Measures.

The SPS Agreement imposes science-based principles to biosecurity risk management, requiring that measures be applied in a way that is transparent, consistent, scientifically based and the least trade restrictive (Goh and Ziegler 2012). The SPS Agreement recognises national sovereignty in determining ALOP, which reflects each country's acceptable level of risk. A country's ALOP is largely a political determination that seeks to balance the economic benefits of trade to the country with the potential biological, economic, and environmental consequences should a pest or disease establish (Goh and Ziegler 2012).

There is no requirement for ALOP to be expressed in quantitative terms, and while measures to ensure achievement of ALOP might differ by import type, member countries must be consistent in their application of measures to avoid discrimination or restrictions on international trade. The vague concepts and language used in the SPS Agreement have aggravated tension between international obligations and national sovereignty in setting ALOP, leading to disputes among member countries (Anderson et al. 2012). When trading partners disagree on the application of measures, they may take their concerns to the WTO's dispute settlement system. Of the cases resolved by a dispute panel process, six have related to biosecurity: salmon (Australia vs Canada); apples (Japan vs USA; Australia vs New Zealand); agricultural products (India vs USA); live pigs, pork and other pig products (Russian Federation vs EU); and animals, meat and other animal products (Argentina vs USA; Box 2.3). For an in-depth analysis of these disputes, see Fraser et al. (2019).

Despite its failings to resolve some disputes, the SPS Agreement remains the cornerstone agreement relating to the protection of national economies from biosecurity risks posed by imports. WTO member countries are obliged to ensure measures imposed on imports are based on risk assessment techniques developed for each sector, broadly defined as plants, animals, food safety, and the environment. This chapter focuses on animals, plants, and the environment, noting the Codex Alimentarius Commission (CAC) develops measures to ensure food is safe and can be traded (FSANZ 2013).

BOX 2.3. MEASURES AFFECTING THE IMPORTATION OF ANIMALS, MEAT, AND OTHER ANIMAL PRODUCTS FROM ARGENTINA TO THE UNITED STATES

In 2012, Argentina brought a dispute resolution to the WTO related to the prohibition by the United States of animals, meat, and meat products from Argentina over concerns about the spread of foot-and-mouth disease (FMD). The dispute dated back to 2001 following an outbreak of FMD in northern Argentina. The ban was still in place more than 10 years later, even though parts of Argentina were determined to be FMD free.

Argentina had repeatedly sought authorisation to send fresh (chilled or frozen) beef into the United States from Patagonia, an FMD-free area. Argentina claimed that the United States was violating multiple SPS Agreement articles, including that the United States had not properly adapted its measures to pest- or disease-free areas (Fraser et al. 2019). For its part, the United States claimed its measures were required to maintain its FMD-free status of 80 years, noting that FMD is one of the most infectious and economically devastating livestock diseases (WTO 2015). The United States used the SPS Agreement to claim that its decisions were based on scientific understanding and the standards of the World Organisation for Animal Health (WOAH).

As is the case with most disputes brought under the SPS Agreement, the dispute settlement panel consulted with scientific experts to assist in the evaluation of scientific evidence in relation to risk assessment techniques, veterinary practices, and surveillance for FMD (WTO 2015). The WOAH was also consulted about its Terrestrial Animal Health Code (WOAH 2022). In 2015, the WTO determined in favour of Argentina, finding that the United States's measures were inconsistent with several provisions of the SPS Agreement (WTO 2015). Exports from FMD-free Argentinian areas to the United States did not immediately resume; trade resumed in 2018 as part of a two-way trade agreement between the two countries.

ANIMAL BIOSECURITY

Under the SPS Agreement, the WOAH (founded as OIE, Office International des Epizooties) is the international standard-setting body for sanitary measures related to animal health and zoonoses (animal diseases that can infect people). The WOAH was established in 1924 and currently comprises 182 member countries. Members are guided by two key documents: the *Terrestrial Animal Health Code* and the *Aquatic Animal Health Code*. Members must establish a Competent Authority (typically the national Veterinary Authority) responsible for ensuring the implementation of animal health and welfare measures. In addition to implementing the Codes, national Veterinary Authorities undertake regular disease reporting and provide veterinary certification for exports (including freedom from pests and diseases of concern).

The WOAH Codes recognise that the importation of animals and their products involves a certain level of disease risk to the importing country (WOAH 2022). Member countries use import risk analysis to assess the disease risks associated with imports and to identify and apply measures to reduce those risks, in line with the SPS Agreement. The four components of import risk analysis are hazard identification, risk assessment, risk management, and risk communication (see Chapter 3. Anticipate). Many diseases have internationally agreed standards concerning risks and, in these cases, only a qualitative assessment may be required (WOAH 2022). Risk management is the process of deciding on, and implementing measures to address risks identified in the risk assessment (in line with WOAH standards) and to ensure negative effects on trade are minimised (WOAH 2022).

PLANT BIOSECURITY

Under the SPS Agreement, the International Plant Protection Convention (IPPC) is the international standard-setting body for phytosanitary measures related to plant health and protection. The IPPC was established by the Food and Agriculture Organisation (FAO) in 1952, with the goal of protecting the world's cultivated and natural plant resources from the introduction and spread of plant pests, while minimising interference with the international movement of people, goods, and conveyances (FAO 2011). The IPPC currently comprises 184 contracting parties.

The IPPC was revised in 1997 and adopted in 2005 to align with the SPS Agreement. One of the key changes under the 1997 revision of the IPPC was the adoption of pest risk analysis as the basis for technically justified measures and improved phytosanitary certificates (see Chapter 3. Anticipate). Other changes included an improved dispute settlement mechanism, improved information sharing, recognition of regulated non-quarantine pests (i.e. pests that are already present in an importing country), and provision for possible electronic certification (Erikson and Griffin 2014).

The IPPC provides a mechanism for developing international standards for phytosanitary measures (ISPMs). As of April 2023, 46 ISPMs have been adopted relating to procedures and references, the application of measures, pest risk analysis, pest surveillance, import regulations, certification, pest management, exotic pest response, and export certification (IPPC 2023). The SPS Agreement allows countries to choose not to use ISPMs, however where phytosanitary measures provide a higher level of protection than ISPMs, these national measures must be justified. The IPPC also requires each signatory country to establish a National Plant Protection Organisation (NPPO), the Competent Authority with the responsibility of protecting the country's plant resources (both natural and cultivated) from the negative impacts of pests (Devorshak 2012). This includes issuing phytosanitary certification for exports; inspecting consignments of plant and plant products; reporting the occurrence, outbreak, and spread of pests; surveillance and maintenance of pest-free areas; and conducting pest risk analysis.

ENVIRONMENTAL BIOSECURITY

WOAH and IPPC have overlapping mandates and cooperate with other international agreements, namely the Convention on Biological Diversity (CBD) and the Cartagena Protocol on Biosafety (an international agreement on biosafety and a supplementary agreement to the CBD). The CBD was established in 1993 to promote the conservation and sustainable use of biological diversity. The CBD recognises invasive species as one of the primary threats to biodiversity, and part of the work of the CBD is directed towards managing the environmental risks associated with invasive species (Devorshak 2012).

International standards and guidelines for plant and animal health developed by the WOAH and IPPC work well when assessing the consequences of horticultural or agricultural threats. However, difficulties arise when the impacts of pests and diseases fall largely on the environment (including the marine environment; see Box 2.4). When pests and diseases affect biodiversity and ecosystem services (i.e. the benefits people obtain from ecosystems, such as pollination), guidance from WOAH and IPPC on risk assessment methods is less clear (Hester and Mayo 2021). Information about threat biology and behaviour is often unavailable, and methods to determine the value of at-risk ecosystem services vary among jurisdictions (see Chapter 3. Anticipate).

In the absence of a consensus methodology for environmental risk assessment, the SPS Agreement allows members to adopt sanitary or phytosanitary measures based on other information, such as relevant environmental information (Fraser et al. 2019). The SPS Agreement provides for the use of provisional measures designed to balance risks in cases where scientific evidence is insufficient or significant uncertainty remains (Erikson and Griffin 2014). A WTO member may provisionally impose measures to protect the environment based on the precautionary principle, although this can only be done on a temporary basis while a more objective risk assessment is undertaken.

BOX 2.4. BIOSECURITY THREATS AND INTERNATIONAL REGULATIONS IN THE MARINE ENVIRONMENT

Marine environments worldwide are under threat from invasive species that can cause significant damage to economic, environmental, and social assets. International shipping is the main pathway for the introduction of exotic marine species via ballast water exchange and vessel biofouling. Ballast water is fresh, or saltwater held in vessel tanks to provide additional stability and manoeuvrability, and the release of ballast water in new locations can release harmful aquatic organisms and pathogens. Since September 2017, all new ships must comply with the Ballast Water Management Convention under the auspices of the International Maritime Organization (IMO). This convention stipulates the use of water management systems to reduce the concentration of organisms and pathogens in ballast water below specific thresholds. Existing ships, prior to having to conform to this standard by 2024, must use a ballast water management plan, a record book, and an international certificate to document their ballast water management methods.

Biofouling is the accumulation of marine aquatic organisms on surfaces and structures exposed to aquatic environments (see Chapter 13. Profiling and Automation). Biofouling on marine vessels navigating international waters can spread non-indigenous marine species, but no international regulations are in place for marine vessel biofouling (Williams et al. 2013). Best-practice management is based on maintaining a biofouling management plan and record book, guided by voluntary biofouling management guidelines (IMO 2011). Biofouling management regulations have been developed at national and regional levels. New Zealand's standard, in force since November 2018, is the most advanced among biofouling requirements—it defines what a clean hull should look like and prescribes biofouling thresholds for short and long visits. The United States has also developed national regulations for international vessels to manage biofouling, and Australia is in the process of doing so. California, two Australian jurisdictions (Western Australia and the Northern Territory), and Ecuador have developed regional-level regulations (Georgiades et al. 2020).

International obligations preclude WTO members from indefinite use of the precautionary principle as grounds for not deciding on quarantine import requests (Nunn 2012). Despite its methodological shortcomings, the introduction of the SPS Agreement is still considered a step forward for incorporating environmental concerns in biosecurity decision making (Haddock 2000).

BIOSECURITY ACTIVITIES

As noted earlier, a biosecurity system encompasses the suite of management activities implemented by a jurisdiction to protect its economy, environment, and human health from the damaging impacts of pests and diseases. Activities occur in pre-border, border, and post-border settings, and involve participants, physical and financial resources, and influencing factors. Pre-border biosecurity activities are informed by international regulations, guidelines, and precedent, whereas border and post-border biosecurity measures are typically the domains of national and local participants (Figure 2.1). This section outlines the main activities, outputs, outcomes, and inputs of biosecurity systems in order to create a holistic understanding of the work national biosecurity agencies undertake, including the main management tools available to risk managers (Table 2.1). Chapters 3–7 expand on the activities listed in Table 2.1 and Figure 2.1.

TABLE 2.1

Biosecurity Activities and Their Expected Outputs and Outcomes. Outcomes Are Wider Goals While Outputs Are the Direct Products, Results, or Services Delivered by an Activity

	Activities What is Being Undertaken	Outcomes Broad Goals	Output Examples Direct Results or Products
Pre-border	**Anticipating** biosecurity risk	The risk profile is identified, assessed, and prioritised	• Intelligence reports • Import risk analyses • Pest risk analyses
	Preventing biosecurity risk material from reaching the border	The number of priority pests and diseases approaching the border is reduced	• International arrangements • Import conditions • Offshore verification and capacity building
Border	**Screening** entry pathways	The number of priority pests and diseases approaching and passing through the border is reduced	• Passengers cleared • Items inspected and cleared • Leakage rate surveys
Post-border	**Detecting** incursions and outbreaks	The time taken to detect incursions of priority pests and diseases is reduced	• Surveillance networks • Hectares surveyed • Diagnostic tests
	Preparing for incursions and outbreaks	Participants in the biosecurity system are ready to respond to incursions and outbreaks	• Response plans • Emergency exercises and farms with biosecurity plans
	Responding to incursions and outbreaks	The number of priority pests and diseases that establish and spread is reduced	• Response plans • Incursions delimited and eradicated
	Recovering from an incursion or outbreak and adapting to new circumstances	The impact of pests and diseases is reduced and disruption to market access is minimised	• National management plans • Markets opened • Community adaptation programs

Source: Adapted from Schneider and Arndt 2020; see Chapter 10, Monitoring, Evaluation, and Reporting for more details.

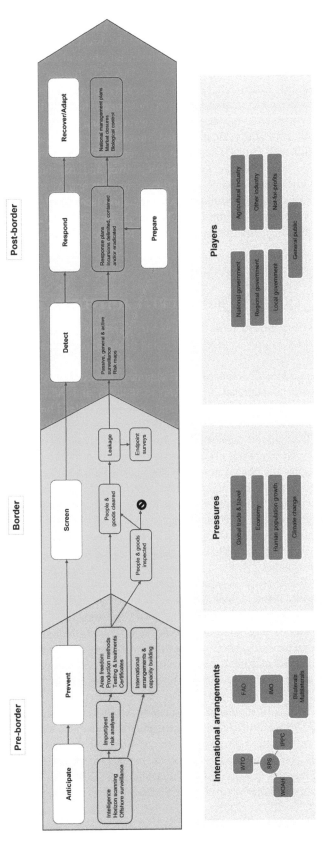

FIGURE 2.1 The Biosecurity Continuum (including pre-border, border, and post-border activities; Bland et al. 2023) and the context in which the system operates (including international arrangements, pressures, and players). FAO: Food and Agriculture Organization. IMO: International Maritime Organization. IPPC: International Plant Protection Convention. SPS Agreement: Agreement on the Application of Sanitary and Phytosanitary Measures. WOAH: World Organisation for Animal Health. WTO: World Trade Organisation.

Biosecurity activities include specific projects and programs undertaken to achieve broad, medium-term goals (biosecurity outcomes), whereas outputs are the direct, tangible products or services delivered by an activity (see Table 2.1). Inputs are the financial, physical, and human resources involved in completing activities and generating outputs (see Box 2.5). In many cases, inputs are limited, and there may be trade-offs in achieving certain biosecurity outcomes. Biosecurity systems also feature influencing and enabling factors that act across the continuum, including strategic planning and policy development; governance; partnerships, engagement, and communications; funding and resource allocation; monitoring and evaluation; and knowledge management, research, and innovation. When viewed in this way—as "a whole of system approach"—biosecurity agencies can identify underperforming areas that do not achieve their objectives and that may increase risk in the overall system or lead to inefficient resource allocation (Schneider and Arndt 2020). As such, a systems approach to biosecurity supports performance evaluation (see Chapter 10. Monitoring, Evaluation, and Reporting), knowledge sharing (see Chapter 11. Research Uptake), and efficient resource allocation across competing activities (see Chapter 9. Resource Allocation).

BOX 2.5. INPUTS OF BIOSECURITY ACTIVITIES

To undertake the activities described in Table 2.1, biosecurity systems rely on system participants (including human resources), physical resources, and financial resources:

- *System participants and human resources*: Participants in biosecurity systems range from government (national, regional, and local) to industry and their representative groups, natural resource managers, users and custodians of land and resources (including farmers), research providers, non-governmental organisations, and the general community. Government agencies operate across the biosecurity continuum, including developing and operating pre-border and border control programs; post-border activities are typically characterised by close partnerships between different levels of government, agricultural industries, and not-for-profit organisations. The general public and community groups contribute to biosecurity effort by complying with biosecurity regulations and by contributing to passive surveillance, citizen science, and other initiatives (Hester and Cacho 2017). Universities, research institutes, and government departments are typically the suppliers of biosecurity research (see Chapter 11. Research Uptake).
- *Physical resources*: Biosecurity system participants rely on extensive physical resources for their work. These include inspection facilities at major points of entry (e.g. airports, seaports, and mail centres), diagnostic facilities (e.g. laboratories, equipment, and taxonomic collections), quarantine facilities, information technology (IT) systems, and office facilities. Many of these resources are government managed and operated, but industries also contribute physical resources, including quarantine premises, facilities, and IT infrastructure operated by customs brokers and freight forwarders.
- *Financial resources*: Financial inputs into biosecurity systems can be significant. In 2015–2016, Australia's expenditure on biosecurity amounted to about AUD 1 billion (USD 0.75 billion), funded through government contributions, industry payments (namely fees, charges, and levies), and in-kind contributions from landholders, community groups, and industry (Craik, Palmer, and Sheldrake 2017). In 2019–2020, New Zealand spent almost NZD 400 million (USD 264 million) on biosecurity measures (MPI 2020).

ANTICIPATING BIOSECURITY RISK

In the pre-border space, international regulations developed by the WTO, IPPC, WOAH, and CAC (food safety) shape the operation of national biosecurity systems. Biosecurity systems start before goods and people reach the border, when current and future biosecurity risks are anticipated. In this phase of the biosecurity system, the risk profile of a jurisdiction is identified, assessed, and prioritised.

Chapter 3. Anticipate introduces the foundational concepts underpinning import risk analyses and pest risk analyses, using qualitative, semi-quantitative, and quantitative methods. Import risk analyses and plant import risk analyses are informed by data gathered through environmental scanning, intelligence forums, and offshore surveillance. Environmental (or horizon) scanning involves examining the external environment to detect emerging biosecurity risks and risks associated with trends in global production, trade, and travel (Camac et al. 2021). Automated open-source search engines are increasingly used to process large amounts of information for environmental scanning. For example, the International Biosecurity Intelligence System (IBIS; Intelliriver) performs analyses to track and forecast animal and plant diseases. The real-time early warning information provided by IBIS allows governments to respond faster to biosecurity threats (Grossel, Lyon, and Nunn 2017). Biosecurity intelligence can also be sourced by participating in intelligence forums, such as the UK's Human Animal Infections and Risk Surveillance group (Wintle, Kennicutt, and Sutherland 2020). Offshore active surveillance relies on animal or plant health surveys conducted in neighbouring countries (in conjunction with local biosecurity authorities) to build capacity for managing emerging threats and to monitor spread of pests and diseases.

PREVENTING BIOSECURITY RISK MATERIAL FROM REACHING THE BORDER

Biosecurity managers can address biosecurity risk offshore by implementing pre-border activities to prevent animal and plant health threats from reaching the border (see Chapter 4. Prevent). The returns on investment for prevention strategies (along with activities to anticipate risk) are commonly considered higher than at later points on the biosecurity continuum (Victorian Government 2009).

Prevention activities encompass pre-border assurance measures designed to reduce the likelihood of pests and diseases being present in imported products, and the development of import protocols (based on import risk analyses). Import conditions can require importers to provide supporting documentation, to organise for the treatment of goods, or to apply for import permits. For low-risk imports, regulators can shift towards requiring offshore certification for sourced products, and treatments, instead of onshore inspections and treatment (Whattam et al. 2014). Offshore activities also include capacity building in neighbouring countries. For example, Australian investments in Indonesia, Papua New Guinea, and Timor-Leste are designed to gain insights into regional biosecurity issues, build networks, and increase capacity to manage risks locally. These efforts include providing technical advice and laboratory services to local authorities to help manage African swine fever (DAWE 2020), for example.

Prevention strategies must be aligned with international regulations. Countries may influence these regulations by participating in WTO meetings and forums, participating in international standard-setting bodies (e.g. WOAH, IPPC, IMO, and CAC), and participating in bilateral or multilateral arrangements. For example, the International Cargo Cooperative Biosecurity Arrangement is a multilateral arrangement for the collaborative development and verification of sanitary and phytosanitary treatment methods. As a further example, the Plant Health Quadrilaterals is a strategic coalition composed of the NPPOs of Australia, Canada, New Zealand, and the United States, which aims to advance the development and implementation of international standards for these countries' shared plant protection priorities and systems.

Screening Entry Pathways

Pre-border activities cannot be completely effective in anticipating and preventing BRM reaching the border (see Chapter 4. Prevent). As such, national governments invest in a range of border interventions to detect incoming risks, including screening, inspection, fumigation, heat treatment, and information provision. Screening activities are conducted on travellers' personal effects, mail, sea and air cargo, vessels, live animals, and plants. Disease vector monitoring within the biosecurity zone at international air and seaports, measures for non-commodity risks (e.g. hitchhiker pests on wooden pallets), and diagnostic testing are other common screening activities at the border.

The major aims of screening are to detect non-compliance with national biosecurity regulations and to prevent pests and diseases from passing through borders (see Chapter 5. Screen). For nearly all types of goods crossing international borders, inspections are based on a selection of items obtained from a consignment, rather than inspections of the full consignment. Statistical frameworks are used to support decision making about whether an import consignment is compliant or not.

Profiling is used to prioritise inspection resources towards the highest-risk passengers or items, where interventions are based on prior experience or beliefs about what makes an entity or pathway risky (see Chapter 13. Profiling and Automation). In many countries, a range of imported goods are cleared for entry upon reception of suitable documentation, that is, without inspection. Profiling can also drive behavioural change within a pathway because knowledge that screening takes place can encourage importers to minimise BRM on a pathway (Sherring 2021). As trade volumes continue to increase, profiling becomes more important to prioritise resources at the border and to determine which goods should be inspected and which goods need not be.

Risk profiling has led to the introduction of risk-based and compliance-based inspection schemes in several countries to improve the efficiency of inspections, where consistently compliant participants are rewarded with fewer biosecurity inspections. The United States has applied risk-based sampling to the import of plant products since the late 1980s and established the National Agriculture Release Program (NARP) in 2007. In Australia, the Compliance-Based Intervention Scheme and the Vessel Compliance Scheme reward consistently compliant behaviour with fewer biosecurity inspections (see Chapter 11. Research Uptake, Chapter 4. Prevent, and Chapter 5. Screen).

Detecting Pest and Disease Incursions or Outbreaks

Detecting an incursion or outbreak early (i.e. before a pest or disease has spread) can improve the outcomes of subsequent emergency response activities. If any notifiable animal diseases and plant pests are found, then they must be reported to WOAH or IPPC, respectively. Surveillance activities, outlined in WOAH and IPPC standards, support the ongoing management of established pests and diseases and provide evidence of pest or disease absence, which underpins claims to pest and disease freedom and facilitates access to international markets (see Chapter 6. Detect).

National and regional governments usually invest in a mixture of surveillance methods to detect incursions, ranging from surveys using visual observations, sophisticated traps, and cameras, to reports from members of the public via hotlines and social media platforms. National Veterinary Authorities, NPPOs, private veterinary practitioners, industries, and non-government organisations also have the capacity to undertake surveillance under different partnership agreements. All these activities must be coordinated by the Competent Authority for these to contribute to the biosecurity system. Citizen science and passive surveillance programmes also feature in national detection activities (see Chapter 6. Detect).

Diagnostic services enable the accurate and early identification of samples and specimens that are notifiable. Veterinary laboratories and plant pest diagnostic facilities are typically operated by national and regional governments, universities, or private entities, and must be regulated by the Competent Authority. Ideally, national networks of diagnosticians underpin both veterinary and plant pest diagnostic facilities by sharing knowledge and diagnostic resources, developing national standards, and providing surge capacity (AHA 2021; PHA 2021). Many countries also operate WOAH reference laboratories that lead research and provide expert advice to other WOAH member countries on priority animal diseases. Apart from supporting early warning, diagnostic services allow biosecurity managers to trace a detected incursion or outbreak back to its source and to trace forward from the source to identify the spread behaviour of a pest or disease (AHA 2021). Both tracing activities assist government agencies with response planning following a detection.

PREPARING FOR AN INCURSION OR OUTBREAK OF PESTS AND DISEASES

To effectively respond to incursions or outbreaks of pests and diseases, biosecurity agencies invest in activities that prepare system participants for incursions. By determining their ALOP, countries acknowledge that biosecurity risks cannot be reduced to zero. As with pre-border and border activities, the return on investment in preparedness activities remains high (Victorian Government 2009). In this part of the biosecurity continuum, responsibility for biosecurity activities mainly shifts from national governments to regional or local governments and private stakeholders. Emergency preparedness is not covered in detail in this book because the measures that biosecurity agencies implement to respond to an incursion or outbreak are specific to the jurisdiction and the invader and therefore difficult to generalise. However, we refer to key components of preparedness in this section and describe emergency response strategies in Chapter 7. Prepare, Respond, and Recover. We also note that preparing for incursions relies on information from the detection and response phases (Figure 2.1 and Table 2.1) and hence is described in a slightly different order to that of Schneider and Arndt (2020; Box 2.2).

A major part of preparedness is to develop and maintain legally binding emergency response agreements and contingency plans. In Australia, three response deeds are used for pests and diseases threatening animals, plants, or the environment and social amenity, respectively (AHA 2019; DAWR 2012; PHA 2020). In New Zealand, all responses are covered under one agreement (GIA 2021). During an emergency animal disease outbreak, the six members of the International Animal Health Emergency Reserve (Australia, Canada, Ireland, New Zealand, the United Kingdom, and the United States) can access additional personnel and resources (AHA 2021).

Training activities and emergency response simulation exercises are undertaken to build the capability and readiness of stakeholders and to test the capacity of the biosecurity system. Farm-level initiatives are also used to encourage the uptake of best practices for priority pests and diseases. For example, in the United States, the "Defend the Flock" program educates poultry owners, growers, and workers on implementing best practices on-farm to reduce the risk of disease outbreaks (APHIS 2021).

Technical tools that assist with preparation for incursions and outbreaks include pest databases, resource tracking systems, livestock identification and tracing systems (see Chapter 6 Detect), case management platforms, disease spread models, and pest risk maps (see Chapter 14. Map). National biosecurity response teams can support emergency responses by deploying specialist government agency personnel. In Australia, the funding of vaccine banks for FMD and anthrax is another preparedness arrangement that can ensure immediate access to vaccines in the event of an outbreak.

RESPONDING TO AN INCURSION OR OUTBREAK OF PESTS AND DISEASES

Biosecurity response activities aim to reduce the number of priority pests and diseases that establish and spread in a jurisdiction. Biosecurity systems that apply a partnership approach to biosecurity risk management coordinate response activities collaboratively; for example, by using a single decision-making body with representatives from different levels of government, industry, and non-government organisations.

If a notifiable pest or disease is detected, a rapid and coordinated "emergency" response can reduce adverse impacts on assets and limit the need for post-incident recovery and adaptation (see Chapter 7. Prepare, Respond, and Recover). Generally, incident management teams follow a sequence of activities prescribed in the emergency response agreements: (1) incident definition (i.e. initial investigation), (2) emergency response, (3) proof of freedom, and where eradication is not feasible, (4) transition to management (AHA 2019; PHA 2020). If response activities have been completed successfully (e.g. see Delpont et al. 2021 for biosecurity measures implemented on French poultry farms in response to outbreaks of Highly Pathogenic Avian Influenza between 2015 and 2017), the proof of freedom phase commences. During this period, research or surveillance is conducted to confirm that a pest or disease has been contained or eradicated.

If the emergency response fails to eradicate the biosecurity threat, managers are faced with four long-term management strategies: 1) continue to pursue eradication, or switch strategies to 2) containment, 3) impact reduction, or 4) mitigation, either separately or in combination. Containment aims to prevent or reduce the likelihood of establishment and reproduction beyond a predefined geographical range (Wittenberg and Cock 2001), whereas impact reduction attempts to reduce threat impacts without necessarily restricting its range (e.g. by suppressing population levels; Grice 2009). Impact reduction methods (e.g. biological control) are usually the only economically feasible, long-term management option for widespread pests and diseases (Cacho and Hester 2022). A fourth course of action, mitigation, does nothing to stop the further spread and establishment of the threat, but rather focuses on mitigating the impacts of the threat on affected assets (e.g. native species; Blackburn et al. 2011; Wittenberg and Cock 2001).

Recovering from an Incursion or Outbreak and Adapting to New Circumstances

Following a biosecurity incident, participants in the biosecurity system implement activities to recover from the immediate impact and to adapt to changed circumstances associated with the ongoing management of established pests and diseases. Recovery and adaptation activities focus on minimising disruptions to market access and reducing the impact of pests and diseases.

The re-opening of markets after a biosecurity incident is supported by surveillance activities to prove area freedom, or in other words, prove that a pest or disease is not present or has not been present for a period of time (see Chapter 6. Detect). Area freedom surveillance underpins export certification when it is a requirement for the export of live animals, animal products, and plant products (Whattam et al. 2014). In some cases, markets can operate where a disease is present, so long as additional measures can be undertaken to reduce risks (e.g. heat treatments). In other cases, markets can operate in zones that are free of pests and disease, even if a country doesn't have national freedom (namely, pest-free areas).

Pests and diseases that circumvent all measures to keep them out and efforts to eradicate or contain them, will have an impact on economic, environmental, and/or social assets. The extent and intensity of the impact depends on the type of pest or disease, the extent of its spread, and the vulnerability of at-risk industries or assets. Long-term management strategies are typically adopted with the aim to reduce adverse impacts of established pests and diseases (see Chapter 7. Prepare, Respond, and Recover).

One way to manage pests and diseases in the long term is to control the movement of goods between or within jurisdictions (e.g. using quarantine zones). For example, all Australian states and territories have established domestic biosecurity or quarantine zones to restrict movement of high-risk material between zones. Domestic trade of plant products across state boundaries can still take place because the Interstate Certification Assurance Scheme allows accredited businesses to issue certificates for their produce and access markets (Australian Interstate Quarantine 2021).

Countries often develop national strategies and frameworks to support and guide biosecurity participants in their efforts to manage already established invasive plant species, invasive terrestrial and freshwater vertebrate animals, and pests and diseases of national significance. Biosecurity New Zealand, for example, manages nine harmful weeds under the National Interest Pest Responses programme, in conjunction with the NZ Department of Conservation and regional councils (MPI

2021). Landholders who fail to control noxious weeds on their property can be directed to do so under legislation. However, it is important that regulations be incentive-compatible if they are to achieve their aim as efficiently as possible (see Chapter 8. Incentives). Industry organisations, producers, and the community are often involved in programs that target established pests and diseases and provide a social benefit (PHA 2021).

BIOSECURITY AS A COMPLEX SYSTEM

Biosecurity systems operate within an environment that is ever-changing and where global and domestic factors put pressure on the system and challenge a jurisdiction's biosecurity status. Biosecurity systems around the world face immense pressure from the growing volume of containerised and non-containerised trade and the increase in international travellers, which are linked to human population growth and increases in consumption (BITRE 2012, 2014). In the future, importers are predicted to source products from a wider range of countries because of changing and expanding transport networks, resulting in more diverse and complex (and potentially higher risk) import pathways (Dodd et al. 2015; IGB 2019). Risks from strategic behaviour are also increasing across pathways as stakeholders continue to find new ways to circumvent biosecurity rules (Pieracci et al. 2021). Adding to this, climate change will offer changing opportunities for exotic pests and diseases to survive in suitable habitats, changing jurisdictions' risk profiles and creating new or unexpected biosecurity risks (see Chapter 14. Map).

Biosecurity systems match the identified characteristics of complex systems such as non-linearity, hierarchical organisation, and feedback loops (Ladyman, Lambert, and Wiesner 2013). Non-linearity suggests that doubling investments in a particular biosecurity activity will not lead to a reduction of risk of the same magnitude because myriad factors influence management outcomes (see Chapter 9. Resource Allocation). Biosecurity systems are usually organised hierarchically, incorporating participants that engage at various levels of authority and creating subsystems (e.g. jurisdictional biosecurity agencies). Importantly, system activities are interconnected, and multiple feedback loops occur among them, which in turn contribute to the overall performance and efficiency of a biosecurity system. For example, pre-border assurance measures (e.g. treatment and processing) and border inspections are interconnected activities, where particular assurance measures may allow fewer or less intense inspections to be conducted at the border (see Chapter 4. Prevent).

An advanced way of dealing with a dense web of interactions and feedback loops is to implement a fully integrated biosecurity framework, featuring legislative authority, a budget, and integrated institutional arrangements. An integrated approach to biosecurity has been implemented by Australia, New Zealand, and Belize (see Box 2.5. and FAO 2016). In an integrated approach, an overarching strategy defining clear outcomes for the biosecurity system provides the basis for a consistent and harmonised biosecurity policy. Underneath the strategy, governance arrangements provide a framework for leadership, management, and defining behaviours and relationships among participants. Effective engagement and communications become crucial to the successful operation of the biosecurity system. Technical capability and capacity for managing, analysing, and sharing information also support system participants. However, in practice, organisational silos and barriers to information sharing can seriously hamper the performance and efficiency of biosecurity systems (see Chapter 10. Monitoring, Evaluation and Reporting and Chapter 11. Research Uptake).

In Australia, key inter-governmental governance arrangements for biosecurity include the Intergovernmental Agreement on Biosecurity (IGAB) and the National Biosecurity Committee (NBC), with its subcommittees and working groups. The IGAB is the primary collaboration mechanism for government, whereas the NBC oversees the identification and implementation of collaborative projects to meet national priorities outlined in the IGAB (Craik, Palmer, and Sheldrake 2016). Collaboration is a principle under the IGAB that sees the Australian government and state and territory jurisdictions committing to sharing responsibility for biosecurity with others (including industries and the community) in a partnership approach (COAG 2019).

Other examples of an integrated approach to biosecurity include New Zealand's Government Industry Agreement for Biosecurity Readiness and Response—a government-industry partnership in which decision making, responsibilities, and costs around biosecurity preparedness and response are shared (GIA 2021). Multiple public-private partnerships have also been formed in the United States in response to the incursion of soybean rust (Markell et al. 2020). Internationally, integrated approaches have been viewed as relatively successful and serve as notable examples of public-private partnerships (de la Cruz 2020; Markell et al. 2020).

Biosecurity risk management not only exhibits features of a complex system (with complex organisation, feedback loops, and uncertain risks) but those of a wicked policy problem, in which values, viewpoints, and objectives vary among stakeholders (DeFries and Nagendra 2017; Head 2008). While some costs of importing goods (e.g. shipping, product spoilage) are the responsibility of the importer, the costs of preventing and responding to pest and disease incursions are transferred to the citizens and the natural environment of the importing country. Current incentives for biosecurity activities do not always encourage the efficient level of effort needed by importers to reduce biosecurity threats and instead might encourage "gaming" of the biosecurity system and strategic behaviour by importers. In Chapter 8. Incentives, we present two options (risk-based insurance premiums and incentive-compatible regulations) to improve the behaviour of biosecurity stakeholders and generate an efficient level of effort across a biosecurity continuum.

IN A NUTSHELL

- The World Trade Organisation's Agreement on the Application of Sanitary and Phytosanitary Measures significantly shapes the activities that governments undertake to reduce biosecurity risks.
- A national biosecurity system consists of interacting components, including activities, participants, resources, and influencing factors.
- The seven key biosecurity activities conducted in pre-border, border, and post-border settings are: anticipating, preventing, screening, preparing, detecting, responding, and recovering/adapting to risks.
- Biosecurity systems match the identified characteristics of complex systems (non-linearity, hierarchical organisation, and feedback loops).
- Conceptualising biosecurity as a complex system can improve coordination across all phases of the system, ensuring high performance and efficient resource allocation.

REFERENCES

AHA. 2019. Government and livestock industry cost sharing deed in respect of emergency animal disease responses. Version No. 19/01 – 06/19. Animal Health Australia, Canberra, Australia.

AHA. 2021. *Animal Health in Australia System Report, First Edition*. Canberra: Animal Health Australia.

Anderson, C., S. Low-Choy, P. Whittle, S. Taylor, C. Gambley, L. Smith, P. Gillespie, H. Löcker, R. Davis, and B. Dominiak. 2017. "Australian Plant Biosecurity Surveillance Systems." *Crop Protection* 100:8–20. https://doi.org/10.1016/j.cropro.2017.05.023.

Anderson, K., C. McRae, and D Wilson. 2012. "Introduction." In *The Economics of Quarantine and the SPS Agreement*, edited by Cheryl McRae, David Wilson and Kym Anderson, 1–6. Adelaide: The University of Adelaide Press.

APHIS. 2021. Plant Health Quadrilaterals Group. Animal and Plant Health Inspection Service. U.S. Department of Agriculture.

Australian Interstate Quarantine. 2021. "Moving plant goods interstate." Australian Interstate Quarantine, accessed 12 January 2023. https://interstatequarantine.org.au/producers/interstate-certification-assurance/.

BITRE. 2012. Air passenger movements through capital and non-capital city airports to 2030-31, Report 133. Canberra: Bureau of Infrastructure, Transport and Regional Economics.

BITRE. 2014. Containerised and non-containerised trade through Australian ports to 2032-33, Report 138. Canberra: Bureau of Infrastructure, Transport and Regional Economics.

Blackburn, T. M., P. Pyšek, S. Bacher, J. T. Carlton, R. P. Duncan, V. Jarošík, J. R. U. Wilson, and D. M. Richardson. 2011. "A Proposed Unified Framework for Biological Invasions." *Trends in Ecology & Evolution* 26 (7):333–339. https://doi.org/10.1016/j.tree.2011.03.023.

Bland, L. M., S. Bau, S. Hester, and E. Arndt. 2023. "Biosecurity Continuum." University of Melbourne. https://doi.org/10.26188/23584230.

Cacho, O. J., and S. M. Hester. 2022. "Modelling Biocontrol of Invasive Insects: An Application to European Wasp (Vespula germanica) in Australia." *Ecological Modelling* 467:109939. https://doi.org/10.1016/j.ecolmodel.2022.109939.

Camac, J., J. Baumgartner, A. Robinson, and T. Kompas. 2021. Estimating trading partner exposure risk to new pests or diseases. Technical report for CEBRA project 190606. Centre of Excellence for Biosecurity Risk Analysis. The University of Melbourne.

COAG. 2019. Intergovernmental Agreement on Biosecurity. Council of Australian Governments, Canberra.

Craik, W., D. Palmer, and R. Sheldrake. 2016. *Is Australia's National Biosecurity System and the Underpinning Intergovernmental Agreement on Biosecurity Fit for the Future?* Canberra: Department of Agriculture and Water Resources.

Craik, W., D. Palmer, and R. Sheldrake. 2017. *Priorities for Australia's Biosecurity System. An Independent Review of the Capacity of the National Biosecurity System and Its Underpinning Intergovernmental Agreement.* Canberra: Department of Agriculture and Water Resources.

DAWE. 2020. *Annual Report 2019-20.* Canberra: Department of Agriculture, Water and the Environment.

DAWR. 2012. *National Environmental Biosecurity Response Agreement.* Version 2012(1). Canberra: Department of Agriculture and Water Resources.

de la Cruz, J. 2020. "Public Private Collaborations Amidst an Emergency Plant Disease Outbreak: The Australian Experience With Biosecurity for Panama Disease." *NJAS - Wageningen Journal of Life Sciences* 92:100316. https://doi.org/10.1016/j.njas.2019.100316.

DeFries, R., and H. Nagendra. 2017. "Ecosystem Management as a Wicked Problem." *Science* 356 (6335):265–270. https://doi.org/10.1016/j.jenvman.2020.111755.

Delpont, M., C. Guinat, J. -L. Guérin, E. Le leu, J. -P. Vaillancourt, and M. C. Paul. 2021. "Biosecurity Measures in French Poultry Farms Are Associated With Farm Type and Location." *Preventive Veterinary Medicine* 195:105466. https://doi.org/10.1016/j.prevetmed.2021.105466.

Devorshak, C. 2012. "International Legal and Regulatory Framework for Risk Analysis." In *Plant Pest Risk Analysis: Concepts and Application*, 29–42. Wallingford: CABI International.

Dodd, A. J., M. A. Burgman, M. A. McCarthy, and N. Ainsworth. 2015. "The Changing Patterns of Plant Naturalization in Australia." *Diversity and Distributions* 21 (9):1038–1050. https://doi.org/10.1111/ddi.12351.

Erikson, L., and R. Griffin. 2014. "The International Regulatory Framework." In *The Handbook of Plant Biosecurity: Principles and Practices for the Identification, Containment and Control of Organisms That Threaten Agriculture and the Environment Globally*, edited by Gordon Gordh and Simon McKirdy, 27–43. Dordrecht: Springer Netherlands.

FAO. 2011. International Plant Protection Convention. Secretariat of the International Plant Protection Convention.

FAO. 2016. *The Biosecurity Approach. A Review and Evaluation of Its Application by FAO, Internationally and in Various Countries.* Rome, Italy: Food and Agriculture Organization of the United Nations.

Fraser, R. W., D. C. Cook, and J. Haddock-Fraser. 2019. *The WTO and Environment-Related International Trade Disputes.* London: World Scientific.

FSANZ. 2013. *Risk Analysis in Food Regulation.* Canberra & Wellington: Food Standards Australia New Zealand.

Georgiades, E., D. Kluza, T. Bates, K. Lubarsky, J. Brunton, A. Growcott, T. Smith, S. McDonald, B. Gould, N. Parker, and A. Bell. 2020. "Regulating Vessel Biofouling to Support New Zealand's Marine Biosecurity System—a Blue Print for Evidence-Based Decision Making." *Frontiers in Marine Science* 7. https://doi.org/10.3389/fmars.2020.00390.

GIA. 2021. About GIA. Government industry agreement for biosecurity readiness and response. Government Industry Agreement for Biosecurity Readiness and Response.

Goh, G., and A. Ziegler. 2012. "Implications of Recent SPS Dispute Settlement Cases." In *The Economics of Quarantine and the SPS Agreement*, edited by Cheryl McRae, David Wilson and Kym Anderson, 75–102. The University of Adelaide Press.

Gordh, G., and S. McKirdy, eds. 2014. *The Handbook of Plant Biosecurity*. Dordrecht, The Netherlands: Springer.

Grice, A. C. 2009. "Principles of Containment and Control of Invasive Species." In *Invasive Species Management: A Handbook of Principles and Techniques*, edited by Mick N. Clout and Peter A. Williams, 61–76. Oxford; New York: Oxford University Press.

Grossel, G., A. Lyon, and M Nunn. 2017. "Open-Source Intelligence Gathering and Open-Analysis Intelligence for Biosecurity." In *Invasive Species: Risk Assessment and Management*, edited by Andrew P. Robinson, Mark A. Burgman, Mike Nunn and Terry Walshe, 84–92. Cambridge: Cambridge University Press.

Haddock, J. 2000. "Rule-Based Decision Making and Its Applicability to Environmental-Related Trade Disputes in the World Trade Organisation (WTO)." *Journal of Environmental Assessment Policy and Management* 02 (01):81–98. https://doi.org/10.1142/S1464333200000060.

Head, B. W. 2008. "Wicked Problems in Public Policy." *Public Policy* 3 (2): 101–118.

Hester, S. M., and O. J. Cacho. 2017. "The Contribution of Passive Surveillance to Invasive Species Management." *Biological Invasions* 19 (3):737–748. https://doi.org/10.1007/s10530-016-1362-4.

Hester, S., and J. Mayo. 2021. Improving the methodology for rapid consequence assessment of amenity and environmental pests. Australian Centre of Excellence for Risk Analysis.

IGB. 2019. *Pest and Disease Interceptions and Incursions in Australia*. Canberra: Inspector-General of Biosecurity. Department of Agriculture and Water Resources.

IMO. 2011. 2011 Guidelines for the Control and Management of ships' Biofouling to Minimize the Transfer of Invasive Aquatic Species. In *Annex 26 Resolution MEPC.207(62)*: Marine Environment Protection Committee, International Maritime Organization.

IPPC. 2019. *Framework for Pest Risk Analysis*. Rome, Italy: Food and Agriculture Organization of the United Nations.

IPPC. 2023. "Adopted standards (ISPMs)." Food and Agriculture Organization of the United Nations, accessed 31 July 2023. https://www.ippc.int/en/core-activities/standards-setting/ispms/.

Ladyman, J., J. Lambert, and K. Wiesner. 2013. "What Is a Complex System?" *European Journal for Philosophy of Science* 3 (1):33–67. https://doi.org/10.1007/s13194-012-0056-8.

Markell, S. G., G. L. Tylka, E. J. Anderson, and H. P. van Esse. 2020. "Developing Public–Private Partnerships in Plant Pathology Extension: Case Studies and Opportunities in the United States." *Annual Review of Phytopathology* 58 (1):161–180. https://doi.org/10.1146/annurev-phyto-030320-041359.

MPI. 2020. *Annual Report 2019/20*. Wellington, New Zealand: Ministry for Primary Industries.

MPI. 2021. *National Interest Pest Responses Programme*. Wellington, New Zealand: Ministry for Primary Industries.

Nunn, M. J. 2012. "The Analytical Foundation of Quarantine Risk Analysis." In *The Economics of Quarantine and the SPS Agreement*, edited by Cheryl McRae, David Wilson and Kym Anderson, 29–52. The University of Adelaide Press.

PHA. 2020. *Government and Plant Industry Cost Sharing Deed in Respect of Emergency Plant Pest Responses*. Canberra: Plant Health Australia.

PHA. 2021. *The National Plant Biosecurity Status Report (2020)*. Canberra: Plant Health Australia.

Pieracci, E. G., C. E. Williams, R. M. Wallace, C. R. Kalapura, and C. M. Brown. 2021. "U.S. Dog Importations During the COVID-19 Pandemic: Do We Have an Erupting Problem?" *PLOS ONE* 16 (9):e0254287. https://doi.org/10.1371/journal.pone.0254287.

Schneider, K., and E. Arndt. 2020. Evaluating the health of Australia's biosecurity system. Final report for CEBRA project 170714. Centre of Excellence for Biosecurity Risk Analysis. The University of Melbourne.

Sherring, P. 2021. "Declare or Dispose: Keeping Biosecurity Threats Out of New Zealand Using Behaviour Change." In *Broadening Cultural Horizons in Social Marketing: Comparing Case Studies from Asia-Pacific*, edited by Rachel Hay, Lynne Eagle and Abhishek Bhati, 211–237. Singapore: Springer Singapore.

Victorian Government. 2009. *Biosecurity Strategy for Victoria*. Melbourne: Victorian Department of Primary Industries.

Whattam, M., G. Clover, M. Firko, and T. Kalaris. 2014. "The Biosecurity continuum and Trade: Border Operations." In *The Handbook of Plant Biosecurity: Principles and Practices for the Identification, Containment and Control of Organisms That Threaten Agriculture and the Environment Globally*, edited by Gordon Gordh and Simon McKirdy, 149–188. Dordrecht: Springer Netherlands.

Williams, S. L., I. C. Davidson, J. R. Pasari, G. V. Ashton, J. T. Carlton, R. E. Crafton, R. E. Fontana, E. D. Grosholz, A. W. Miller, G. M. Ruiz, and C. J. Zabin. 2013. "Managing Multiple Vectors for Marine Invasions in an Increasingly Connected World." *BioScience* 63 (12):952–966. https://doi.org/10.1525/bio.2013.63.12.8.

Wintle, B. C., M. C. Kennicutt, and W. J. Sutherland. 2020. "Scanning Horizons in Research, Policy and Practice." In *Conservation Research, Policy and Practice*, edited by Juliet A. Vickery, Nancy Ockendon, Nathalie Pettorelli, Peter N. M. Brotherton, William J. Sutherland and Zoe G. Davies, 29–47. Cambridge: Cambridge University Press.

Wittenberg, R., and M. J. W. Cock, eds. 2001. *Invasive Alien Species: A Toolkit of Best Prevention and Management Practices*. Wallingford, Oxon, UK: CAB International.

WOAH. 2018. "Import risk analysis." World Organisation for Animal Health, accessed 31 July 2023. https://www.woah.org/fileadmin/Home/eng/Health_standards/tahc/2018/en_chapitre_import_risk_analysis.htm.

WOAH. 2022. "Terrestrial Animal Health Code." World Organisation for Animal Health, accessed 13 January 2023. https://www.woah.org/en/what-we-do/standards/codes-and-manuals/terrestrial-code-online-access/.

WOAH. 2023. "Old Classification of Diseases Notifiable to the OIE – List A." World Organisation for Animal Health, accessed 13 January 2023. https://www.woah.org/en/what-we-do/animal-health-and-welfare/animal-diseases/old-classification-of-diseases-notifiable-to-the-oie-list-a/.

WTO. 1995. "The WTO agreement on the application of Sanitary and Phytosanitary Measures (SPS Agreement)." World Trade Organization, accessed 13 January 2023. https://www.wto.org/english/tratop_e/sps_e/spsagr_e.htm.

WTO. 2015. "United States — measures affecting the importation of animals, meat and other animal products from Argentina." World Trade Organization, accessed 13 January 2023. https://www.wto.org/english/tratop_e/dispu_e/cases_e/ds447_e.htm.

WTO. 2021. *WTO Annual Report 2021*. Geneva: World Trade Organization.

Section I

Pre-Border

3 Anticipate
Assessing Biosecurity Risks Pre-Border

Susan M. Hester, Sana Bau, and Lucie M. Bland

ABSTRACT

Risk is defined as the product of the likelihood and consequences of a threat (which may be economic, social, or environmental). Before biosecurity risks can be managed, they must be anticipated through a process known as risk assessment. Biosecurity risk assessments aimed at informing international import requirements are based on the principles and processes set out by international organisations (e.g. the International Plant Protection Convention and the World Organisation for Animal Health). Biosecurity risk assessments are also important within national borders, where they inform regulation, efficient resource allocation, and the identification of research priorities. In this chapter, we introduce the foundational concepts underpinning risk assessment for international trade and outline common methods for qualitative, semi-quantitative, and quantitative risk assessment. Regardless of the context or methods used, anticipating risks is difficult because it involves making judgements around the likelihood and consequences of events that are yet to occur. As such, poorly designed risk assessment protocols may be prone to logical or implementation errors. Anticipating risks should not be regarded as a "set and forget" process—biosecurity agencies need to be aware of unexpected, emerging and changing biosecurity risks, and risks from strategic behaviour should not be overlooked by biosecurity agencies.

GLOSSARY

SPS Agreement The Agreement on the Application of Sanitary and Phytosanitary Measures sets out the measures that may be applied by signatory countries (countries who are members of the World Trade Organisation) to protect themselves from risks to animal, plant, or human health arising from the entry, establishment, or spread of pests and diseases via trade.

Appropriate level of protection The SPS Agreement defines appropriate level of protection (ALOP) as "the level of protection deemed appropriate by the Member establishing a sanitary or phytosanitary measure to protect human, animal or plant life or health within its territory" (Annex A, paragraph 5).

Biosecurity risk material (BRM) Living organisms exotic to a country, or biological material carrying an exotic pest or disease. Biosecurity risk material includes live animals and plants, animal and plant material (e.g. fur, feathers, pollen, and flowers), food scraps, and soil.

Risk analysis The broad term used to refer to the practice of assessing risk, managing risk, and communicating about risk. Under the SPS Agreement, risk analysis is an important part of determining whether WTO member countries are able to regulate imports associated with a pest or disease and the strength of the measures to be applied.

Risk assessment protocol A formal method for describing and characterising risks.

DOI: 10.1201/9781003253204-4

Risk (or expected consequence) The product of the likelihood and consequence of an event (risk = likelihood × consequence).

Likelihood The chance of an event occurring, usually expressed as a probability (ranging between 0 and 1) or frequency (e.g. 1 in a 100-year event).

Consequence The magnitude of an adverse event if it occurs, which can include the extent, intensity, and/or severity of the impact.

Qualitative risk assessment The use of subjective information and judgements based on evidence, experience, and expert judgement to provide a description of the risk (Devorshak 2012b).

Quantitative risk assessment The use of numerical values derived from data or expert opinion to provide a numerical estimate of risk (Griffin 2012b). Numerical estimates of risk are often derived from mathematical formulae or quantitative models.

Quarantine pest A pest of potential economic importance to an area where it is not yet present, or where it is present but not widely distributed and is being officially controlled (IPPC 2022).

Interception data Data on the detection of pests and diseases collected during inspection or testing of an imported consignment.

Scenario tree A way of identifying and examining a chronological series of significant events or consequences and relating them in a linear fashion.

Non-market valuation methods Describe the methods used to find the economic value of goods and services that are not traded in markets (i.e. environmental goods and services) where a market price is not readily available. For these goods it becomes necessary to estimate the willingness to pay for these non-market goods indirectly. Methods can be classified as either revealed preference techniques or stated preference techniques.

Benefit transfer Where existing estimates of non-market values are transferred to the present study if the study sites are considered broadly similar.

Social amenity Any tangible or intangible resources developed or provided by humans or nature such as dwellings and parks, or views and outlooks (DAWE 2021b).

INTRODUCTION

Biosecurity systems are fundamentally about anticipating risks and managing them. Substantial focus is given to anticipating risks posed by goods, parcels, and people before they reach national borders (Camac et al. 2021; Constable, Kelly, and Dall 2021; Robinson, Burgman, and Cannon 2011), where risks are more cost-effectively managed and prevented (Leung et al. 2014). Pre-border biosecurity decisions are informed by international regulations, guidelines, and precedent (see Chapter 2. Biosecurity Systems and International Regulations). Decisions to allow or disallow the import of a commodity rely on a credible characterisation of risk that satisfies the expectations of the Agreement on the Application of Sanitary and Phytosanitary Measures (SPS Agreement). Assessments of biosecurity risks are also important domestically to inform biosecurity regulations, resource allocation, and research priorities. While anticipating risks is important across the biosecurity continuum, in this chapter, we focus on the pre-border setting, although methods for assessing risks can be applied at different stages of biosecurity systems.

RISK ASSESSMENT AND RISK ANALYSIS

Risk assessment refers to the scientific part of a risk analysis, where experimental and other data are used to reach a conclusion about the likelihood and consequence of a threat (also called an adverse event or hazard). In the biosecurity context, the threat will usually be the introduction into a country or region of a harmful pest, disease, or weed—also known as biosecurity risk material (BRM).

BOX 3.1. A SIMPLE ESTIMATION OF EXPECTED CONSEQUENCES

The conventions of risk assessment encourage decision making based on the "expected consequence" of a threat, defined as risk = likelihood × consequence. Likelihood is the chance of an event occurring, such as seeds of an exotic weed attached to a traveller's shoes passing through the border. Consequence is the potential magnitude of an impact if the event occurs, such as the economic costs of the weed establishing, the extent of its spread, or the number of native species that could be threatened. For example, for a species assessed to have an invasion likelihood of 0.30 (i.e. it has a 30% chance of invading a novel environment within a given timeframe) and a cost of $60M from potential invasion, the expected consequence is calculated as 0.30 × $60M = $18M.

A key challenge in anticipating biosecurity risks is that it is difficult to judge the likelihood of events that are yet to occur. This difficulty can overwhelm decision makers, who may choose to only focus on the severity of impacts and not consider their likelihood. Ignoring the likelihood of a threat means decisions are based on worst-case outcomes: when ranking scenarios whose risks (i.e. expected consequences) are equal, most people overestimate the risk of low-likelihood causes of death (e.g. living near a nuclear reactor) and underestimate high-likelihood ones (e.g. riding a bicycle; Burgman 2005). This can lead to resources being wasted on severe threats that are not very likely, at the expense of threats that are less severe but more common.

Broadly, risk assessment involves estimating: (1) the likelihood of entry, establishment, and spread of a threat and (2) its economic, social, and/or environmental consequences were entry, establishment, and spread to occur. Risk is formally defined as the "expected consequence" of a threat, expressed as the product of the likelihood and consequence of a threat (risk = likelihood × consequence; Box 3.1).

Biosecurity agencies aim to identify and reduce, eliminate, or avoid the risks associated with biosecurity threats. During risk analysis (Figure 3.1), biosecurity agencies generally follow a process of:

- *Threat identification*: identifying the hazard (e.g. the specific organism) and its potential adverse impacts.
- *Risk assessment*: estimating the risks (expected consequences) associated with a threat.

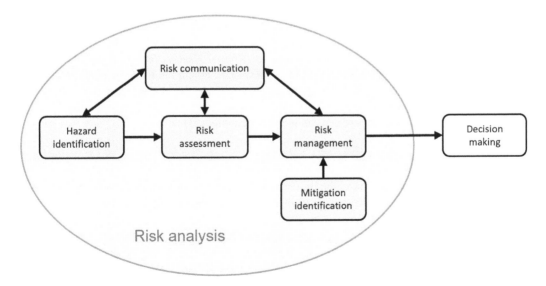

FIGURE 3.1   Components of risk analysis. Definitions of risk analysis vary among sources, but generally encompass the components listed here (Refer FSANZ 2013; Griffin 2012a; IPPC 2019a; WOAH 2022b).

- *Risk management*: deciding which measures to implement to mitigate risks.
- *Risk communication*: communicating with interested and/or impacted parties about risks and mitigation measures.
- *Decision making*: where political, social, and other factors are considered along with risk analysis results to inform decisions.

RISK ANALYSIS AND INTERNATIONAL TRADE

Under the World Trade Organisation (WTO) Agreement on the Application of Sanitary and Phytosanitary Measures (SPS Agreement), signatory countries are entitled to set an appropriate level of protection (ALOP) to protect their economy, society, and environment from risks posed by pests and diseases. When biosecurity risks associated with international trade are identified, countries may impose measures to reduce risks, but need to do so in accordance with WTO rules (see Chapter 2. Biosecurity Systems and International Regulations for a more detailed exploration). Underestimating biosecurity risk may lead to a jurisdiction exceeding its ALOP and incurring damages to domestic production and costs from controlling and eradicating incursions. On the other hand, overestimating risk may unnecessarily restrict trade, breach WTO rules, and limit social welfare gains both domestically and internationally (i.e. the benefits to the global economy of entering into trade agreements; Cook and Fraser 2008). In some cases, the stakes are very high if the import is highly valuable or when biosecurity decisions are under high scrutiny (e.g. contested views on the import of New Zealand apples to Australia; see Higgins and Dibden 2011).

Biosecurity decisions rely on appropriate evaluations of expected consequences to achieve biosecurity objectives. Specifically, a risk assessment may be required when:

- A new pest is identified on an import pathway.
- The susceptibility of a plant or animal to a pest or disease has (or is suspected to have) changed.
- A trading partner plans to start exporting a commodity to an importing country.
- Existing phytosanitary or sanitary measures or policies are being reviewed.
- Domestic programmes are implemented in response to an outbreak.
- Horizon scanning is undertaken to identify new or emerging biosecurity threats.
- Determining whether a particular organism is a pest.

Guidelines for risk analysis differ depending on whether the risks relate to trade in animals and animal products or to trade in plants and plant products. Animal disease guidelines for import risk analysis (IRA) are developed by the World Organisation for Animal Health (WOAH, founded as OIE), and plant guidelines for pest risk analysis (PRA) are developed by the International Plant Protection Convention (IPPC). For both plants and animals, member countries are able to use both *quantitative* and *qualitative* risk assessments to identify potential risks under the SPS Agreement—a risk assessment does not necessarily have to arrive at a numerical (quantitative) result.

RISK ASSESSMENT FOR ANIMALS AND ANIMAL PRODUCTS

The focus of risk assessments around the imports of animals and animal products is on diseases. Biosecurity agencies are tasked with identifying which diseases pose the highest risks to their domestic economies based on:

- The species or product being imported.
- Whether the disease is present in the exporting and/or importing country.
- Available veterinary services.
- Surveillance and control programmes in the exporting country.

Prior to undertaking any risk assessment, threat identification involves identifying the causative agent(s) (i.e. pathogen(s)) of a disease that could generate adverse consequences when importing a commodity (Figure 3.1). For example, *Hendra henipavirus* is the causative agent of Hendra, a disease that can affect many mammal species (including horses and humans). Causative threats are classified dichotomously as threats or not. If no threats are identified within an import, a risk assessment is not conducted (WOAH 2022b). If a risk assessment is required, agencies follow the guidelines provided in the *Terrestrial Animal Health Code* (WOAH 2022b) or the *Aquatic Animal Health Code* (WOAH 2022a).

Risk assessments conducted under WOAH guidelines should:

- Accommodate a variety of commodities and multiple threats within commodities.
- Be based on the best available scientific information.
- Be transparent, particularly in the use of data and judgement.
- Document assumptions and uncertainty properly.
- Be amenable to updating when new information becomes available.

Depending on the threat, risk assessments may be qualitative or quantitative. For many diseases for which there is broad international agreement on risks, only a qualitative assessment is required. Both qualitative and quantitative risk assessments under the SPS Agreement rely on the following steps (WOAH 2022b).

Entry Assessment

For a threat to materialise, there must be a potential pathway for exposure—a series of events that must occur, each with a probability of occurrence (Griffin 2012b). Entry assessments identify the pathways necessary for the entry of a threat, the probability of a threat occurring at different stages of a pathway, and any actions that may affect this probability of entry. Entry assessments for animal diseases typically rely on information about biological factors (e.g. species, treatment methods, and diagnostic testing), country factors (e.g. incidence or prevalence), and commodity factors (e.g. quantities imported, processing, and contamination). If the entry assessment does not find a significant risk, the risk assessment does not proceed further (WOAH 2022b).

Exposure Assessment

Exposure assessment involves describing how animals and humans in the importing country might become exposed to a disease and the probability of this occurring. Information about properties of the disease, importing country factors (e.g. presence of vectors, human and animal demographics, and environmental characteristics), and commodity factors (e.g. quantities to be imported and intended use) are used to assess exposure.

Consequence Assessment

A consequence assessment describes the potential consequences of exposure. Consequences may be direct (e.g. animal infection, production losses, and public health consequences), indirect (e.g. compensation costs, potential trade losses, and environmental impacts), and/or related to biosecurity interventions (e.g. surveillance and control costs). More details on consequence assessment are provided later in the chapter.

Risk Estimation

In this step, the results of the entry, exposure, and consequence assessments are combined to produce an overall measure of the risk associated with a threat. Risk estimation may be qualitative or quantitative. In quantitative estimation, various outputs may be provided, including estimated numbers of herds, animals, or people likely to experience health impacts over time, probability distributions or confidence intervals to express uncertainty, and sensitivity analyses. Following risk estimation, risks may be prevented, reduced, or mitigated using risk management (WOAH 2022b).

RISK ASSESSMENT FOR PLANTS AND PLANT PRODUCTS

The term "pest" refers to any organism that is injurious to plants or plant health, including plant pathogens (fungi, viruses, bacteria, and nematodes), arthropods (including insects), other types of invertebrates (e.g. molluscs), and weeds (Devorshak 2012a; IPPC 2022). The steps involved in risk assessment for plant pests are described in "international standards for phytosanitary measures" (ISPMs) provided by the IPPC, the international body recognised under the SPS Agreement as responsible for setting phytosanitary measures related to plant health and protection (see Chapter 2. Biosecurity Systems and International Regulations).

For international trade in plants and plant products, a PRA involves evaluating the probability of the introduction and spread of a pest and the magnitude of the potential economic consequences of this introduction. Two ISPMs developed by the IPPC relate specifically to risk assessment: ISPM No 2. (*Framework for pest risk analysis*; IPPC 2019a) and ISPM No. 11 (*Pest Risk Analysis for Quarantine Pests*; IPPC 2019b). Typically, data are qualitative, and a narrative approach or a risk-rating system is used to describe risk (see Box 3.3). In some cases, quantitative data may be available (e.g. on the prevalence of the pest in the country of origin or its ability to survive treatment). Regardless of approach, the assessment must meet the requirements of the SPS Agreement: it must be transparent, apply the criteria used in the analysis consistently, the evidence should be sound and scientifically robust, and the analysis clearly documented (Nunn 2012).

The *initiation stage* of the risk assessment identifies the pest(s) and pathway(s) that may be considered for pest risk assessment and the scope of the PRA, including its area of focus (Devorshak and Neeley 2012; see Australian Government Department of Agriculture 2019).

Pest Categorisation

Once a risk assessment is deemed necessary, the first step is to determine whether the pest meets the criteria for a *quarantine pest* (a pest that does not already occur in a country or region, but may cause economic harm if it did) or a *regulated non-quarantine pest* (pests that affect plants for planting and have an unacceptable economic impact; IPPC 2022). This requires examining evidence about whether the pest already occurs in the PRA area, whether it could survive in the area, and whether it has potential to cause economic harm (Devorshak and Neeley 2012).

Pest categorisation is summarised via the following elements (IPPC 2019b):

- Identity of the pest.
- Presence or absence in the PRA area.
- Regulatory status.
- Potential for establishment and spread in the PRA area.
- Potential for economic consequences (including environmental consequences) in the PRA area.

If no harm is likely or if the pest could not survive, the PRA concludes here. Otherwise, identified quarantine pests progress through the risk assessment process. A useful example of the pest categorisation process is provided in the *Final pest risk analysis for brown marmorated stink bug* (*Halyomorpha halys*) undertaken by the Australian Government Department of Agriculture (2019).

Probability of Entry, Establishment, and Spread

The process involved in assessing introduction (entry and establishment) and spread is outlined in ISPM 11 (IPPC 2019b). For a pest to be introduced and establish in a new area, a sequence of events must occur with a reasonable probability (Devorshak and Neeley 2012). Visual descriptions of export pathways, as part of a systems approach, can be used to depict the events that could cause introduction and spread to take place. ISPM 11 provides a detailed description of the elements that should be analysed in determining the probability of introduction and spread (IPPC 2019b). These include the probabilities that the pest:

- Occurs in a life stage that would be associated with the commodity.
- Survives transport or storage.
- Escapes detection.
- Survives existing pest management procedures.
- Finds a favourable location and/or suitable host material in the PRA area.
- Is able to reproduce and spread.

Potential Economic Consequences

Under ISPM 11, the potential "economic consequences" of pests may impact agriculture, the environment, or society. Impacts may be *direct* (e.g. impacts on plant life or health, or other aspects of the environment) or *indirect* (e.g. via effects on feasibility and cost of eradication or containment; domestic trade, international trade, or tourism). Understanding potential impacts requires selecting a time horizon and estimating the spread potential of the pest over that time period (Devorshak and Neeley 2012). More details of impact assessment are given below.

QUALITATIVE AND SEMI-QUANTITATIVE RISK ASSESSMENT PROTOCOLS

When data are unavailable or when risks are well known, qualitative risk assessments may be warranted (Devorshak 2012b). In qualitative risk assessments, subjective information and judgements based on evidence and expert judgement are used to describe risk (Box 3.2). Qualitative methods include rating and scoring systems, which often become "semi-quantitative" when ratings are converted to numerical values.

BOX 3.2. PEST RISK ANALYSIS IN AUSTRALIA

Australia's guidelines for undertaking biosecurity import risk analysis (BIRA) provide a formal methodology for characterising biosecurity risks that may be associated with the importation of goods into Australia (Biosecurity Australia 2009). Within PRA in Australia, risk is assessed in qualitative terms using a risk assessment matrix. The risk assessment matrix combines the likelihood of a disease or pest entering, establishing, and spreading in Australia with its potential consequences, should those events occur. In the risk assessment matrix, risk is the product of likelihood and consequence, both of which are described using six-point ordinal classes (Table 3.1). Likelihood is the product of judgements of the probability of entry, establishment, and spread, expressed as ordinal classes on the vertical axis. A corresponding numerical scale is given to avoid biased interpretations of the language (e.g. a "negligible" likelihood corresponds to a probability from 0 to 10^{-6}; Table 3.2).

A plant commodity is allowed for import if it satisfies Australia's ALOP, equivalent to a verbal descriptor of "very low" risk. The Australian Government Department of Agriculture, Fisheries and Forestry also uses this matrix to determine whether risk management measures are required to achieve ALOP. To reflect Australia's approach to biosecurity risk, the risk assessment matrix is not symmetrical. A "low" likelihood combined with "high" consequences is not the same as a "high" likelihood combined with "low" consequences (Table 3.1).

TABLE 3.1

The Risk Estimation Matrix Used for Pest Risk Analysis (PRA) in Australia.

		Negligible	Very low	Low	Moderate	High	Extreme
	High	Negligible risk	Very low risk	Low risk	Moderate risk	High risk	Extreme risk
	Moderate	Negligible risk	Very low risk	Low risk	Moderate risk	High risk	Extreme risk
	Low	Negligible risk	Negligible risk	Very low risk	Low risk	Moderate risk	High risk
	Very low	Negligible risk	Negligible risk	Negligible risk	Very low risk	Low risk	Moderate risk
	Extremely low	Negligible risk	Negligible risk	Negligible risk	Negligible risk	Very low risk	Low risk
	Negligible	Negligible risk	Negligible risk	Negligible risk	Negligible risk	Negligible risk	Very low risk

Likelihood of entry and exposure (vertical axis)

Consequences of pest entry, establishment, and spread (horizontal axis)

Note: The upper limit of Australia's appropriate level of protection (ALOP) is a "Very Low" Risk.
Source: Taken from DAWR (2016).

TABLE 3.2

Verbal Descriptors of Likelihoods and Associated Probability Intervals Used by Australia in Pest Risk Analysis (PRA; DAWE 2021a).

Likelihood Descriptor	Probability Interval
High	$0.7 < \text{to} \leq 1$
Moderate	$0.3 < \text{to} \leq 0.7$
Low	$0.05 < \text{to} \leq 0.3$
Very low	$0.001 < \text{to} \leq 0.05$
Extremely low	$10^{-6} < \text{to} \leq 0.001$
Negligible	$0 < \text{to} \leq 10^{-6}$

Many qualitative and semi-quantitative methods have been developed (see Hester and Mayo 2021 for a review), including to inform quarantine screening for potential pests before entry into a jurisdiction (see Box 3.3 for a weed risk assessment protocol, and Bomford (2008) for vertebrate species), to combine likelihoods and consequences of introduction to justify risk mitigation measures under the SPS Agreement (Department of Agriculture 2014), to determine the extent of cost-sharing between public and private parties once a pest enters or establishes (GIA 2023; PHA 2021), or to prioritise the management of exotic species with potential environmental or economic impacts (Raphael et al. 2009).

COMMON ISSUES WITH QUALITATIVE AND SEMI-QUANTITATIVE RISK ASSESSMENT PROTOCOLS

Risk assessment protocols assist decision making by breaking down different aspects of a problem, and they encourage decision makers to be clear about their judgements and assumptions (Bedford and Cooke 2001). However, risk assessment protocols can vary in quality and may be prone to logical or implementation errors. Five limitations are commonly identified in qualitative and semi-quantitative risk assessment protocols—these are summarised below, along with rules of thumb to ensure these limitations are avoided.

Vague Formulation of the Risk Assessment Problem or Context

Issues with problem formulation can compromise predictions of expected consequence, for example in the weed risk assessment protocol, "weediness" is not clearly defined (Pheloung et al. 1999).

BOX 3.3. WEED RISK ASSESSMENT PROTOCOL

Pheloung, Williams, and Halloy (1999) developed a point-scoring protocol to inform border screening for taxa likely to become environmental or agricultural weeds in Australia, and the protocol has been used, with and without modification, in other countries (Daehler et al. 2004; He et al. 2018; Koop et al. 2012). The point-scoring protocol aims to produce a more objective risk assessment compared to asking experts in an unstructured manner. The protocol includes 49 questions in 8 categories covering biogeography, undesirable attributes, and the biology and ecology of the species (Table 3.3). Answers contribute points to an overall score, where a high score indicates that a species has a high risk of becoming a weed. If a species assessment exceeds a specified threshold, the species is denied entry.

The protocol includes questions on agricultural and environmental impacts based on the subjective judgement of experts, but within the protocol, the total number of points associated with agriculture and environment differ (perhaps representing a deliberate weighting in favour of environmental impacts or an arbitrary categorisation of cues). The protocol explicitly considers the logical relationship between "climate and distribution" and "weed elsewhere", such that species that are climatically suited to Australia AND have been recorded as environmental or agricultural weeds elsewhere are assigned higher scores. However, all other cues (e.g. undesirable traits, dispersal mechanisms, and persistence attributes) are simply added, implying an additive relationship rather than conditional or more complex relationships among cues. In the case of the weed risk assessment protocol, summation still seems to provide reasonable discrimination between weeds and non-weeds (Gordon et al. 2008).

TABLE 3.3
Weed Risk Assessment Questionnaire

Category	Question	Impact Type
1. Domestication/ cultivation	1.01 Is the species highly domesticated	A
	1.02 Has the species become naturalised where grown	C
	1.03 Does the species have weedy races	C
2. Climate and distribution	2.01 Species suited to Australian climates	C
	2.02 Quality of climate match data	
	2.03 Broad climate suitability (environmental versatility)	C
	2.04 Native or naturalised in regions with extended dry periods	C
	2.05 Does the species have a history of repeated introductions outside its natural range	
3. Weed elsewhere	3.01 Naturalised beyond native range	C
	3.02 Garden/amenity/disturbance weed	E
	3.03 Weed of agriculture/horticulture/forestry	A
	3.04 Environmental weed	E
	3.05 Congeneric weed	
4. Undesirable traits	4.01 Produces spines, thorns or burrs	A
	4.02 Allelopathic	C
	4.03 Parasitic	C
	4.04 Unpalatable to grazing animals	A
	4.05 Toxic to animals	C
	4.06 Host for recognised pests and pathogens	C
	4.07 Causes allergies or is otherwise toxic to humans	C
	4.08 Creates a fire hazard in natural ecosystems	E

(Continued)

TABLE 3.3 *(Continued)*

Weed Risk Assessment Questionnaire

Category	Question	Impact Type
	4.09 Is a shade tolerant plant at some stage of its life cycle	E
	4.10 Grows on infertile soils	E
	4.11 Climbing or smothering growth habit	E
	4.12 Forms dense thickets	E
5. Plant type	5.01 Aquatic	E
	5.02 Grass	C
	5.03 Nitrogen fixing woody plant	E
	5.04 Geophyte	C
6. Reproduction	6.01 Evidence of substantial reproductive failure in native habitat	C
	6.02 Produces viable seed	C
	6.03 Hybridises naturally	C
	6.04 Self-fertilisation	C
	6.05 Requires specialist pollinators	C
	6.06 Reproduction by vegetative propagation	C
	6.07 Minimum generative time	C
7. Dispersal mechanisms	7.01 Propagules likely to be dispersed unintentionally	A
	7.02 Propagules dispersed intentionally by people	C
	7.03 Propagules likely to disperse as a produce contaminant	A
	7.04 Propagules adapted to wind dispersal	C
	7.05 Propagules buoyant	E
	7.06 Propagules bird dispersed	E
	7.07 Propagules dispersed by other animals (externally)	C
	7.08 Propagules dispersed by other animals (internally)	C
8. Persistence attributes	8.01 Prolific seed production	C
	8.02 Evidence that a persistent propagule bank is formed (>1 year)	A
	8.03 Well controlled by herbicides	A
	8.04 Tolerates or benefits from mutilation, cultivation or fire	C
	8.05 Effective natural enemies present in Australia	E

Note: Questions Relevant to: A: agricultural impacts, E: environmental impacts, and C: agriculture and environment combined.

Source: Pheloung, Williams, and Halloy (1999).

If a risk assessment protocol does not specify the timeframe over which a threat is expected to occur, estimated likelihoods become uninterpretable. Estimates of expected consequence derived from protocols that do not specify the basic elements or context of the risk assessment are unlikely to be reliable for informing decision making.

Issues with Subjective Scales and Risk Matrices

Matrices based on subjective scales are commonly used in risk assessment (Table 3.1). Ordinal categories for likelihoods and consequences (defined by verbal descriptors and a corresponding numerical scale) are multiplied to calculate an overall risk score or risk rating. Risk matrices collapse a set of probabilistic judgements about complex events (e.g. probabilities, post-impact recovery times, eradication costs) into a limited number of ordinal categories, which may be vaguely defined. This loss of precision leads to range compression, where quantitatively different risks result in the same score and relative changes are lost (Hubbard 2020). The sharp thresholds delineating ordinal levels can lead to illogical scenarios—small changes in likelihood- and consequence-scale values can translate into large jumps in risk scores when multiplied, and risks with the same numerical expected consequence may be assigned different risk scores depending on the score on which likelihood and consequence estimates land.

Poor Estimations of Likelihood

By definition, likelihood is a necessary component of the estimation of risk, so any protocol that only estimates consequences (and not likelihoods) is in fact an impact assessment, not a risk assessment. When likelihoods are subjectively estimated, people (including experts) tend to overestimate the likelihood of rare events and underestimate the likelihood of common events (Burgman 2005). Using qualitative, verbal descriptors of likelihood (e.g. "negligible") can exacerbate this problem (Cox, Babayev, and Huber 2005).

Ignoring Uncertainty

Estimates of likelihood and consequence always involve uncertainty, but people often struggle with accounting for uncertainty or may find thinking about it uncomfortable. Not reporting uncertainty imposes a risk-neutral attitude on decision makers, when in fact a decision maker may prefer to consider worst-case outcomes (i.e. a risk-averse attitude) or the potential payoff a decision could yield (i.e. a risk-seeking attitude).

Vague Language

Linguistic uncertainty is the uncertainty that arises due to different use and interpretation of language (Regan, Colyvan, and Burgman 2002). Many protocols specify categorical or ordinal responses to questions or criteria (e.g. "low" or "very low"), but linguistic uncertainty can generate problems with interpretation. Where possible, category thresholds should be clearly defined or defined numerically (e.g. Table 3.2). Language may also influence decision making, in that a species described as "weedy" may be viewed as undesirable and may be more likely to be controlled or eradicated than a species described as "exotic".

CREATING SOUND QUALITATIVE AND SEMI-QUANTITATIVE RISK ASSESSMENTS

Qualitative or semi-quantitative methods are often chosen because they are relatively easy to apply (Devorshak 2012b), but to avoid the problems listed above, risk analysts need to ensure that:

- Judgements, criteria, and decisions are applied consistently.
- Assumptions are clearly documented.
- Uncertainty is clearly described and communicated.

Structured decision making can be used to design sound risk assessment protocols that meet these criteria (Walshe et al. 2012). Structured decision making is a systematic, deliberative process for making decisions based on both scientific facts and value judgements (Gregory et al. 2012).

QUANTITATIVE ESTIMATION OF LIKELIHOODS AND CONSEQUENCES

Quantitative methods use numerical values from empirical data or expert elicitation to provide an estimate of risk, elements of risk, and/or strategies to mitigate risk. Quantitative methods such as probabilistic scenario analyses (e.g. decision trees), Monte Carlo methods within simulation models, Frequentist or Bayesian models, and benefit-cost analyses may all be used in biosecurity risk assessments. For example, a biosecurity agency might use historical interception data to calculate the likelihood a pest enters a country, and then use a partial or full benefit-cost analysis (BCA) to understand the consequences of potential establishment and spread. Quantitative methods can be time- and resource-intensive—often requiring specialised expertise, data, and computing resources—and because of this, may not always be feasible (EFSA Panel on Plant Health 2012; Griffin 2012b).

QUANTITATIVE ESTIMATION OF LIKELIHOODS

Quantitatively estimating the likelihood of a threat entering a country or arriving at a point in the landscape relies on: (1) data (e.g. historical *interception data* from border inspections or data generated by expert opinion) and (2) a formula or model that can be used to quantitatively characterise the likelihood. Interception data can be used to model how an invasive species threat is likely to arrive (e.g. imported goods, people, wind) and to determine the likelihood of establishment in particular locations of interest (Camac et al. 2021). These likelihood models are commonly referred to as "pathway models" as they are often focused on estimating contamination rates occurring at particular points in the landscape (e.g. ports). These models can be incorporated with other data and models to develop maps of establishment potential (see Chapter 14. Map).

Biosecurity agencies are also interested in changes in the likelihood a biosecurity threat is still present on a pathway, in spite of the application of various treatments and processing activities along the import supply chain. For example, to reduce the likelihood that particular forest pests will enter Australia on imported timber products, various physical and/or chemical processing and/or treatments are required prior to arrival in Australia (see Chapter 4. Prevent).

A *scenario tree* allows an analyst to develop a quantitative estimate of the conditional likelihood a biosecurity threat (1) occurs on a consignment, (2) avoids border screening; (3) survives transport; and (4) comes into contact with a suitable environment or host material. Developing a linear event tree relies on an understanding of the import scenario, the initial volume and pest load, and an estimate of the effect of specific treatments or events on pest load (Griffin 2012b). Figure 3.2 shows a hypothetical scenario tree for an imported product and how likelihoods of particular pre-border

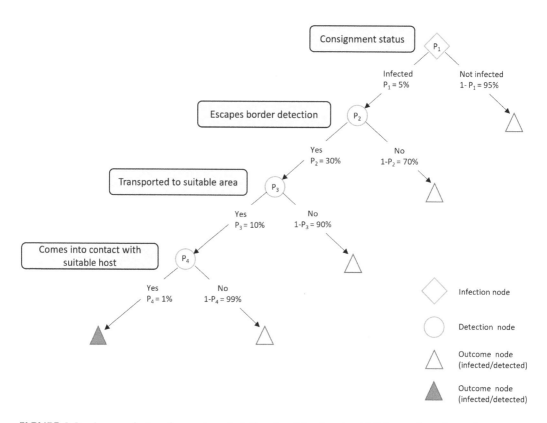

FIGURE 3.2 A scenario tree for use in calculating the risk of pest establishment in a domestic economy caused by imported fruit containing BRM. (Adapted from Arndt et al. 2022.)

phytosanitary measures on the pathway can be combined to predict the likelihood of an incursion event and the number of such events per year.

A starting point for a scenario tree might be the likelihood that a consignment, for example of oranges, is infected with a pest upon arrival at its import-country destination, P_1—the biosecurity agency has ascertained that this value is 5%. Assume also that there is a 30% chance that contaminated fruit will escape detection at the destination country's border (P_2), and 10% of this will go to areas deemed suitable for establishment (P_3), with a tiny portion (1%) of these coming into contact with a suitable host plant (P_4) (i.e. almost all of the oranges are consumed). The likelihood of an outbreak can be expressed as follows:

$$Likelihood = (P_1) \times (P_2) \times (P_3) \times (P_4)$$
$$= 0.05 \times 0.3 \times 0.1 \times 0.01$$
$$= 0.000015$$

If 50,000 oranges were imported during a particular time period, we could use the likelihood value to calculate that there would be 0.75 incursions per time period (0. 000015 × 50,000). The risks outlined in the above example could be mitigated by pre-border treatments (additional nodes added in Figure 3.2), or additional inspection effort at the border (to reduce the value of P_2). The type and level of intervention effort should be informed by the cost of risk-reduction options (see Chapter 9. Resource Allocation).

A mix of qualitative and quantitative data can be used to inform likelihoods at the various nodes in scenario trees. For example, interception data and or census data (data obtained from checking each item rather than taking a sample of items) are often available for understanding infection status of consignments and level of detection at the border (Chapter 5. Screen), while qualitative data from expert opinion might only be available to understand remaining nodes on the pathway.

Other relevant examples of scenario tree-usage in biosecurity include for understanding the likelihood a pest will arrive via multiple suitable pathways (Chapter 14. Map), and understanding domestic surveillance systems (Arndt et al. 2022).

Quantitative Estimation of Consequences

Calculating consequences of invasive species entry, establishment, and spread can be challenging where the monetary value of avoided damages are not easy to obtain. BCA is the standard method of evaluating the consequences of biosecurity threats should they establish (see Chapter 9. Resource Allocation). BCA involves comparing the costs of responding to an outbreak and benefits (avoided impacts) of managing the outbreak. Since impacts of a particular pest will accrue over time, a key part of a BCA is predicting the extent of each impact and the future time periods in which they are likely to occur. This means taking account of social, political, and climatic uncertainty that could affect impacts in the future, in addition to predicting the spread of the pest over time. Information on spread may not exist, so models will need to be developed if a quantitative approach is required. Using the spread models to derive impact will require an understanding of how the invasive species might cause damage.

Impacts can be categorised in different ways, but for the purposes of incorporating impacts into SPS-compliant risk assessments, impacts are often described as being direct or indirect. Direct impacts affect host plants or animals (e.g. animal infection, production losses, public health consequences, direct impacts on flora and fauna). Indirect impacts result from the presence of the pest and extend beyond the specific pest-host dynamics (e.g. via the loss of environmental amenity, changes in relative prices of inputs, compensation costs). Below, we discuss impacts in terms of the broad sectors affected: business activity, people and the environment. Ideally, all impacts would be amenable to calculation in dollar terms (monetised), but this may not be feasible for some impacts that affect people or the environment (Summerson, Hester, and Graham 2018). Where monetisation of impacts is not possible, impacts may be: (1) quantified but not monetised (e.g. the number of species threatened by the pest) or (2) described qualitatively (i.e. not quantified or monetised, the least preferred option).

Business Activity

Anticipating biosecurity risks is a key concern for agriculture, fisheries, and forestry industries. Consequences relating to production losses and management costs are readily measured in monetary units (e.g. dollars), which makes calculations possible using a range of techniques and methodologies (e.g. farm budgeting, mathematical models, partial equilibrium models, computable general equilibrium (CGE) models; Jones 2000; Welsh et al. 2021).

- **Farm budgeting** for example, gross margin and whole-farm budgets—provide a very simple approach to understanding the impact of a pest, disease, or weed on a particular agricultural enterprise. Gross margin budgets could be used to show the impact of an outbreak on input costs (e.g. chemicals might be purchased to control a weed), yield (a pest outbreak may reduce the amount of fruit produced) or product quality (despite control, a pest may have left fruit with some lasting visual damage which results in a lowered unit price). Whole farm budgets could be used to show farm-wide impacts of pest, diseases, or weeds. Budgeting approaches are often the building blocks for more complex economic analysis (e.g. Jones and Vere 1998).
- **BCA** involves identifying all the costs and benefits from controlling a pest, disease, or weed and valuing them. Because benefits and costs accrue at different points in time they must be compared in "present value" terms via a process known as discounting (Chapter 9. Resource allocation). Useful metrics derived via BCA include net present value and benefit-cost ratio, which can be used to compare and rank alternative control scenarios. BCA can be incorporated into techniques used to understand impact.
- **Simulation models** are simplified representations of natural systems and can be used to understand pest, disease, and weed spread overtime, and the performance of various strategies used to manage the spread. Bioeconomic models consist of a biological model describing a production system and an economic model relating the production system to prices and resource constraints (see Cacho 1997) and can be used to understand impact of pests, disease, and weeds in a range of settings (e.g. Cacho et al. 2008; Choquenot and Hone 2000).
- **Partial equilibrium models** analyse relationships within a particular market or sector of the economy in isolation from the broader national economy. These models provide an estimate of the change in economic welfare due to, for example a pest incursion (Sinden et al. 2004; Soliman et al. 2012; Welsh et al. 2021). Partial equilibrium approaches can be employed where the impacts of a pest are likely to be contained within a sector that does not have significant linkages to other sectors (e.g. horticulture or floriculture; Cook et al. 2012). For larger agricultural industries (e.g. grain crops), several other industries may be reliant on a product impacted by a pest, including other agricultural industries that use those products as inputs, wholesalers, packaging and distribution firms, retailers, and exporters.
- **CGE models** account for flow-on effects to the rest of the economy and may be more appropriate for pests and diseases whose consequences are unlikely to be contained within a specific industry. If prevention measures are modelled, indirect or flow-on benefits accrue to "downstream" industries. CGE models incorporate all available economic sectors, can reflect responses over a long time horizon, but can be difficult to build, use and validate due to requirements for comprehensive databases of economic activity by sector (Welsh et al. 2021). Specialised CGE software packages exist (e.g. Global Trade Analysis Project (GTAP) model; Hertel 1997) or analysts can use mathematical programming software

to build and implement CGE models. Examples of CGE models in biosecurity contexts include McDermott, Finnoff, and Shogren (2013) and Do and Vanzetti (2019).

Environmental Impacts

Monetising the impacts of biosecurity threats on the environment is usually more difficult than monetising impacts on business activity, as environmental goods and services are typically not traded in markets, and their value is not easily observable. The Total Economic Value (TEV) framework (Barbier, Acreman, and Knowler 1997; Emerton and Howard 2008; Pearce 1993; Tait and Rutherford 2018) is commonly used where monetisation is not straightforward (see Summerson et al. 2018 for more details). Under the TEV approach, environmental assets (goods and services) are classified into use and non-use values to apply non-market valuation methods (NMV). In these cases, rather than directly observing an individual's willingness to pay for more of the environmental good or service, it becomes necessary to estimate a good's value indirectly.

Two broad groups of NMV methods have been developed for this purpose:

- **Stated preference techniques** involve directly surveying people's hypothetical behaviour in constructed markets for the environmental good or service in question (OBPR 2020). The two main stated preference techniques are *choice modelling* and the *contingent valuation*. Using contingent valuation, people are asked directly about their willingness to pay for improvements in particular environmental goods and services. This involves asking people via a survey, to compare a business-as-usual scenario (no extra cost) with an improvement scenario (extra payment) (see, e.g. Rolfe and Windle 2017). In choice modelling studies, respondents are presented with a number of alternatives (including a business as usual option) and asked to choose between them. Choice modelling differs from contingent valuation in that it describes the situation of interest in terms of attributes; it varies improvement options over different levels; and respondents are asked to complete a series of trade-offs via surveys (Rolfe and Windle 2014).
- **Revealed preference techniques** (also known as surrogate market approaches) seek to elicit peoples' willingness to pay for a good or service by observing their actual behaviour in real, related markets (OBPR 2020). Revealed preference techniques are used to value impacts on non-extractive, direct uses of the environment, such as tourism, recreation, and aesthetic or cultural impacts. These methods infer value of the environment from observed behaviour and market interactions, where market "proxies" provide information on the value of a non-market good (OBPR 2020). The two main revealed preference techniques are the *travel cost method* (typically used for recreation impacts) and *hedonic pricing* (typically used for housing and lifestyle impacts). The travel cost method involves collecting data on the costs incurred by each individual in travelling to a recreational site or amenity, and is based on the premise that the time and travel expenses incurred to visit a particular site represent people's willingness to pay. When recreational sites are impacted by a biosecurity threat, the travel cost method can be used to understand the changes in tourism and recreational earnings that make the site less attractive to visitors (Emerton and Howard 2008). The hedonic pricing method also involves collection of market prices (in this case, property prices) and in the biosecurity context these are used to examine differences in property prices between locations with different environmental qualities or landscape values (Emerton and Howard 2008). Emerton and Howard (2008) describe in detail the steps involved in applying travel cost methods and hedonic pricing techniques to invasive species outbreaks.

Non-market valuation techniques are not simple to apply and in many cases the time and resources required for primary data collection may not be available in a biosecurity context. Often the only appropriate course of action to value environmental impacts is *benefit transfer,* where estimates of non-market values for one study site are transferred to a comparable study site (Tait and Rutherford 2018). If undertaking benefit-transfer appears to be an option, an economist with expertise in this process should assist with this task. Examples where benefit transfer has been used to value non-market impacts of invasive species include Cacho, Hester, and Tait (2021) and Dodd et al. (2020).

People

Decision makers increasingly recognise the importance of considering the impacts of pests and diseases on people, specifically on human health (e.g. via stinging, poisoning, and disease), infrastructure (e.g. buildings, power plants, and electrical grids), and social and cultural amenity. Social impact assessment (Vanclay 2002) offers an overarching framework within which to understand various social and cultural consequences of invasive species (Crowley, Hinchliffe, and McDonald 2017).

Where invasive species cause negative impacts on human health, the value of these impacts can be calculated using information on average medical costs involved in treatment, the number of days off work required for recovery, and average wages (Summerson et al. 2018). When an invasive species damages infrastructure, consequences can be calculated as the expenditure required to refurbish the infrastructure to a pre-invasion standard. There may also be losses incurred as a result of shutdowns required for clean-up and repair of equipment. Connelly et al. (2007) list 12 expenditure categories for controlling zebra mussel in drinking water treatment facilities, including: lost production and revenues, prevention efforts, construction of new intakes, physical removal of mussel accumulations, and/or chemical treatments of affected intake components. Calculation of consequences on infrastructure is reasonably straightforward because market prices can be used (e.g. labour costs, equipment purchase costs, revenue lost).

Social amenity refers to the values associated with natural environments (e.g. parks, reserves, and ecosystems). Impacts on social amenity are difficult to quantify when it comes to estimating expected consequences, but non-market valuation techniques can also be applied in the social amenity context. For example, in the case of an invasive algae that has covered a beach normally used for recreation, revealed preference techniques would be an appropriate valuation tool because recreation is a direct, non-extractive use. Where resources are unavailable to undertake the primary research required for non-market valuation, benefit transfer may also be used, as discussed above.

Cultural amenity refers to cultural sites, customs and practices, heritage, and a sense of place. Cultural assets can be material, emotional, or spiritual, such as special tracts of the natural environment that encourage particular types of cultural activities (Rolfe and Windle 2003). For example, Australia's Aboriginal and Torres Strait Islander peoples maintain strong cultural, social and spiritual values with their land and sea country, many of which have been impacted by invasive species (Bangalang et al. 2022). Traditional approaches to valuing these impacts, which consider only the one-way flow of ecosystems' benefits to people, are fundamentally incompatible with Indigenous values, concepts and relationships with Country (Jarvis et al. 2022). A promising approach is to work with Indigenous peoples to develop an alternative, parallel "valuation" model and a generic process and set of indicators to enable cultural values, concepts, and connections to be accounted for (e.g. Jarvis et al. 2022).

DISCUSSION

Accurately anticipating biosecurity risks is crucial to protecting countries and regions from the damage caused by incursions of pests, diseases, and weeds. National biosecurity agencies put substantial resources into anticipating risks by developing import conditions based on the result of risk assessments and undertaking border inspections to check that importers have adhered to import conditions. Anticipating risks accurately is also crucial for efficiently allocating resources to mitigate and respond to risks (see Chapter 9. Resource allocation and Chapter 7. Prepare, Respond, and Recover).

Typically, when a country is known to have a biosecurity threat, imports from that country are the subject of higher rates of border inspection and may have to meet additional pre-border obligations for treatment and processing. While this approach can work well for biosecurity threats that spread slowly, it is problematic for emerging threats that spread quickly and that are not immediately detected in exporting countries. For such threats, incorporating information about exposure risk for different trading countries based on border interception data is likely to improve the accuracy of risk assessments. Camac et al. (2021) outline a method that integrates border interceptions, international trade flow data, pest occurrence records, and climate suitability models to estimate the exposure risk of potential and current trading partners obtaining an established population of a new high threat pest or disease as a function of what they import and from whom. The method is designed to be implemented by biosecurity practitioners and is illustrated using the brown marmorated stink bug (*H. halys*; a "hitchhiker" pest; i.e. a pest that is transported with non-commodity goods) and Australian interception data (Camac et al. 2021). Results highlight which countries are most exposed to the incursion and establishment of brown marmorated stink bug, and vulnerable tariff codes that are likely to introduce the greatest number of the hitchhiker pest into Australia. Coupled with estimated consequences from a brown marmorated stink bug incursion, information on the likelihood of entry can then be used to prioritise resources at the border (Chapter 9. Resource Allocation).

Biosecurity agency staff must also be aware of unexpected biosecurity risks—those that relate to the presence of any BRM on a good, conveyance, or pathway upon which it is not reasonably expected or known to occur (Hoffmann and Robinson 2019). Potential unexpected biosecurity risks include: "hitchhiker" pests (i.e. pests associated with non-commodity goods); goods or people "exposed" to contamination, infestation or infection; risks from strategic behaviour (e.g. fraudulent documents); and changing and emerging risks, for example, goods from zones where a change in pest status has yet to be detected, or diseases that have not been found in a region or are yet to be scientifically classified.

Anticipating risks should not be regarded as a "set and forget" process. Models can assist with forecasting changing and emerging risks, but their results should be used in combination with other information, including information derived from horizon scanning and intelligence-related activities. Open source intelligence tools such as Epiwatch (EPIWATCH 2023), EIOS (Epidemic Intelligence from Open Sources; WHO 2023), and IntelliRiver (Intelliriver Systems 2023) are useful in this regard. The online system IntelliRiver (originally IBIS; International Biosecurity Intelligence System) tracks and forecasts terrestrial animal, aquatic animal, and plant diseases (Grossel, Lyon, and Nunn 2017). Useful information on emerging risks might be held by different departments within a biosecurity agency, in different biosecurity-focused agencies within a country, or in international fora. Having the tools and capacity required to ensure awareness of changing and emerging risks can therefore be challenging for biosecurity agencies (Hoffmann and Robinson 2019), and as such, it is important that biosecurity agencies are open to collaboration and knowledge sharing (see Chapter 11. Research Uptake).

Risks from strategic behaviour should not be overlooked by biosecurity agencies. Strategic behaviour occurs where entities make choices strategically to meet their own objectives, and these choices might not necessarily align with the regulator's objective. In a biosecurity context, such risks emerge whenever a biosecurity agency delegates tasks to a third party (e.g. a Competent Authority, National Plant Protection Organisation, or third-party certifier) or when incentives inherent in biosecurity rules have not been considered in their design (Campbell et al. 2021). Economic theories such as contract theory and incentive theory provide methods for identifying risks associated with strategic behaviour and tools for developing incentive-compatible rules and regulations (see Chapter 8. Incentives).

Indeed, taking a multidisciplinary approach to anticipating biosecurity risks improves the efficiency and effectiveness with which risk assessment is carried out, with flow-on benefits for all the other biosecurity activities that rely on accurate and timely anticipation of risks.

IN A NUTSHELL

- ALOP is a political construct, risk estimation is a scientific process.
- Both qualitative and quantitative assessments are allowed under the SPS Agreement.
- Qualitative decision-support protocols can be limited by vague formulation of the decision problem, poor estimation of likelihood, reluctance to include uncertainty, and vague use of language in regard to estimating risk.
- Multiple quantitative techniques are available to estimate biosecurity consequences for primary production, social amenity, and the environment.

REFERENCES

Arndt, E., L. Rumpff, S. Lane, S. Bau, M. Mebalds, and T. Kompas. 2022. "Estimating Probability of Visual Detection of Exotic Pests and Diseases in the Grains Industry—An Expert Elicitation Approach." *Frontiers in Ecology and Evolution* 10. https://doi.org/10.3389/fevo.2022.968436.

Bangalang, N.-g., J. Nadji, A. Nayinggul, S. Nadji, A. Nayinggul, S. Dempsey, K. Mangiru, J. Dempsey, S. McCartney, J. Mairi Macdonald, and C. J. Robinson. 2022. "Understanding Indigenous Values and Priorities for Wetlands to Guide Weed Management Actions: Lessons from the Nardab Floodplain in Northern Australia's Kakadu National Park." *Ecological Management & Restoration* 23 (S1):105–116. https://doi.org/10.1111/emr.12542.

Barbier, E. B., M. Acreman, and D. Knowler. 1997. *Economic Valuation of Wetlands: A Guide for Policy Makers and Planners.* Gland, Switzerland: Ramsar Convention Bureau.

Bedford, T., and R. Cooke. 2001. *Probabilistic Risk Analysis: Foundations and Methods.* Cambridge: Cambridge University Press.

Biosecurity Australia. 2009. *Final Import Risk Analysis Report for Fresh Unshu Mandarin Fruit from Shizuoka Prefecture in Japan.* Canberra: Biosecurity Australia.

Bomford, M. 2008. *Risk Assessment Models for Establishment of Exotic Vertebrates in Australia and New Zealand: A Report Produced for the Invasive Animals Cooperative Research Centre.* Canberra: Invasive Animals Cooperative Research Centre.

Burgman, M. 2005. "Values, History and Perception." In *Risks and Decisions for Conservation and Environmental Management*, edited by Mark Burgman, 1–25. Cambridge: Cambridge University Press.

Cacho, O. J. 1997. "Systems Modelling and Bioeconomic Modelling in Aquaculture." *Aquaculture Economics & Management* 1 (1-2):45–64. https://doi.org/10.1080/13657309709380202.

Cacho, O., S. M. Hester, and P. Tait. 2021. *Re-Evaluating Management of Established Pests Including the European Wasp, Vespula germanica Using Biocontrol Agents.* Centre of Excellence for Biosecurity Risk Analysis.

Cacho, O. J., R. M. Wise, S. M. Hester, and J. A. Sinden. 2008. "Bioeconomic Modeling for Control of Weeds in Natural Environments." *Ecological Economics* 65 (3):559–568. https://doi.org/10.1016/j.ecolecon.2007.08.006.

Camac, J., J. Baumgartner, A. Robinson, and T. Kompas. 2021. Estimating trading partner exposure risk to new pests or diseases. Technical report for CEBRA project 190606. Centre of Excellence for Biosecurity Risk Analysis.

Campbell, A., F. Mody, A. Mooney, J. Whyte, and S. Hester. 2021. *Increasing Confidence in Pre-border Risk Management.* Centre of Excellence for Biosecurity Risk Analysis.

Choquenot, D., and J. Hone. 2000. "Using bioeconomic models to maximize benefits from vertebrate pest control: Lamb predation by feral pigs." Human conflicts with wildlife: economic considerations. Proceedings of the Third NWRC Special Symposium, Fort Collins, CO.

Connelly, N. A., C. R. O'Neill, B. A. Knuth, and T. L. Brown. 2007. "Economic Impacts of zebra Mussels on Drinking Water Treatment and Electric Power Generation Facilities." *Environmental Management* 40 (1):105–112. https://doi.org/10.1007/s00267-006-0296-5.

Constable, F., G. Kelly, and D. Dall. 2021. "Viruses in Cucurbit Seeds from on-Line Mail-Order Providers." *Australasian Plant Disease Notes* 16 (1):10. https://doi.org/10.1007/s13314-021-00423-1.

Cook, D. C., and R. W. Fraser. 2008. "Trade and Invasive Species Risk Mitigation: Reconciling WTO Compliance With Maximising the Gains from Trade." *Food Policy* 33 (2):176–184. https://doi.org/10.1016/j.foodpol.2007.07.001.

Cook, D. C., G. Long, H. P. Possingham, L. Failing, M. Burgman, R. Gregory, R. Estevez, and T. Walshe. 2012. Potential methods and tools for estimating biosecurity consequences for primary production, amenity, and the environment. Australian Centre of Excellence for Risk Analysis.

Cox, L. A., D. Babayev, and W. Huber. 2005. "Some Limitations of Qualitative Risk Rating Systems." *Risk Analysis* 25 (3):651–662. https://doi.org/10.1111/j.1539-6924.2005.00615.x.

Crowley, S. L., S. Hinchliffe, and R. A. McDonald. 2017. "Invasive Species Management Will Benefit from Social Impact Assessment." *Journal of Applied Ecology* 54 (2):351–357. https://doi.org/10.1111/1365-2664.12817.

Daehler, C. C., J. S. Denslow, S. Ansari, and H.-C. Kuo. 2004. "A Risk-Assessment System for Screening Out Invasive Pest Plants from Hawaii and Other Pacific Islands." *Conservation Biology* 18 (2):360–368. https://doi.org/10.1111/j.1523-1739.2004.00066.x.

DAWE. 2021a. *Final Group Pest Risk Analysis for Soft and Hard Scale Insects on Fresh Fruit, Vegetable, Cut-Flower and Foliage Imports.* Canberra: Department of Agriculture, Water and the Environment.

DAWE. 2021b. National Environmental Biosecurity Response Agreement - 2021. edited by Department of Agriculture Water and the Environment. Canberra.

DAWR. 2016. *Biosecurity Import Risk Analysis Guidelines 2016: Managing Biosecurity Risks for Imports into Australia.* Canberra: Department of Agriculture and Water Resources.

Department of Agriculture. 2014. *Final Report for the non-Regulated Analysis of Existing Policy for Table Grapes from Japan.* Canberra: Australian Government Department of Agriculture.

Department of Agriculture. 2019. *Final Pest Risk Analysis for Brown Marmorated Stink Bug (Halyomorpha halys).* Canberra: Australian Government Department of Agriculture.

Devorshak, C. 2012a. "International Legal and Regulatory Framework for Risk Analysis." In *Plant Pest Risk Analysis: Concepts and Application*, edited by C. Devorshak, 29–42. Wallingford: CABI International.

Devorshak, C. 2012b. "Qualitative Methods." In *Plant Pest Risk Analysis: Concepts and Application*, edited by C. Devorshak, 97–117. Wallingford: CABI International.

Devorshak, C., and A. Neeley. 2012. "Pest Risk Assessment." In *Plant Pest Risk Analysis: Concepts and Application*, edited by C. Devorshak, 135–150. Wallingford: CABI International.

Dodd, A., N. Stoeckl, J. B. Baumgartner, and T. Kompas. 2020. Key Result Summary: Valuing Australia's Biosecurity System, CEBRA Project 170713. Centre of Excellence for Biosecurity Risk Analysis.

Do, H. L., and D. Vanzetti. 2019. "Possible Effects of the Recent Australian Ban on Prawn Imports." *Aquaculture Economics & Management* 23 (2):218–231. https://doi.org/10.1080/13657305.2018.1469682.

EFSA Panel on Plant Health. 2012. "Guidance on Methodology for Evaluation of the Effectiveness of Options for Reducing the Risk of Introduction and Spread of Organisms Harmful to Plant Health in the EU Territory." *EFSA Journal* 10 (6):2755. https://doi.org/10.2903/j.efsa.2012.2755.

Emerton, L., and G. Howard. 2008. *A Toolkit for the Economic Analysis of Invasive Species.* Nairobi: Global Invasive Species Programme.

EPIWATCH. 2023. "EPIWATCH." accessed 28 July 2023. https://www.epiwatch.org/.

FSANZ. 2013. *Risk Analysis in Food Regulation.* Canberra & Wellington: Food Standards Australia New Zealand.

GIA. 2023. "Government Industry Agreement (GIA)." accessed 3 February 2023. https://www.gia.org.nz/.

Gordon, D. R., D. A. Onderdonk, A. M. Fox, and R. K. Stocker. 2008. "Consistent Accuracy of the Australian Weed Risk Assessment System Across Varied Geographies." *Diversity and Distributions* 14 (2):234–242. https://doi.org/10.1111/j.1472-4642.2007.00460.x.

Gregory, R., L. Failing, M. Harstone, G. Long, T. McDaniels, and D. Ohlson. 2012. *Structured Decision Making: A Practical Guide to Environmental Management Choices.* Chichester, West Sussex; Hoboken, N.J: Wiley-Blackwell.

Griffin, R. 2012a. "Basic Concepts in Risk Analysis." In *Plant Pest Risk Analysis: Concepts and Application*, edited by C. Devorshak, 7–18. Wallingford: CABI International.

Griffin, R. 2012b. "Quantitative Methods." In *Plant Pest Risk Analysis: Concepts and Application*, edited by C. Devorshak, 119–134. Wallingford: CABI International.

Grossel, G., A. Lyon, and M. Nunn. 2017. "Open-Source Intelligence Gathering and Open-Analysis Intelligence for Biosecurity." In *Invasive Species: Risk Assessment and Management*, edited by Andrew P. Robinson, Mark A. Burgman, Mike Nunn and Terry Walshe, 84–92. Cambridge: Cambridge University Press.

Hertel, T. W., ed. 1997. *Global Trade Analysis: Modeling and Applications.* Cambridge Cambridge University Press.

Hester, S. M., and J. Mayo. 2021. Improving the methodology for rapid consequence assessment of amenity and environmental pests. Centre of Excellence for Biosecurity Risk Analysis.

He, S., L. Yin, J. Wen, and Y. Liang. 2018. "A Test of the Australian Weed Risk Assessment System in China." *Biological Invasions* 20 (8):2061–2076. https://doi.org/10.1007/s10530-018-1680-9.

Higgins, V., and J. Dibden. 2011. "Biosecurity, Trade Liberalisation, and the (anti)politics of Risk Analysis: The Australia-New Zealand Apples Dispute." *Environment and Planning A: Economy and Space* 43 (2):393–409. https://doi.org/10.1068/a43289·

Hoffmann, M., and A. Robinson. 2019. Identifying unexpected biosecurity risks. Centre of Excellence for Biosecurity Risk Analysis.

Hubbard, D. W. 2020. "Worse than Useless: The Most Popular Risk Assessment Method and Why It doesn't Work." In *The Failure of Risk Management: Why It's Broken and How to Fix It*, 163–192. Hoboken, New Jersey: Wiley.

Intelliriver Systems. 2023. "Intelliriver Source." accessed 28 July 2023. https://ibis.intelliriver.systems/source/home.

IPPC. 2019a. International standard for phytosanitary measures. ISPM No. 2: Framework for pest risk analysis. Food and Agriculture Organization of the United Nations (FAO).

IPPC. 2019b. International standard for phytosanitary measures. ISPM No. 11: Pest risk analysis for quarantine pests. Food and Agriculture Organization of the United Nations (FAO).

IPPC. 2022. International standard for phytosanitary measures. ISPM No. 5: Glossary of phytosanitary terms. Food and Agriculture Organization of the United Nations (FAO).

Jarvis, D., N. Stoeckl, M. Douglas, D. Grainger, S. Larson, G. Finau, A. Larson, R. Barrowei, B. Coleman, D. Groves, J. Hunter, M. Lee, and M. Markham. 2022. *Valuing Indigenous Cultural Connections*. Cairns: James Cook University.

Jones, R. E. 2000. "The Economics of Weed Control." In *Australian Weed Management Systems*, edited by B. M. Sindel, 267–283. Meredith, Australia: R.G. and F.J. Richardson.

Jones, R. E., and D. T. Vere. 1998. "The Economics of Serrated Tussock in New South Wales." *Plant Protection Quarterly* 13 (2): 70–76.

Koop, A. L., L. Fowler, L. P. Newton, and B. P. Caton. 2012. "Development and Validation of a Weed Screening Tool for the United States." *Biological Invasions* 14 (2):273–294. https://doi.org/10.1007/s10530-011-0061-4.

Leung, B., M. R. Springborn, J. A. Turner, and E. G. Brockerhoff. 2014. "Pathway-Level Risk Analysis: The Net Present Value of an Invasive Species Policy in the US." *Frontiers in Ecology and the Environment* 12 (5):273–279. https://doi.org/10.1890/130311.

McDermott, S. M., D. C. Finnoff, and J. F. Shogren. 2013. "The Welfare Impacts of an Invasive Species: Endogenous vs. Exogenous Price Models." *Ecological Economics* 85:43–49. https://doi.org/10.1016/j.ecolecon.2012.08.020.

Nunn, M. J. 2012. "The Analytical Foundation of Quarantine Risk Analysis." In *The Economics of Quarantine and the SPS Agreement*, edited by Cheryl McRae, David Wilson and Kym Anderson, 29–52. The University of Adelaide Press.

OBPR. 2020. *Cost–benefit Analysis*. Canberra: Office of Best Practice Regulation.

Pearce, D. W. 1993. *Economic Values and the Natural World*. Cambridge, MA: The MIT Press.

PHA. 2021. "Pest categorisation process." accessed 1 March 2021. https://www.planthealthaustralia.com.au/biosecurity/emergency-plant-pests/pest-categorisation-process/.

Pheloung, P. C., P. A. Williams, and S. R. Halloy. 1999. "A Weed Risk Assessment Model for Use as a Biosecurity Tool Evaluating Plant Introductions." *Journal of Environmental Management* 57 (4):239–251. https://doi.org/10.1006/jema.1999.0297.

Raphael, B., J. Lizzio, J. Wright, L. Richmond, and J. Baker. 2009. *Establishing a List of Nationally Significant Environmental Invasive Pathogens and Invertebrates*. Canberra: Bureau of Rural Sciences.

Regan, H. M., M. Colyvan, and M. A. Burgman. 2002. "A Taxonomy and Treatment of Uncertainty for Ecology and Conservation Biology." *Ecological Applications* 12 (2):618–628. https://doi.org/10.1890/1051-0761(2002)012[0618:ATATOU]2.0.CO;2

Robinson, A., M. A. Burgman, and R. Cannon. 2011. "Allocating Surveillance Resources to Reduce Ecological Invasions: Maximizing Detections and Information About the Threat." *Ecological Applications* 21 (4):1410–1417. https://doi.org/10.1890/10-0195.1.

Rolfe, J., and J. Windle. 2003. "Valuing the Protection of Aboriginal Cultural Heritage Sites." *Economic Record* 79 (Special Issue):S85–S95. https://doi.org/10.1111/1475-4932.00094.

Rolfe, J., and J. Windle. 2014. "Public Preferences for Controlling an Invasive Species in Public and Private Spaces." *Land Use Policy* 41:1–10. https://doi.org/10.1016/j.landusepol.2014.04.013.

Rolfe, J., and J. Windle. 2017. "Valuing Protection Against Invasive Species Using Contingent Valuation." In *Invasive Species: Risk Assessment and Management*, edited by Andrew P. Robinson, Mark A. Burgman, Mike Nunn and Terry Walshe, 252–265. Cambridge: Cambridge University Press.

Sinden, J., R. Jones, S. Hester, D. Odom, C. Kalisch, R. James, and O. Cacho. 2004. The Economic Impact of Weeds in Australia. Glen Osmond, SA: CRC for Australian Weed Management.

Soliman, T., M. C. M. Mourits, W. van der Werf, G. M. Hengeveld, C. Robinet, and M. O. Lansink. 2012. "Framework for Modelling Economic Impacts of Invasive Species, Applied to Pine Wood Nematode in Europe." *PLOS ONE* 7 (9):e45505. https://doi.org/10.1371/journal.pone.0045505.

Summerson, R., S. Hester, and S. Graham. 2018. Methodology to Guide Responses to Marine Pest Incursions Under the National Environmental Biosecurity Response Agreement. Centre of Excellence for Biosecurity Risk Analysis.

Tait, P., and P. Rutherford. 2018. Non-Market Valuation of Environmental Impacts for Biosecurity Incursion Cost Benefit Analysis: A Guidance Manual for Public Policy. Agribusiness and Economics Research Unit.

Vanclay, F. 2002. "Conceptualising Social Impacts." *Environmental Impact Assessment Review* 22 (3):183–211.: https://doi.org/10.1016/S0195-9255(01)00105-6.

Walshe, T., M. Cole, N. Grant, L. Failing, G. Long, and R. Gregory. 2012. A review of current methods and tools used by biosecurity agencies to estimate consequence impacts on primary production, amenity, and the environment. Australian Centre of Excellence for Risk Analysis.

Welsh, M. J., J. A. Turner, R. S. Epanchin-Niell, J. J. Monge, T. Soliman, A. P. Robinson, J. M. Kean, C. Phillips, L. D. Stringer, J. Vereijssen, A. M. Liebhold, T. Kompas, M. Ormsby, and E. G. Brockerhoff. 2021. "Approaches for Estimating Benefits and Costs of Interventions in Plant Biosecurity Across Invasion Phases." *Ecological Applications* 31 (5):e02319.: https://doi.org/10.1002/eap.2319.

WHO. 2023. "Epidemic Intelligence from Open Sources (EIOS)." accessed 28 July 2023. https://www.who.int/initiatives/eios.

WOAH. 2022a. "Aquatic Animal Health Code." World Organisation for Animal Health, accessed 13 January 2023. https://www.woah.org/en/what-we-do/standards/codes-and-manuals/aquatic-code-online-access/.

WOAH. 2022b. "Terrestrial Animal Health Code." World Organisation for Animal Health, accessed 13 January 2023. https://www.woah.org/en/what-we-do/standards/codes-and-manuals/terrestrial-code-online-access/.

4 Prevent
Aligning Border Surveillance with Pre-Border Intervention

Anthony Rossiter and Susan M. Hester

ABSTRACT

Governments often require specific processing and treatments to be applied to imported products for those products to be allowed into a country or region. These pre-border *interventions* are thought to reduce the biosecurity risks of the imported products to which they are applied. When pre-border interventions are in place, consideration should be given whether border intervention regimes should be adapted in response. Since any intervention imposes costs on import-supply chain participants and consumers, the question of "which pre-border intervention measures to impose" is as much an economic question as a statistical and scientific one.

In this chapter, we discuss the issue of mitigating biosecurity risks pre-border, the various regulatory frameworks that can be used, and their implications for efficiency. We outline a tool that allows estimates of the effectiveness of interventions to be incorporated into an existing border inspection arrangement, known as the Compliance-Based Intervention Scheme, currently in operation at the Australian border. This framework underscores the importance of taking a multidisciplinary approach to determine appropriate pre-border and border interventions as part of a biosecurity risk mitigation strategy for imported goods.

GLOSSARY

Border intervention Includes measures and actions implemented by a biosecurity agency or requested of importers (e.g. screening or inspection, fumigation, heat treatment, information provision); also known as *assurance measures.*

Biosecurity risk material (BRM) Living organisms exotic to a country, or biological material carrying an exotic pest or disease. Biosecurity risk material includes live animals and plants, animal and plant material (e.g. fur, feathers, pollen, and flowers), food scraps, and soil.

Traditional command-and-control regulation Application of one intervention regime to all products in a biosecurity category, regardless of the risk mitigation measures undertaken or compliance history.

Risk-based regulation Intervention resources are applied according to an assessment of the risks posed by an import (i.e. accounting for both the likelihood and consequences of the potential entry, establishment, and spread of a pest or disease).

Incentive regulation Considers the behavioural response of regulated entities in designing interventions. The regulator may use rewards and/or punishments to induce behaviours by regulated entities that are consistent with the regulator's objective.

Regulatory performance An assessment of how a regulator and/or the interventions for which they are responsible are meeting their objectives. For example, specific requirements

DOI: 10.1201/9781003253204-5

may be assessed on the extent to which they reduce biosecurity risks and impose regulatory burden on regulated entities.

Appropriate level of protection The Agreement on the Application of Sanitary and Phytosanitary Measures (SPS Agreement) defines the appropriate level of protection (ALOP) as "the level of protection deemed appropriate by the Member establishing a sanitary or phytosanitary measure to protect human, animal or plant life or health within its territory" (Annex A, paragraph 5).

Approach rate An estimate of the likelihood of entry of pests and diseases, calculated using border inspection data.

Continuous sampling plan (CSP) A class of technical rules for determining whether to inspect a consignment, based on the recent intervention history of the pathway (Dodge 1943; Dodge and Torrey 1951).

Clearance number A key parameter of the CSP algorithms. It represents the number of consecutive clean imports that must be reached before a target's goods can be switched to a compliance-based rate of intervention in monitoring mode.

Monitoring fraction A parameter in the CSP algorithms used to determine the frequency of intervention once an importer has demonstrated sufficient compliance with biosecurity requirements in the monitoring mode of the CSP algorithm. This parameter governs the compliance-based rate of intervention (MF) to be applied that enables intervention on less than 100% of consignments imported.

Compliance-Based Intervention Scheme (CBIS) An intervention scheme offered by the Australian Department of Agriculture, Fisheries and Forestry for selected products to automate the implementation of CSP algorithms.

Leakage Occurs when a consignment containing biosecurity risk material that would or should have been detected by an intervention and treated to ameliorate biosecurity risks but was not. This results in a disease or pest entering Australia, with the possibility it becomes established and spreads post-border.

Leakage (post-intervention non-compliance) rate An estimate of the proportion of consignments on a pathway containing biosecurity risk material that cross Australia's international borders after undergoing any required interventions.

Markov chain A model describing a sequence of possible events in which the probability of each event depends only on the state attained in the previous event (Grimmett and Stirzaker 2020). This means that, conditional on the present state of the system, its future and past states are independent.

Menu of (regulatory) contracts From the economic theory of contracts, a menu of regulatory contracts is approach to regulation where the regulator offers the regulated entity a suite of options (the menu) as to how it can meet requirements. A well-designed menu of regulatory contracts encourages the regulated entity to reveal information to the regulator through its menu choice, under the assumption that the regulated entity has chosen the scheme that is "optimal" for them.

INTRODUCTION

Governments must balance the benefits of international trade against the likelihood that damaging pests and diseases enter their country via imported products. Concerns about biosecurity risks motivate governments to regulate the importing process by prescribing interventions. Interventions include border inspections as well as pre-border assurance measures that prescribe specific actions or treatments that must be completed before goods can enter a country or region.

Pre-border assurance measures are designed to reduce the likelihood of pests and diseases being present in imported products. These include activities related to production (e.g. crop rotation, pest exclusion, in-field chemical treatment, and biological control), harvest and packing (e.g. good sanitation and hygiene, pest exclusion), processing (e.g. cooking and washing), pre-export activities (e.g. heat and cold treatments, irradiation, and fumigation), and other testing. Pre-border interventions imposed by national biosecurity agencies can increase costs on various import-supply chain participants. These may be direct financial costs or costs associated with delays in getting products into domestic markets. Some of the costs imposed by these interventions (e.g. inspections) will be passed on to domestic businesses and consumers through higher costs of imported products or more limited access to certain goods.

Although border interventions and pre-border assurance measures are often considered separately, they are intertwined and form part of the importing country's overall biosecurity risk mitigation strategy. Pre-border assurance measures are implemented to reduce (mitigate) biosecurity risks at the border, because particular pre-border measures may allow fewer or less intense interventions to be conducted at the border. While it is important to understand the connection between pre-border and border measures from a risk perspective, it is also important to understand the budgetary implications of alternative intervention strategies. The challenge for governments is to design interventions that make the best use of the limited resources that can be devoted to biosecurity, given the cost of risk reduction and its potential benefits (see Chapter 9. Resource Allocation).

Designing appropriate intervention strategies requires a balance between the often-competing objectives of government (trade benefits versus protection from biosecurity risks) and understanding the relative benefits of intervening in the import-supply chain. This involves understanding the trade-offs involved in the biosecurity assurance system as a whole, including how entities in the import-supply chain will respond to different regulatory requirements.

Given the finite nature of resources available to undertake border interventions, it is important to implement an appropriate mix of pre-border and border interventions that reflect an efficient allocation of limited government resources. In this chapter, we outline the different models of regulation and how these are improved by incorporating knowledge from economics around efficiency and incentives. We focus on how border inspection regimes should be adapted in response to pre-border intervention measures, using Australia's Compliance-Based Intervention Scheme (CBIS) as an example.

REGULATORY MODELS FOR BIOSECURITY INTERVENTIONS

TRADITIONAL COMMAND-AND-CONTROL REGULATION

Many biosecurity intervention regimes are currently based on, or have evolved from, a traditional command-and-control regulatory model. Under this model, every consignment on a pathway is treated as having the same biosecurity risk potential and is subject to mandatory interventions (e.g. inspections) at the border. For some pathways, consignments may also be subject to multiple pre-border interventions, which may involve import-supply chain participants needing to apply specific biosecurity measures at different stages along the value chain for the imported product. Under this regulatory approach, the same types of interventions are applied to all consignments on the pathway regardless of the risks that a particular individual consignment may pose to maintaining a country's biosecurity status.

This traditional regulatory model is a "one-size-fits-all" approach with respect to product pathways and to participants in the import sector. In applying the same intervention approach to all products and stakeholders in a biosecurity category, irrespective of investments in pre-emptive biosecurity measures or compliance history, some consignments may be subject to more intervention than necessary while others may be subject to less intervention than required to appropriately manage biosecurity risks (Berg 1998; Black and Baldwin 2010).

RISK-BASED REGULATION

Risk-based regulation has been proposed as a strategy to reform regulatory systems (Black and Baldwin 2010), and it is increasingly taken up by governments for biosecurity management (Beale et al. 2008; Lunn 2014). Risk-based regulation involves applying inspection resources according to an assessment of the risks posed by an import. This risk assessment involves an understanding of both the likelihood of biosecurity risk material (BRM) entering on a particular pathway and the consequences of the leakage through the border of this BRM.

Risk-based regulation involves a regulator selecting the risks it seeks to control and monitor and determining its tolerance for certain risks. When biosecurity risks associated with international trade are identified, countries may impose measures to reduce risks, but need to do so in accordance with World Trade Organisation (WTO) rules contained in the Application of Sanitary and Phytosanitary Measures (SPS Agreement). Under the SPS Agreement, WTO members are entitled to maintain a level of protection they consider appropriate to protect life or health within their territory (see Chapter 3. Anticipate). For example, Australia's Appropriate Level of Protection (ALOP) is expressed as providing a high level of sanitary and phytosanitary protection aimed at reducing risk to a very low level, but not to zero (Biosecurity Act 2015, Section 5).

Risk-based approaches to regulation are characterised by a focus on the regulator's objectives and display a greater focus on outcomes. The focus of risk-based regulation is on mitigating risks rather than prescribing particular rules for enforcement. Better targeting the level and frequency of interventions allow the regulator to devote appropriate resources to imports that pose different levels of risk (i.e. resource-efficient regulation). In the biosecurity context, this would imply structuring interventions to reward importers for "good" importing behaviour (e.g. complying with rules and bringing in clean consignments) relative to those who demonstrate a lower level of compliance with biosecurity requirements.

The CBIS is an example of a risk-based mechanism, adopted by the Australian national biosecurity regulator (Robinson et al. 2012, Rossiter and Hester 2017). CBIS is based on continuous sampling plan (CSP) algorithms (Dodge 1943; Dodge and Torrey 1951), which introduced lower intervention frequencies on pathways with relatively low failure rates for border inspections (see Box 4.2). These protocols assume that the likelihood of BRM being present in a consignment is related to past compliance, with lower intervention frequencies afforded to entities with a strong compliance record.

Risk-based regulatory frameworks may involve some elements of command-and-control regulation, where the nature of the risks to be managed warrants it. This might reflect that the entry of some pests or diseases (e.g. foot-and-mouth disease) could have significant consequences for agricultural industries and the local environment were they to establish.

INCENTIVE REGULATION

Incentive regulation provides a further layer of sophistication, compared to risk-based regulation (QCA 2014). In risk-based regulation, a regulator often does not consider the feedback loop from the way regulations are set to actions taken by regulated entities (e.g. importers). Such responses will reflect the regulated entity's own motivations, and their actions may or may not be in line with the government's regulatory objective (Lunn 2014). The presence of this "alignment problem" means there are potential gains to the regulator and regulated entity from improving the incentives of the latter for undertaking the task (see Chapter 8. Incentives; Laffont and Tirole 1993). In biosecurity, this would involve designing interventions aimed at better aligning the goals of all parties. This is the premise of incentive regulation.

Incentive regulation considers the behavioural response of regulated entities, with the regulator using both rewards and punishments to induce behaviours consistent with the regulator's objective.

The regulator seeks to design an incentive scheme which allows the regulated entity's objective (which is often to maximise profits) to be aligned with the regulator's objectives. If incentives are not explicitly considered when regulations are designed, the regulator's objectives are unlikely to be met. In some cases, the behavioural response by the regulated entity may be the opposite to what was intended by the regulator, resulting in significant costs incurred in checking compliance, managing trade relationships, or managing incursions. Appropriately incorporating incentives into biosecurity regulations becomes a problem of *mechanism design* (Roth 2002), where the regulator seeks to choose the rules for the strategic interaction to maximise their own objective (see Chapter 8. Incentives).

For example, while the CBIS algorithms were implemented as risk-based regulation, they have inherent incentive properties because they may provide cost and/or time advantages to importers and/or suppliers with a good compliance record. The algorithms also benefit the regulator through reducing the costs of administering the biosecurity assurance system. The challenge with incentive regulation is to fine-tune the intuitive incentive structures that currently exist in the CSP algorithms to utilise the information and expertise of system participants in ways that make the biosecurity system more effective and efficient.

ECONOMIC IMPLICATIONS OF REGULATORY MODELS

In terms of its use of intervention resources and benefits of intervention outcomes, the traditional command-and-control regulatory model is the least efficient of all models. Traditional regulatory models increase the costs to import-supply chain participants of achieving biosecurity objectives, regardless of the threat posed by the imported product. For importers, traditional regulatory models reduce or eliminate the cost advantage of sourcing goods from suppliers that have appropriately processed their exports to ensure they are less likely to contain BRM. Prescriptive interventions may therefore act as a disincentive for entities in the product's value chain to explore alternative ways to reduce biosecurity risks—importers and suppliers are not rewarded for developing effective mitigation options achieved at lower cost, stifling innovation, and potentially increasing the costs of supply. Consumers may also forgo some of the advantages of international trade, such as access to higher-quality and/or lower-cost goods or a more diverse range of products than can be sourced locally.

Efficiency is improved under risk-based regulation, although constructing rules based on the regulator's objectives alone will not guarantee a system that fosters behaviours focused on compliance. For this to happen, the objectives of regulators *and* stakeholders need to be taken into account, which occurs in incentive regulation. This form of regulation leads to the creation of incentive structures that encourage importers to choose suppliers with better biosecurity compliance records or lead importers to try and influence suppliers to improve their mitigation efforts. Whether such a biosecurity assurance system affects import-supply chain behaviour depends on the value of the cost advantages induced by the system. In situations where mitigation involves large fixed costs, ongoing benefits would need to be realised to ensure that firms in the import-supply chain would adopt this technology or process; otherwise, it would be in their best interests to maximise expected profits to use the less effective existing mitigation strategies (Rossiter and Hester 2017).

Incentive regulation also brings further improvements in efficiency because it involves regulators focusing their activities on those that are of greatest net benefit to their regulatory objectives. Offering a suite of options to the importer—a *"menu of contracts"*—as to how biosecurity assurance is provided for their consignment, can deliver a more tailored, lower-cost method for providing assurance that biosecurity risks are being appropriately managed (Rossiter et al. 2016, Rossiter et al. 2018, Sappington 1994). A more familiar application of this idea can be illustrated with motor car insurance. The buyer of insurance is asked to choose between a menu of contracts including at one extreme a high excess but low premium, and at the other a low excess with high premium. Confronted with these options, the driver accounts for private information about their driving habits and capabilities (i.e. the factors influencing their probability of making a claim) and the premium. The driver then selects the excess and premium combination that maximises their wellbeing.

In a biosecurity context, it should be possible to design a reward structure that provides increasing benefits to an importer for higher levels of biosecurity compliance, with options offered as part of the menu being interdependent in terms of trade-offs to induce better behaviour. The menu of contracts idea was developed to allow market participants to reveal information that can assist the market to work efficiently.

While understanding the economic implications of regulatory models allows improvement in efficiency, embedding economic concepts in a multidisciplinary approach to regulatory design (Box 4.1) should ensure intervention rules are effective, efficient, and implementable.

BOX 4.1. A MULTI-DISCIPLINARY APPROACH TO DESIGNING AND IMPLEMENTING ASSURANCE MEASURES

Designing effective, efficient, and implementable assurance measures requires a multidisciplinary approach.

- **Scientific knowledge** is required to inform estimates of the likelihood and consequences of entry, establishment, and spread of pests and diseases post-border, and understand the efficacy of a range of pre-border interventions in reducing the likelihood that pests and diseases will be present in imported products (and/or the prevalence of pests and diseases). It remains important to independently verify (e.g. through technology, audit or certification) that certain risk-reducing interventions have taken place pre-export.
- **Statistical rules that are risk based** provide the basis for implementing schemes that have known properties that can be easily "encoded" into automated systems. With increasing volumes of trade, these systems have become integral to directing cargo for specific border interventions. Automated schemes allow decisions around how much or how often to intervene (inspect) to be easily changed depending on regulatory performance (see Chapter 5 Screen).
- **Economics** is required to understand consequences of BRM establishing within a country. Economics is also required to understand the in-built incentives for compliance inherent in statistical rules, including how stakeholders (suppliers/producers and importers) will respond to different rules (i.e. change their tendency to comply). This requires an understanding of production processes, the costs of supply, and associated interventions to importers. Knowledge of economics is also required to assess and quantify the wider (social) costs and benefits of intervening to ensure the proposed regulatory approach adequately accommodates the trade-offs involved. Quantification, as opposed to more qualitative risk-rating approaches (e.g. risk matrices covering likelihood and consequences; Chapter 3. Anticipate), is required to discriminate between alternative settings for the regulatory rules.
- **Technology** plays a role in verification of pre-export production processes and treatments and can assist in the design of assurance measures that are "incentive-compatible" (Chapter 8. Incentives).

OPERATIONALISING RISK-BASED AND INCENTIVE REGULATION

A key challenge in designing effective, efficient, and implementable assurance measures is understanding the import-supply chain characteristics that may increase or decrease biosecurity risks, compliance rates, and thus *approach rates*—the likely entry rate of BRM—at national borders. Characteristics that may influence biosecurity risks and the approach rate of BRM include, for example:

- whether offshore certification has taken place.
- the level and type of processing that has occurred.
- physical treatments that have occurred (e.g. irradiation, cold disinfestation, heat treatments).
- the production standards in the country of origin.

Biosecurity regulators in several countries, including Australia, are adopting risk-based and incentive regulation frameworks as they recognise the inefficiency of a traditional one-size-fits-all approach (Rossiter and Hester 2017, Rossiter et al. 2016). This reflects a desire to remove unnecessary regulatory burden on business and individuals and to create an environment that fosters productivity and innovation. The CBIS, introduced by Australia's Department of Agriculture, Fisheries and Forestry (DAFF) in 2013, is an example of a risk-based approach that can be augmented with additional information to form an incentive regulation scheme.

The CBIS adaptive sampling algorithm is a mechanism for guiding biosecurity inspection resource allocation for specific commodities (Chapter 11. Research Uptake). For most applications adopted by DAFF, the CBIS uses CSP algorithms (CSP-1 and CSP-3; Dodge 1943; Dodge and Torrey 1951) to determine whether a consignment requires inspection (Box 4.2). While the CBIS was brought in as a risk-based regulation tool, the algorithm was demonstrated to have inherent incentive properties, which became important to consider during pathway, rule, and parameter selection (Rossiter et al. 2016, 2018).

BOX 4.2. CSP ALGORITHMS APPLIED TO INSPECTIONS

CSP-1: When a new importer starts on the CSP-1, they are usually subject to mandatory inspections in "census mode"—all consignments are inspected in this mode, until they build up a good compliance record (Figure 4.1). The two key parameters that need to be chosen in this rule are:

- *Clearance number (CN)*: the number of successive consignments that must pass inspection in census mode for the importer to be eligible for a reduced inspection frequency.

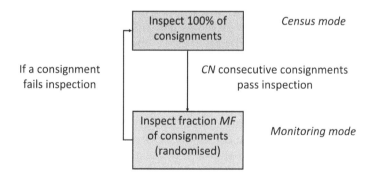

FIGURE 4.1   Schematic representation of the CSP-1 algorithm. (Reproduced from Rossiter and Hester 2017). *Abbreviations*: CSP, continuous sampling plan; CN, clearance number; MF, monitoring fraction.

- *Monitoring fraction (MF)*: the reduced inspection frequency and probability that a given consignment is inspected in "monitoring mode".

If an importer's consignment fails inspection when the importer is in monitoring mode, their subsequent consignments become subject once again to mandatory inspection in census mode (Figure 4.1). The importer only receives the reduced inspection frequency again after another *CN* successive consignments pass inspection.

CSP-3: The CSP-3 algorithm differs from the CSP-1 rule in terms of what happens to an importer following a failed inspection in "monitoring mode" (Figure 4.2). This rule has less severe consequences for occasional non-compliance when an importer is on "monitoring mode", but at the cost of a more complex penalty mechanism. The additional parameters that need to be chosen in this rule are:

- *Tight census number (TC)*: if an importer's consignment fails inspection in monitoring mode, the next *TC* consignments are subject to mandatory inspection in "tight census mode". *TC* is typically set to 4 as suggested in the original Dodge and Torrey (1951) algorithm and in the statistical analysis of Robinson et al. (2012).
- *k*, which is the minimum permitted spacing of two detected contaminated consignments (usually the same as *CN*, as shown in Figure 4.2). In other words, if a subsequent failure is detected within *k* inspections of any other fail (*TC* of which will be in tight census) then the pathway moves to census mode.

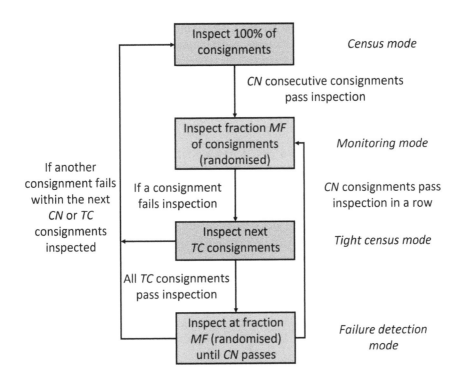

FIGURE 4.2 Schematic representation of the CSP-3 algorithm. (Reproduced from Rossiter and Hester 2017). *Abbreviations*: CSP: Continuous sampling plan. CN: Clearance number. MF: Monitoring fraction. TC: Tight Census number.

This rule is designed to protect the regulator against a sudden systematic (long-lasting) problem that would significantly raise the likelihood of a consignment failing inspection. If the next *TC* consignments following a failure pass inspection, the importer's consignments go back to being inspected at the reduced rate (*MF*) while the regulator keeps track of the number of inspections passed since the last failure. This part of the algorithm is usually referred to as "failure detection mode". Provided the importer passes inspection *k* times since their last failure (set as *CN* in Figure 4.2), the importer remains eligible to be inspected at the reduced rate of inspection; otherwise, on recording another failure within *k* inspections of the previous one, the importer's consignments revert to mandatory inspection until they pass inspection *CN* times in a row.

Pathway Selection

CBIS was first introduced as a trial on four pathways (dried apricots; dates, green coffee beans, hulled sesame seeds); by August 2021, 68 low-risk plant products and one animal product were subject to CBIS (DAFF 2022) with a total of around 350 import cases on which it could be applied. Implementation of the algorithm to a wider range of commodities was hindered by challenges in assessing suitability of pathway and parameter selection (Box 4.3). The department's approach to identifying suitable rule parameters relied on analysing historical pathway data using simulation techniques. However, it was difficult to apply CBIS rules where import data were unavailable or unsuitable. In addition, implementation of CBIS was hampered by a limited understanding of import-supply chain characteristics and how these may affect approach rates at the Australian border.

Understanding the suitability of pathways for CBIS involves more than a statistical analysis of intervention outcomes. Rather, it requires a more in-depth understanding of pathway cost structures, mitigation options, and consequences of leakage. This brings what can be thought of as "non-statistical intelligence"—the use of economics, scientific knowledge and technology—to consider how best to design biosecurity assurance protocols across a variety of circumstances.

BOX 4.3. KEY CONCEPTS UNDERPINNING BIOSECURITY ASSURANCE MEASURES

Pre-intervention rate of non-compliance: The (incoming) rate of non-compliance prior to completing the interventions (heat treatment, processing, etc.) specified by the importing nation. In the CBIS Sensitivity Module, this is also referred to as the **approach rate** (see Glossary).

Post-intervention rate of non-compliance: The (outgoing) rate of non-compliance after completing the interventions (heat treatment, processing, etc.) specified by the importing nation.

Sampling plan: The method used to determine whether or not to inspect a consignment. For example, the CBIS inspects using the CSP family of rules (Dodge 1943, Rossiter and Hester 2017).

Failure at inspection: Occurs when a non-compliance is detected at inspection. Non-compliance might be due to the incorrect declaration of goods, packaging failures and/or evidence suggesting the possible presence of biosecurity risk material.

Quarantine failure: Not all failures at inspection are due to the discovery of pests and diseases that are not already present in in a country. Those failures due to the discovery of BRM are known as quarantine failures.

Average outgoing quality (AOQ): Describes the modelled relationship between (i) the incoming (pre-intervention) rate of non-compliance, measured by the approach rate of the relevant "failure" concept (failure at inspection or quarantine failure), and (ii) the outgoing (post-intervention) rate of non-compliance (i.e. BRM leakage) after completing the specified intervention protocols.

Average outgoing quality curve (AOQ curve): Shows how the post-intervention rate of non-compliance (vertical axis) depends on the pre-intervention rate of non-compliance (horizontal axis) (e.g. Figure 4.3). The maximum value of the AOQ curve is the average outgoing quality limit.

Average outgoing quality limit (AOQL): Provides a "worst-case" scenario for the post-intervention rate of non-compliance for a given sampling plan.

RULE SELECTION (CSP-1 OR CSP-3)

For most pathways, the CBIS uses the CSP-3 algorithm to determine inspections, as CSP-3 provides less of a "cost" to the importer if recording a failure in one inspection does not increase the probability that future consignments will be more likely to fail. CSP-3 was adopted following recommendations in Robinson et al. (2012) based on a statistical analysis of the department's administrative data for several plant product pathways. However, subsequent analysis of the CSP rules (Rossiter and Hester 2017) suggested that the CSP-1 algorithm would be preferable from the regulator's perspective, particularly where the consequences of BRM leakage are perceived to be large. The CSP-1 algorithm is simpler and more easily communicated to stakeholders, which is likely to result in a clearer understanding of the incentive properties of the inspection rule.

RULE PARAMETERS

The difficulty for technical officers in choosing CSP rule parameters CN and MF is the need to understand the potential trade-offs associated with different parameter choices. Different rules (CSP-1 or CSP-3) and combinations of rule parameters could be suitable for managing a given pathway or assurance mechanism. A key element of this approach to parameter selection requires quantifying the biosecurity agency's maximum tolerance for *post-intervention non-compliance* that is consistent with ALOP – its "tolerance for leakage" (Box 4.3). This quantification departs from the usual practice in conducting biosecurity import risk analyses (*Biosecurity Act 2015*), such as using risk estimation matrices (Chapter 3. Anticipate). To select CBIS rule parameters, more fine-scale differentiation through quantifiable benchmarks is required to distinguish which parameter combinations offer "acceptable" risk profiles. Applied to the Australian context, this provided new challenges for technical officers responsible for selecting parameters and implementing the algorithm.

CBIS SENSITIVITY TESTING

To improve implementation of CBIS (pathway, rule, and parameter selection), an Excel spreadsheet-based tool, the *CBIS Sensitivity Module* (Rossiter 2021) was developed to assist staff with selecting the two key parameters: CN (clearance number) and MF (monitoring fraction), with TC remaining at 4 and k equal to CN (Hester et al. 2020). The tool was developed based on modelling the long-run properties of the CSP-1 and CSP-3 rules as *Markov chains*, which obviated the need to analyse pathway intervention data (see Hester et al. 2020 for more details). This allowed for rules of thumb, drawing on concepts from the statistical quality control literature (Dodge 1943; Stephens 2001; Box 4.4), to be developed for CN and MF and for parameters under the CSP-1 and CSP-3 rules.

To quantify the biosecurity agency's maximum tolerance for post-intervention non-compliance that is consistent with its ALOP, the CBIS Sensitivity Module requires information on the amount saved by importers when an intervention is not required, the likelihood a pest will establish if it escapes border detection, establishes and spreads, the cost of eradication and the value of damages caused by the pest (see Box 4.4).

The Markov-chain approach does not provide a definitive answer on which CSP rule parameters should be adopted on a pathway. Instead, it provides an implementable framework for considering

BOX 4.4. EXAMPLE CALCULATION OF TOLERANCE FOR LEAKAGE

On a given pathway where the CBIS is to apply to inspections, it is estimated that not having an inspection will save $760 on average for importers, comprising:

- $200 in inspection fees, given a typical inspection takes between 45 minutes and 1 hour.
- $100 in inspection attendance costs, given the customs broker typically attends the full inspection.
- $20 in inspection booking opportunity costs, given it can take up to 15 minutes of a broker's time to book in the inspection.
- No costs from destroyed or unsaleable product, given the inspection process does not damage the goods.
- $240 in saved storage costs, based on an average saving of two days' storage at $120 per day. and
- $200 in transport cost savings, based on reduced transhipment requirements.

Further, from the perspective of the expected cost of leakage, it is estimated that:

- if a pest known to be found on the pathway makes its way past the Australian border, then there is a 0.5% chance the pest will establish in a small local area. Eradication of the pest from the local area is expected to cost around $5 million; and
- if the pest establishes, there is a 1% likelihood that it will spread beyond the initial establishment site, costing an additional $200 million in damages and eradication costs.

On the basis of this, the candidate threshold for post-intervention non-compliance would be estimated as:

$$\text{Leakage tolerance threshold} = \frac{200 + 100 + 20 + 0 + 240 + 200 + 0}{0.005 * (5,000,000 + 0.01 * 200,000,000)} \times 100\% \approx 2.2\%.$$

and quantifying potential trade-offs associated with different parameter choices. Specifically, it allows comparison of two measures of interest for selecting rule parameters:

- How the long-run share of "consignments saving inspection" changes for importers with different failure rates.
- How the rate of stakeholder-specific BRM leakage varies according to pre-intervention failure rates.

Taken together, these measures demonstrate the trade-offs associated with choosing different CSP algorithms and/or values for CN and MF. The first metric documents the relative strength of incentives, in the form of rewards, experienced by biosecurity system stakeholders with different failure rates, while the second assesses the risk to the biosecurity agency's overarching objective of managing the potential leakage of BRM into Australia.

Each combination of CN and MF for a CSP algorithm produces an *average outgoing quality* (AOQ) *curve* that describes how the rate of post-intervention non-compliance varies with the pre-intervention rate of non-compliance (approach rate; Figure 4.3). A key measure that describes the potential risk to the biosecurity agency's objective is the *average outgoing quality limit* (AOQL), which is the maximum of the AOQ curve and represents the worst-case scenario for modelled importer-level non-compliance.

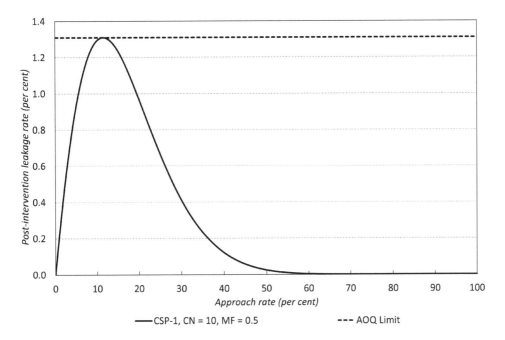

FIGURE 4.3 Average outgoing quality (AOQ) curve and limit for the CSP-1 algorithm with $CN = 10$ and $MF = 0.5$. (From Hester et al. 2020.)

Both concepts are shown in Figure 4.3 for the CSP-1 rule with $CN = 10$ and $MF = 0.5$, which highlights the typical "hump shape" of the AOQ curve. When incoming consignments are highly compliant (i.e. have low approach rates), the rate of post-intervention non-compliance will also be low. If the incoming approach rate is high, most consignments will fail the intervention and undergo some form of treatment to rectify the presence of BRM, thereby ensuring a low rate of post-intervention non-compliance. As a result, the AOQL (worst-case scenario) typically occurs at an intermediate failure rate, thereby generating a peaked profile as shown in Figure 4.3.

One way of establishing which rules could be suitable for adoption on a pathway is to assess which parameter combinations yield an AOQL that is no higher than the biosecurity agency's threshold tolerance for post-intervention non-compliance (Box 4.4). Once the appropriate cost and likelihood estimates are available, such a procedure is easily implementable, given it does not rely on detailed historical pathway data, and provides a high degree of assurance to the biosecurity agency that, at an importer or pathway level, the overall rate of post-intervention non-compliance would be anticipated to be no higher than that threshold. Obtaining likelihood and cost estimates related to establishment and spread will be non-trivial for a range of pests and diseases; however, by adopting an approach that seeks to account for these uncertainties through scenario analysis on key parameters, threshold tolerances for post-intervention non-compliance can still be developed to inform the biosecurity agency's regulatory settings.

This Markov-chain modelling highlighted that, all other things being equal, increasing CN or MF will reduce the AOQL. Two other considerations are relevant for parameter selection (Hester et al. 2020):

1. There is a trade-off between the value of CN and MF that will meet the threshold leakage tolerance, in that a higher CN will be required if the MF is lowered.
2. Because increasing the CN will lower the AOQL, for any given value of MF, it will be of most interest to find the minimum CN for a given monitoring fraction for which the threshold tolerance is satisfied.

The CBIS Sensitivity Module uses these two characteristics to recommend the minimum clearance number that can be combined with a given monitoring fraction so that the maximum leakage tolerance is not exceeded. For a range of potential *MF* values that have a natural (fraction) interpretation for communication with stakeholders, the module:

- Assesses whether a clearance number of no more than 20 would meet the tolerance requirements.
- Reports the minimum *CN* and associated maximum modelled post-intervention non-compliance rate.

Exploration with the CBIS Sensitivity Module suggests that if the tolerance for leakage on a pathway is 1% or below, there will be few if any circumstances under which a CSP algorithm on its own can meet the required maximum leakage tolerance. In these circumstances, changes to the import conditions, such as specific offshore treatments that lower the maximum feasible approach rate, may be needed before even a subset of the pathway may be suitable for management through the CBIS.

Table 4.1 shows the *MF* and *CN* combinations for the CSP-3 algorithm that meet the maximum tolerance threshold calculated in Box 4.4 (2.2%). Any of these parameter combinations, or combinations where the *CN* is above the designated minimum, will not exceed the tolerance determined for the inspection or assurance measure on the pathway.

Considering Incentives in Selection of Parameter Combinations

Once the suite of rules that satisfy the tolerance for leakage has been determined, recommendations around which particular *CN* and *MF* combinations are likely to be *most* suitable for a given set of circumstances rests with consideration of the incentives for compliance—how highly compliant importers might be "rewarded" relative to less-compliant importers via saved inspections or interventions. Figure 4.4 illustrates the pattern for three of the eligible rules shown in Table 4.1, namely those corresponding to *MF* values of 0.2 (Rule 1, black), 1/3 (Rule 2, orange), and 0.5 (Rule 3, red). These illustrate the general features of how these eligible rules vary with changes in the monitoring fraction and clearance number. In particular, the rules with the highest clearance number (and lowest monitoring fraction) offer highly compliant importers with the greatest share of saved inspections.

The rewards for compliance also tend to decay more quickly for rules with a higher *CN*, for modest to high degrees of non-compliance with requirements, allowing greater separation of the benefits for importers with low approach rates compared to those with higher approach rates. Ideally, such separation of rewards can help encourage system stakeholders to change their behaviour, such as through switching to more compliant suppliers or modifying processes to better manage the biosecurity status of their

---

TABLE 4.1

MF and Minimum *CN* Combinations for the CSP-3 Algorithm that Yield a Modelled Maximum Importer Leakage Rate of No More Than 2.2%.

Monitoring Fraction (MF)	Minimum Clearance Number (CN)	Modelled Maximum Importer Leakage (%)
0.2	20	2.17
0.25	17	2.14
1/3	13	2.10
0.4	10	2.15
0.5	7	2.10

Source: Hester et al. (2020.)
Note: *MF* Options where the *CN* required would exceed 20 have been omitted.

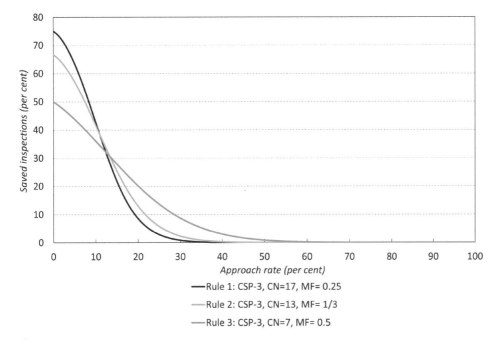

FIGURE 4.4 Comparison of the saved inspection fraction for three candidate CSP-3 parameter combinations with maximum post-intervention leakage rates less than 2.2%. (From Hester et al. 2020).

goods. As such, a default position could be to offer something akin to Rule 1, being the eligible rule with the lowest monitoring fraction to provide the greatest expected rewards to importers with a strong compliance record.

However, there are countervailing factors that could result in a biosecurity agency preferring a rule with a higher monitoring fraction. Primarily these focus on:

- The potential for strategic behaviour (gaming) of the rules, where parties in the supply chain put in less effort with respect to maintaining census-mode levels of BRM.
- The throughput on the pathway, which affects how quickly importers will "feel" the rewards of holding a good compliance record.

On the first issue, the game-theoretic model by Rossiter and Hester (2017) showed that there were stronger incentives for importers to "cheat" and reduce the effort they deployed to ensure consignments were free from BRM when subject to a lower monitoring fraction—typically below MF = 0.25. The potential to game the rules largely depends on how easy it is for the importer or their agents in the supply chain to avoid (costly) risk-reducing processes. In many cases, strategic behaviour by import-supply chain participants may not be feasible; in others, processes integral to reducing the inherent biosecurity risks in products are readily verifiable on a consignment-by-consignment basis, thereby mitigating cheating potential. As such, the extent to which gaming is a credible risk can be assessed based on knowledge of the production processes for the goods. Where there is evidence of strong incentives and an ability to game the rules, the biosecurity agency could then recommend eligible rules with a higher monitoring fraction to limit the scope for strategic behaviour.

The biosecurity agency may also wish to recommend an eligible rule with a lower clearance number and higher monitoring fraction on pathways where the volume of imports is lower. On these pathways, it may take a considerable amount of time for importers to build up the required number of consecutive passes to qualify for monitoring mode, thereby blunting the incentives for compliance from the importer's perspective. More immediate rewards would be available if a rule with a lower clearance number was adopted instead.

Differentiated Rules Based on Pre-border Assurance Measures and Information

The discussion to date has focused on applying CBIS across a whole pathway, accounting for the full range of approach rates possible for a pathway. However, appropriate CBIS rules for consignments that have completed independently verifiable pre-border assurance measures—known to reduce the likelihood of BRM being present in consignments—may not need to take account of all potential approach rates. Completing a given treatment or undergoing a specific production process, for example, could be effective in limiting the approach rate of consignments to below a pre-specified threshold. This implies the eligibility criterion could be based on the maximum post-intervention non-compliance rate taken over the relevant (feasible) part of the AOQ curve, rather than the entire curve.

As part of a pragmatic approach to regulatory risk management, estimates of a candidate maximum approach rate should be based on available scientific evidence that can be independently verified. This will ensure credibility of the arrangements and help preserve a high biosecurity status. The effectiveness of pre-border assurance or other control measures associated with a proposed maximum approach rate may need to be reliably evaluated before adoption by a biosecurity agency. Where there are sufficient data captured to support a rigorous statistical analysis, it may be possible to use hypothesis tests to assess the effectiveness of some pre-border measures.

This approach reflects the principle of a "menu of regulatory contracts" (Rossiter et al. 2016) and enables a biosecurity agency to better target its interventions on pathways. In some cases, an agency may allow consignments undergoing certain additional pre-border assurance measures to be eligible for more generous intervention rules. Where the biosecurity agency has only a very low tolerance for leakage, consignments undergoing specific and verifiable pre-border treatments may be eligible for CBIS-type rules, while others face mandatory interventions.

In the CBIS Sensitivity Module, users have the option to provide an estimate of the maximum approach rate under additional intervention or assurance measures and test the sensitivity of potential rules to alternative assumptions about this maximum approach rate. The Module provides alternative minimum clearance number estimates for consignments subject to the risk-reducing intervention. Most commonly, the additional measure will enable a lower minimum clearance number to be used for meeting the threshold tolerance for the rate of post-intervention non-compliance; in some circumstances, the additional measure may even enable rule combinations with lower monitoring fractions to be applied.

Table 4.2 and Figure 4.5 compare the minimum clearance numbers for the pathway as a whole (second column of Table 4.2 and Figure 4.5a) with the situation where an additional pre-border

TABLE 4.2

Monitoring Fraction and Minimum Clearance Number Combinations for the CSP-3 Algorithm with and without Additional Assurance Measures That Guarantee an Approach Rate of No Higher Than 8%.

	Minimum Clearance Number	
Monitoring Fraction	Applicable to Whole Pathway	Applicable to Importers with the Additional Process or Intervention
0.2	20	20
0.25	17	16
1/3	13	9
0.4	10	5
0.5	7	5

Source: Hester et al. (2020).

assurance measure is effective in preventing the failure rate at inspection going higher than 8% (third column of Table 4.2 and Figure 4.5b) under the assumption that the intervention is always effective.

In most cases, the minimum clearance number required to meet the 2.2% post-intervention non-compliance threshold is lower for consignments completing the additional pre-border assurance measure.

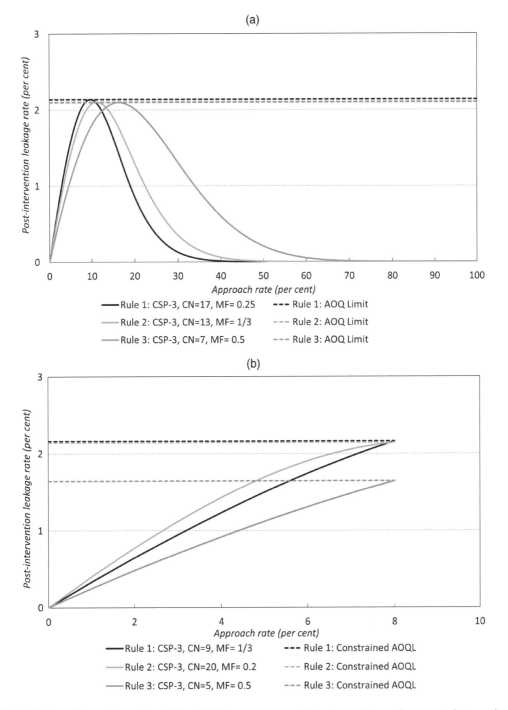

FIGURE 4.5 Comparison of candidate CSP-3 parameter combinations with maximum post-intervention leakage rates less than 2.2% across the whole pathway (a) or a subset with an additional pre-border assurance measure effective at preventing the failure rate at inspection going higher than 8% (b). (From Hester et al. 2020.)

Given biosecurity risk identification reflects evolving scientific knowledge, there may be circumstances in which new risks are identified or instances where some pre-border measures are not being effectively implemented. This may require the introduction of new interventions for risk management or verification purposes, imposing additional regulatory burden on stakeholders.

Where a risk has been identified, there may be limited evidence available to establish whether non-compliances are being detected. As part of establishing new interventions at the border in situations where the tolerance for non-compliance is above a certain threshold, the biosecurity agency could introduce the additional intervention as part of a two-step process.

1. The biosecurity agency could communicate the newly identified risk to stakeholders and explain that it wishes to build an evidence base to establish whether this risk is present in consignments landing in the domestic economy. Randomised testing or inspection (which may be stratified by country, if appropriate) at a fixed probability of referral could be carried out for a given time period (see Chapter 5. Screen). Data collected through this process can be used to determine whether the arrival rate of the identified risk likely exceeds the agency's tolerance for non-compliance.
2. If the tolerance is exceeded, either across the pathway as a whole or by certain countries of origin, then the biosecurity agency can introduce an appropriate CBIS rule to provide assurance that post-intervention non-compliance is managed consistent with its tolerance for leakage. If the current approach rate of the risk is below the pre-determined tolerance level, then the randomised testing or inspection could be discontinued.

As part of establishing suitable parameters for the two-step process, a biosecurity agency could use the CBIS Sensitivity Module to choose values for CN and MF for the second stage that are consistent with managing the risks below an agreed post-intervention non-compliance threshold. The value for MF determined in this stage can then be adopted as the referral rate for intelligence-gathering purposes in the first stage.

DISCUSSION

The need to move from a one-size-fits all regulatory approach to risk-based or incentive regulation should allow biosecurity agencies to manage their budgets and address the biosecurity risks from the growing movement of goods, passengers, and mail across the globe. In this process, regulators need to take into account the inherent incentive properties of pre-border and border assurance requirements they implement. Considering both sets of assurance measures together and incorporating the impact on stakeholder behavioural responses is critical to ensure a well-functioning biosecurity management system.

Adaptive intervention approaches, such as Australia's CBIS, offer the opportunity to target biosecurity agency resources towards products and import-supply chain stakeholders who present a greater threat to national biosecurity objectives. Successful implementation requires a multidisciplinary approach that incorporates knowledge and insights from science, statistics, and economics and leverages appropriate technologies as part of the assurance framework.

The approach outlined in this chapter reflects an evolution of the existing regulatory framework that underpins biosecurity assurance activities around the world. As shown in Chapter 8. Incentives and Stoneham et al. (2021), one option is to redesign the risk-sharing system to follow an insurance model. Whether the insurance approach is more appropriate for the management of biosecurity risks in the long term will depend on whether certain pre-conditions, such as traceability, are met and the net benefits of the different frameworks, which may differ depending on specific circumstances. Biosecurity regulatory frameworks that leverage elements of risk-based and incentive regulation are therefore likely to have application in the long term.

IN A NUTSHELL

- A key challenge in designing effective, efficient, and implementable assurance measures is understanding how particular pre-border interventions may decrease the prevalence of biosecurity risks and thus the likelihood of pests and diseases arriving at national borders.
- Biosecurity agencies should consider the design of pre-border and border interventions together to ensure they provide appropriate incentives for compliance and innovation in biosecurity risk management.
- Adaptive intervention regimes, such as Australia's CBIS, offer a way of efficiently allocating government resources while maintaining existing biosecurity risk mitigation requirements.
- Successful implementation of risk-based and incentive regulatory schemes requires a multidisciplinary approach, drawing on knowledge and insights from science, statistics, and economics and leveraging appropriate technologies as part of the assurance framework.
- Statistical modelling approaches, such as the CBIS Sensitivity Module, allow biosecurity agencies to test different combinations of key rule parameters to understand the potential likelihood of leakage and the rewards for compliant importers relative to those who are non-compliant.
- Uncertainties associated with inputs to the rule's parameter-selection process mean sensitivity analysis and professional judgement play a distinct role in the choice of appropriate intervention arrangements.

REFERENCES

Beale, R., J. Fairbrother, A. Inglis, and D. Trebeck. 2008. One Biosecurity. A working partnership. The independent review of Australia's quarantine and biosecurity arrangements. Report to the Australian Government. Quarantine and Biosecurity Review Panel, Department of Agriculture, Fisheries and Forestry (Australia).

Berg, S. V. 1998. "Introduction to the Fundamentals of Incentive Regulation." In *Infrastructure Regulation and Market Reform: Principles and Practice*, edited by Margaret Arblaster and Mark Jamison, 37–45. Melbourne: Australian Competition and Consumer Commission and Public Utility Research Centre.

Biosecurity Act 2015 (Cth). https://www.legislation.gov.au/C2015A00061/latest/text

Black, J., and R. Baldwin. 2010. "Really Responsive Risk-Based Regulation." *Law & Policy* 32 (2):181–213. https://doi.org/10.1111/j.1467-9930.2010.00318.x.

DAFF. 2022. "Compliance-Based Intervention Scheme (CBIS)." accessed 13 March 2023. https://www.agriculture.gov.au/biosecurity-trade/import/goods/plant-products/risk-return.

Dodge, H. F. 1943. "A Sampling Inspection Plan for Continuous Production." *The Annals of Mathematical Statistics* 14 (3): 264–279.

Dodge, H. F., and M. N. Torrey. 1951. "Additional Continuous Sampling Inspection Plans." *Industrial Quality Control* 7 (5): 7–12.

Grimmett, G., and D. Stirzaker. 2020. *Probability and Random Processes.* 4th ed. Oxford: Oxford University Press.

Hester, S. M., A. Rossiter, A. Robinson, J. Sibley, B. Woolcott, C. Aston, and A. Hanea. 2020. CBIS/CSP sensitivity: Incorporating pre-border information analysis. Centre of Excellence for Biosecurity Risk Analysis.

Laffont, J. -J., and J. Tirole. 1993. *A Theory of Incentives in Procurement and Regulation.* Cambridge, Massachusetts: MIT Press.

Lunn, P. 2014. *Regulatory Policy and Behavioural Economics.* Paris: OECD Publishing.

QCA. 2014. *Incentive Regulation: Theory and Practice.* Queensland Competition Authority.

Robinson, A., J. Bell, B. Woolcott, and E. Perotti. 2012. AQIS quarantine operations risk return: Imported plant-product pathways. Final Report for ACERA 1001 Study J. Australian Centre of Excellence for Risk Analysis, University of Melbourne.

Rossiter, A. 2021. CBIS Sensitivity Module Version 2. University of Melbourne. https://doi.org/10.26188/23734218.v1.

Rossiter, A., and S. M. Hester. 2017. "Designing Biosecurity Inspection Regimes to Account for Stakeholder Incentives: An Inspection Game Approach." *Economic Record* 93 (301):277–301. https://doi.org/10.1111/1475-4932.12315.

Rossiter, A., S. M. Hester, C. Aston, J. Sibley, G. Stoneham, and F. Woodhams. 2016. Incentives for importer choices. Final Report for CEBRA 1304C, Centre of Excellence for Biosecurity Risk Analysis, University of Melbourne.

Rossiter, A., A. Leibbrandt, B. Wang, F. Woodhams, and S. M. Hester. 2018. Testing compliance-based inspection protocols. Final Report for CEBRA 1404C, Centre of Excellence for Biosecurity Risk Analysis, University of Melbourne.

Roth, A. E. 2002. "The Economist as Engineer: Game Theory, Experimentation, and Computation as Tools for Design Economics." *Econometrica* 70 (4):1341–1378. https://doi.org/10.1111/1468-0262.00335.

Sappington, D. E. M. 1994. "Designing Incentive Regulation." *Review of Industrial Organization* 9 (3):245–272. https://doi.org/10.1007/BF01025724.

Stephens, K. S. 2001. *The Handbook of Applied Acceptance Sampling: Plans, Principles, and Procedures.* Milwaukee: ASQ Quality Press.

Stoneham, G., S. M. Hester, J. S.-H. Li, R. Zhou, and A. Chaudhry. 2021. "The Boundary of the Market for Biosecurity Risk." *Risk Analysis* 41 (8):1447–1462. https://doi.org/10.1111/risa.13620.

Section II

Border

5 Screen
Designing Sampling Schemes for Border Inspection

Raphaël Trouvé, Lucie M. Bland, Andrew P. Robinson, Mark J. Ducey, and Susan M. Hester

ABSTRACT

Effective border inspections allow regulators to monitor the presence of pests and diseases at the border, reduce the biosecurity risk at point of entry, and deter wrongdoing. Despite the values at stake, biosecurity inspections are conducted under tight budget constraints and so for nearly all types of goods crossing international borders, inspections are based on a selection of items—a sample—obtained from a consignment. The usual goal of sampling is to make a statistical statement about a collection of items. This chapter presents a range of statistical frameworks that can support decision making about whether an import consignment is compliant or not and whether a pathway is compliant with biosecurity policies or not. We review two commonly used inference frameworks: design-based and model-based inference. We also review adaptive inspection schemes, where some consignments are inspected, and others are not, based on the recent inspection history of the pathway. We conclude the chapter by summarising the pros and cons of using alternative inspection frameworks.

GLOSSARY

Biosecurity risk material (BRM) Living organisms exotic to a country, or biological material carrying an exotic pest or disease. Biosecurity risk material includes live animals and plants, animal and plant material (e.g. fur, feathers, pollen, and flowers), food scraps, and soil.

Sample The portion of a population (the *sampling frame*) that is selected for analysis.

Consignment An import of goods lodged in a single phytosanitary certificate or import declaration. In general, a consignment consists of all the goods for a single consignee that arrives on the same voyage of a vessel. A single consignment may be composed of one or more commodities or lots (lines).

Pathway An import supply chain, often associated with a particular commodity and country of origin; that allows the entry or spread of BRM (passengers, mail, cargo, individual commodities).

Leakage Occurs when a consignment containing BRM that would or should have been detected by an intervention and treated to ameliorate biosecurity risks, but was not. This results in a disease or pest entering a country, with the possibility it becomes established and spreads post-border.

Leakage (post-intervention non-compliance) rate An estimate of the proportion of consignments on a pathway containing BRM that cross international borders after undergoing any required interventions.

DOI: 10.1201/9781003253204-7

Detectability The detectability (or efficacy) of a test or associated with a single unit inspection is the probability of detecting BRM when inspecting a contaminated unit. Please note that detectability is equivalent to the term sensitivity as used in Chapter 6. Detect, whereas sensitivity in the border inspection setting has a different meaning (see below).

Prevalence The prevalence (sometimes called contamination rate, infestation rate, or approach rate) is the proportion of units that contain BRM in a consignment. In a border inspection setting, the numerical attribute studied is typically the prevalence of incoming consignments.

Design prevalence The design prevalence (sometimes called risk cut-off or detection level) is the lower limit of the prevalence we want to detect with a given sensitivity. In the "600-unit samples" rule, the design prevalence is 0.5% (or 1 in 200 units contaminated).

Sensitivity The sensitivity of an inspection (sometimes called the confidence level) is the minimum probability with which we wish to detect at least one instance of BRM in the inspected sample, given that the baseline prevalence is at the design prevalence or higher. In the "600-unit samples" rule, the sensitivity S is 95%.

Sample size The number of units that will be inspected, typically chosen to provide a given sensitivity to detect a given design prevalence. Sample size is usually denoted by the letter n.

Sampling frame A list of numbers corresponding to each unit in the population (e.g. a list of airplane passengers). Sampling is done by randomly selecting identification numbers from the frame and measuring the units that correspond to the selected identifiers (much like pulling numbers out of a hat).

Simple random sampling Is easy to define and difficult to implement. A simple random sample of size n is the outcome of a process for which every possible combination of n units from the population has the same probability of being selected.

Binomial sampling Simple random sampling *with* replacement (where units can be selected more than once) is referred to as binomial sampling. More complex formulae for sensitivity and sample size are available for simple random sampling *without* replacement (also called hypergeometric sampling, see Venette, Moon, and Hutchison 2002). Fortunately, the binomial distribution is conservative in that the required sample size n to yield a desired sensitivity is slightly larger than that computed from the hypergeometric distribution. Binomial sampling also approximates well sampling without replacement when the sample size is much smaller than the population of interest (i.e. when $n \ll$ all the units contained in a consignment).

Cluster sampling With cluster sampling, the population is divided into groups (or clusters). Clusters are randomly selected from the population, and several units are sampled per cluster. For example, we might select individual fruits within selected crates within a container, rather than selecting individual fruits completely at random from the entire container.

Intra-cluster correlation The intra-cluster correlation coefficient (ICC, often written as ρ in equations) characterises the degree of similarity among units within a cluster. An ICC of 1 indicates that all units from the same cluster are identical (i.e. if 1 unit is infested, all other units from the same cluster are infested, and vice versa). An ICC of 0 indicates that two units sampled from the same cluster are no more similar than two units sampled from different clusters.

INTRODUCTION

Invasive species (e.g. arthropods, plants, fungi, and microbial pathogens) pose a significant and growing risk to agricultural systems and native ecosystems worldwide. The impacts of an incursion and the costs of control escalate dramatically once an invader has become established, so effective surveillance at the border is critical in national efforts to reduce the social and ecological costs of invasive species (Quinlan, Stanaway, and Mengersen 2015; Whattam et al. 2014). Despite the values at stake, biosecurity inspections are conducted under tight budget constraints. For nearly all types of goods crossing international borders, inspections are based on a selection of items obtained from a consignment, rather than inspections of the full consignment (IPPC 2008). Indeed, the infrastructure, time, and budget required to inspect everything exhaustively would bring the global economy to a halt. As Beale et al. (2008, ix) stated, "Zero biosecurity risk is unattainable and undesirable".

The goal of sampling schemes is to make a statistical statement about a collection of items. For example, we may wish to estimate the proportion of contaminated fruit in a consignment of oranges that has arrived at the national border. Many methods can be used to detect biosecurity risk material (BRM) in imported goods. Some of the methods are simple (e.g. opening an apple with a knife to check the presence of insect larvae) and others are more complex (e.g. using molecular biology on crushed seeds to detect the presence of viruses; Constable et al. 2019). From a risk-based perspective, the *efficacy* or *detectability* of the test are important (i.e. whether BRM is detected if it is present in the sampled unit) as well as both its cost, which might limit the number of samples that can be inspected, and timeliness, which might also limit the number of samples but also the latency of decision making. When inspection is destructive, such as for the testing of seeds, sampling is the only option to avoid destroying all imports.

At the border, inspection of a sample of a whole consignment is performed to determine whether the consignment is likely to be compliant with biosecurity regulations. If the sample is non-compliant, then further action is taken (e.g. treatment, re-export, or destruction), whereas if no non-compliance is detected, then the consignment is typically released. When deciding whether a consignment is compliant based on the inspection of a sample, a decision maker is typically looking for sufficient assurance that the proportion of BRM in the accepted consignment is below a certain threshold (this threshold is typically above zero which means that some leakage in the accepted consignment might remain). Sampling and inspection for biosecurity compliance are therefore a little different to traditional sampling applications, because the intended outcome is a decision about compliance rather than an estimate of an unknown quantity (e.g. the contamination rate).

Different levels of sampling are implemented depending on the pathway risk and budget available (see Chapter 3. Anticipate and Chapter 14. Map). Sampling can be conducted at the consignment level (i.e. inspecting *all* consignments arriving at the border, but only inspecting a sample from each consignment), or sampling can be conducted at the pathway level (i.e. inspecting only a certain proportion of consignments, and then inspecting all or a selection of units from within the selected consignments; Figure 5.1). The first approach (inspecting all consignments) might be used for pathways that have the potential to carry high-priority quarantine pests and thus is for the purpose of risk mitigation, whereas the latter might be chosen for less risky pathways and so is typically for monitoring (also see Chapter 6. Detect).

Implementing a sampling regime on pathways or consignments prompts an important question, namely: how large a sample should be taken to adequately manage the biosecurity risk of a pathway? Existing international agreements, such as the ISPM-6 and ISPM-31 guidelines of the International Plant Protection Convention (IPPC 2008, 2018; see Chapter 2. Biosecurity Systems and International Regulations), recommend that biosecurity decisions be based on scientifically and statistically sound procedures. However, international agreements provide little guidance on the procedures to be used. A complete answer would require detailed information on all the pests that could be carried by pathways, including count, status, invasiveness, and impact (e.g. economic, environmental, social, and human health), as well as the effectiveness of any other measures that are deployed to manage the risk. This is clearly an impossible burden.

Perhaps as a result, a wide range of applications rely on a small set of sampling and statistical approaches that have become accepted as common practice. One example is the 600-unit sample "rule", which has been adopted as standard practice in a range of situations (Ransom 2017; Rossiter and Hester 2017). In this approach, a sample size of 600 units is nominated based on historical familiarity and stakeholder acceptance rather than based on specific knowledge about the threats that create the risk that is being managed, which often would present an unreasonable burden of expectations on the decision maker.

Unfortunately, in biosecurity as in other areas, the conceptual basis for drawing inferences from sample data is rarely specified (see Gregoire 1998 for a useful discussion). This lacuna can lead to confusing or even contradictory interpretations of key statistical terms, with important consequences for what is considered an acceptable sampling design or what is valid inference once data have been collected. A poor understanding of sampling design and inference frameworks might cause the inspection system to be less efficient than intended and allow more leakage of BRM than expected, with negative consequences for the management of risk across the biosecurity continuum.

The goal of this chapter is to provide guidance on the level of sampling effort required to reach a degree of assurance about the prevalence of BRM in consignments that are accepted by a jurisdiction. We discuss the *design-based* approach to sampling, where inferences about a population are based on the characteristics and design of the sampling process. This is the approach most commonly used by biosecurity agencies to assess biosecurity risks at the border. *Model-based* inference is an alternative approach, where a model is postulated as generating the data; statistical inferences are then made based on that model (Dumelle et al. 2022; Gregoire 1998).

We also clarify some misconceptions about different inference frameworks, focusing primarily on design-based and model-based inference and comparing those with the Bayesian inferential framework—a form of model-based inference in which prior beliefs about uncertain quantities or parameters are updated after observing the data. Finally, we review adaptive inspection schemes, a group of methods where some consignments are inspected, and others are not, based on the recent inspection history of the pathway.

DESIGN-BASED INFERENCE

Design-based inference is one of several frameworks that can be used to develop assurance about regulatory compliance of consignments, and is the main type of inference used for border biosecurity inspections (Trouvé et al. 2020). Design-based inference is based on the premise that conclusions can be drawn about the population from the sample, because we know exactly how the sample was collected. The sample is collected using one of several sample designs—commonly, simple random sampling, systematic sampling, cluster sampling, or two-stage sampling. These are discussed below, firstly in terms of possible border inspection scenarios (Figure 5.1)—depending on whether the purpose of inspection is to mitigate risk (consignment-level inspection) or to undertake monitoring (pathway-level inspection)—and secondly, in terms of sample size formulas. All *consignments*—which may be composed of one or more commodities or lots (lines)—arriving at the border are assumed to be inspected via random sampling of a certain proportion of units or goods within each consignment.

Consignment-Level Inspections

The purpose of most consignment-level inspections is risk mitigation—to determine whether the amount of a pest or disease in the consignment (prevalence; p) exceeds a maximum allowable amount (design prevalence, p^*; left side of Figure 5.1a). All consignments arriving at the border will be inspected via a sample from each consignment (following, for example, MIL-STD-1916; see Department of Defense 1996, 1999). The chosen sampling design will depend on whether the units within the consignment are "clustered" in groups of similar or related units within the population (Figure 5.1b). For example, fruits packed within cartons, seeds within sacks, or even passengers travelling within family groups can all be considered as clusters of sampling units. If no clustering exists, a random sample should be taken; specifically, a simple random sample or a systematic random sample. For large consignments (relative to sample size) the binomial formula should be used to determine sample size (Eq. 5.1, Box 5.1); for small populations relative to sample size (e.g. sample size >10% of consignment size) the hypergeometric formula for sample size may be used and may lead to some savings (this approach is omitted here for clarity). If clustering does exist—that is, the units are packed in boxes or bags—but there is no further clustering within those boxes or bags, then cluster sampling or two-stage sampling should be undertaken. Where clustering does exist within the boxes or bags, model-based estimation should be used to obtain the correct sensitivity and the required sample size based on the estimated clustering of BRM.

Pathway-Level Inspections

Sample-based pathway-level inspections are usually applied to less risky pathways and so are typically used for monitoring pathway compliance (Figure 5.1, right side of Panel A). Under this approach, only a sample of consignments is chosen for inspection. When no auxiliary information is available about each consignment (e.g. its country of origin or supplier) a *systematic* or *simple*

BOX 5.1. SAMPLE SIZE USING BINOMIAL SAMPLING

Under *binomial sampling*, given a BRM rate (prevalence) p, a probability of detection (d), the probability of observing compliance after inspecting just 1 unit that was randomly sampled from the consignment is $1 - dp$. Then the probability of observing compliance after inspecting n units is $(1 - dp)^n$, and the probability of non-compliance S (namely, the sensitivity of the inspection) follows from the laws of probability:

$$S = 1 - (1 - dp)^n \qquad \text{(Eq. 5.1)}$$

We can rearrange Eq. 5.1 to solve for n as a function of S and p^*, which is the design prevalence we are willing to accept. The sample size is as follows:

$$n = \frac{\ln(1 - S)}{\ln(1 - dp^*)} \qquad \text{(Eq. 5.2)}$$

Under the *binomial* assumptions (namely, (1) that sampling is undertaken with replacement; (2) sampling of each unit is independent and random; and (3) the sample size is fixed and known in advance) and with $p^* = 0.005$ and $S = 0.95$, Eq. 5.2 gives a required sample size of $n = 598$, which is often rounded to $n = 600$, hence the "600-unit samples" rule. In an ideal world, the design prevalence (p^*) and sensitivity (S) would depend on the likely prevalence of risk material within a particular type of incoming material and the risk associated with an incursion. However, the "600-unit sample" and its underlying $p^* = 0.005$ and $S = 0.95$ appear to have become a default for historical reasons, much in the same way that Fisher's tentative suggestion of a p-value of 0.05 (about one time in twenty, Fisher 1925) has evolved into a de facto scientific standard.

random sample should be taken. When auxiliary information is available, the pathway may be stratified and different cohorts of consignments should be sampled at different rates using profiling (see
Chapter 13. Profiling and Automation for a more complete description), and a simple or a systematic
random sample of consignments should be taken within each stratum. Where no intervention data
are available, consignments should be selected for inspection via *CSP-based algorithms* (see below
and Chapter 4. Prevent).

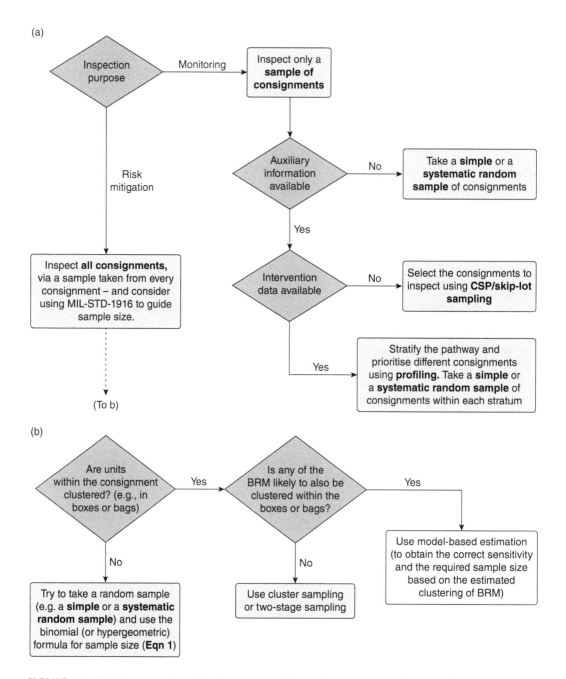

FIGURE 5.1   Decision trees describing the purpose of inspection and some of the sampling methods used.
Panel (a) details the sampling decisions where monitoring is the purpose. Panel (b) describes sampling decisions where risk mitigation is the purpose.

Sampling Methodologies

We now briefly introduce the most commonly used sampling methodologies.

Simple Random Sampling

Most presentations of simple random sampling follow a simple setup. Based on a list (or frame) spanning the population of interest, simple random sampling proceeds by selecting n (the sample size) units from the list in such a way that all possible combinations of n units have the same probability of being selected. For example, the passenger list of an inbound aircraft is a frame from which a simple random sample of passengers can be drawn (Lane et al. 2017).

When the number of incoming sample units is small (e.g. shipments of large or high-value items such as tropical logs), it may be possible to construct a list on arrival. When the number of incoming units is great or incoming units arrive in bulk, a list or other convenient frame may not be available, but the absence of a list-based frame does not prevent the use of simple random sampling. It is possible to physically mix some commodities, akin to shuffling a deck of cards, so that the selection of sample units is uniform and random. Many nuts and seeds can be handled in this way, although large shipments may already be subdivided into sacks or other containers, which would suggest a cluster sampling approach. For other commodities, the geometry of packing facilitates simple random sampling. If nursery seedlings are packed in flats (trays) in which individuals are laid out with rectangular or hexagonal spacing, and the flats themselves are arranged in a regular fashion, then the location of a flat, combined with a row and column number for a position within a flat, constitutes a form of "address" that can be sampled randomly if the number and configuration of flats are known.

Importantly, one needs to be aware of sampling contexts where it might appear that simple random sampling is taking place, but it isn't. For example, Barron (2006) reports that bananas are typically packed in cartons containing on average 16 bunches (actual numbers may vary), and a consignment might consist of 20,000 or more rectangular cartons. Although it is possible to sample the cartons by simple random sampling, if one then selects a single bunch of bananas from a chosen carton by simple random sampling, then bunches from cartons containing larger numbers of bunches have a lower probability of selection than those containing fewer bunches, which is not simple random sampling. Instead, this approach is two-stage sampling (see below).

Systematic Sampling

Systematic sampling is a potentially efficient alternative to simple random sampling that is used when the sampling units have some spatial or temporal structure, such as units in a container, or on a conveyer belt. A systematic sample is taken by selecting a random starting point and then inspecting every kth unit, where k is set to achieve the desired sample size n. For example, if we know the consignment has 100 units and we require a 9-unit sample (to keep the example simple) then we would start randomly somewhere between units 1 and 10 inclusive and take every 10th unit.

Stratified Sampling

Stratified sampling is a way of organising the sampling units before they are selected. Briefly, all the units are classified into cohorts (strata) and then each stratum is sampled as though it were a unique process. Stratified sampling therefore requires extra information about each unit, which is used to allocate the unit into the appropriate stratum; this is called auxiliary information. Splitting the process up this way can have benefits in efficiency and simplicity although it creates additional administrative overheads in managing the extra information needed.

Cluster and Two-Stage Sampling

Clustering of goods to be inspected arises naturally in a variety of biosecurity contexts (Barron 2006; Hughes, Madden, and Munkvold 1996). For example, fruits packed within cartons, seeds within sacks or passengers travelling in family groups can all be considered clusters of sampling units. Clustering reduces inspection sensitivity because it reduces the chances of finding infested units in an inspection. If the data come in clusters and we can conduct simple random sampling (i.e. each unit is equally likely to be sampled), then we can ignore the clustered nature of the data and use formulae for non-clustered data (Eqs. 5.1 and 5.2). However, in some cases, it may be wiser to take advantage of the clustered nature of the population. Two-stage sampling allows sampling of more than 1 unit per cluster.

Cluster and two-stage sampling recognise the hierarchical structure in the population by designating primary and secondary sampling units. Suppose a consignment consists of fruits packed within cartons, then our first step is to select a predetermined number of cartons at random, followed by a selection of fruits within the chosen cartons. In cluster sampling, once a carton is selected, all fruits in that carton are included in the sample; the chosen fruits form a cluster. In two-stage sampling, the fruits within a carton are chosen by a further subsampling approach (e.g. simple random sampling of fruits within that carton). Compared to our banana example above, two-stage sampling allows sampling of more than 1 unit per cluster, which introduces a complication in determining the number of cartons and/or fruits that should be sampled. This complication is beyond the remit of the current text but see Cochran (1977) for technical details.

For cluster or two-stage sampling to be attractive, it must offer benefits in terms of cost, practicality, or the opportunity to inspect other aspects of the consignment (e.g. packing materials). Design-based inference avoids unnecessary assumptions about the sampled population, but it does so at the cost of ignoring background information (e.g. clustering in past consignments). Using past information to formulate more efficient inference is a strength of model-based and Bayesian inferences.

DISCUSSION OF DESIGN-BASED INFERENCE

If contamination is detected in any unit in the inspected sample, then the consignment is rejected; if we detect no contamination, we accept the consignment. Formally, this sampling plan is an *acceptance sampling plan* (Stephens 2001). In acceptance sampling plans, a sample is drawn at random from the lot, and based on the sample information, we decide on the fate of the lot (i.e. we either accept or reject the lot).

Typically, biosecurity agencies like to be very confident that even a very small amount of a pest or disease, if present, would have been detected. We might naively prefer to be able to say that a consignment has zero prevalence (i.e. $p^* = 0$) but proving this would require 100% inspection of the consignment, which can rarely be justified (recall Beale et al. 2008s dictum against zero risk!). Instead, a more useful goal is to aim to state that $p < p^*$ with a high level of confidence where p^* is a very small value. A common choice in biosecurity is a design prevalence of $p^* = 0.005$ (1 unit contaminated per 200 units) and a *sensitivity* (*S*) of 95% (IPPC 2008; Venette et al. 2002). A sensitivity of 95% means that, at the design prevalence, only 5% of consignments are accepted when they should have been rejected; these are also called false negatives. When these values of sensitivity and design prevalence are used in the binomial formula for sample size (Box 5.1), the result is a required sample size (*n*) of 598, which is often rounded to $n = 600$—the 600-unit sample rule.

Although in theory, setting the design prevalence and level of confidence of an inspection allows the required sample size for an inspection to be determined (Eq. 5.2), in practice, the reverse process often prevails: the regulator inspects 600 units per consignment because it seems an appropriate amount of effort, and reports the corresponding 95% confidence level and 0.5% design prevalence. A more in-depth analysis is needed, however, to determine the optimal sample size, along with biological and economic information. Determining an acceptable *leakage rate* for a pathway—estimating the proportion of consignments containing BRM that enter a country after undergoing any required interventions or considering the economic costs of eradication—can help calculate inspection sample size. As an example of this kind of thinking, Lane et al. (2018) used maximum average leakage rate as a basis for computing sample size. This method suggests using a higher sample size for larger consignments than for smaller consignments—for example, if two consignments having the same prevalence are accepted into a country, the larger consignment will contain more BRM and thus pose a higher biosecurity risk.

ENHANCED SAMPLING METHODOLOGIES

Although the previously described sampling methods are widely embedded and accepted by biosecurity agencies, there are other techniques that are more sophisticated in their approach to the use of sample data. While such sophistication has advantages in efficiency, the techniques are typically unavailable to biosecurity agencies because in practice the data required to implement them is seldom collected. These approaches are summarised below.

MODEL-BASED INFERENCE

Model-based inference may be used to develop assurance about regulatory compliance of consignments, although its use for this purpose is proven challenging because of its technical requirements (Trouvé et al. 2020). In model-based inference, a probabilistic model provides the foundation to extrapolate from the sample to the population—we postulate a model that might have generated the inspection data (e.g. a binomial model), check the assumptions of the model and then make inferences about the contamination rate of the consignment. One of the advantages of model-based inference is that it allows the use of external information (e.g. external predictors, or information about the clustering of the data taken from, e.g. the published literature). When data come from simple random sampling, model-based inference postulated to have been generated by a binomial model gives the same sensitivity and sample size as design-based inference (Eq. 5.2).

The key difference between design-based and model-based inference is how inference from the sample is conceptually connected to the target process. In design-based inference, we can draw conclusions about the target process from the sample because we know exactly how the sample was collected. The population from which we sample is viewed as fixed; randomness arises through the selection of sample units. In model-based inference, the variable that we observe on the sample units is treated as a random outcome from an unknown process. Within a model-based framework, we are concerned with how well the data and the mechanisms that generated the data conform to our assumed model. If that conformity is poor, then the inferences from our survey will be fragile. However, certain aspects of the design (e.g. the use of subjective sampling) may not create serious problems.

Recent texts outlining a model-based perspective include Valliant, Dorfman, and Royall (2000) and Chambers and Clark (2012). Model-based inference is often considered to involve a hypothetical "superpopulation", since the actual population can be considered as just one random realisation of the possible populations generated by the underlying model. Inference is based not on the randomisation distribution created through the inclusion of units under a sampling design but on the distributions assumed in the model itself. Such a shift means that exactness in the specification of the design is less important than the reasonableness of the model.

BAYESIAN INFERENCE

Bayesian inference is rarely used in regulatory settings as far as we are aware, so we mention it only briefly. Bayesian inference is in essence a form of model-based inference with its origins in the work of Bayes (1763) and Laplace (1812). Central to Bayesian inference is the use of Bayes' Theorem to update prior knowledge or belief about unknown quantities after observing the data. The workflow proposes the potential values that the parameter of interest might take (e.g. prior information on potential prevalence in incoming consignments derived from past data or expert elicitation, referred to as the *prior distribution*) as well as a model that might have generated the inspection data (e.g. a binomial model, referred to as the *likelihood function* of the model). The prior distribution and the model with the inspection data are then combined to make an inference on the parameter of interest (e.g. the prevalence of the consignment after inspection, referred to as the *posterior distribution*). In Bayesian inference, external sources of information other than past data can be used as a prior. In the absence of past data on a pathway, structured expert elicitation (see Chapter 12. Elicit) can be a good starting point.

ADAPTIVE INSPECTION SCHEMES

Adaptive inspection schemes provide methods to choose to inspect or not inspect consignments based on the recent inspection history of the pathway. In a sense, adaptive inspection schemes use recent history to provide assurance about the pathway and the next consignment.

The oldest such scheme is the continuous sampling plan (CSP) family of rules. In the biosecurity context, a CSP is a rule for determining whether to inspect a consignment based on the recent inspection history of the pathway and parameters set by the pathway manager, thereby rewarding consistently compliant importers with, for example, reduced inspection rates. The simplest instance is CSP-1, which requires the user to choose:

- the clearance number, *CN*—the number of consecutive consignments that must pass inspection for the importer to be eligible for a reduced inspection frequency; and
- the monitoring fraction, *MF*—the reduced inspection frequency; that is, the probability a given consignment is inspected in "monitoring mode".

Imagine that the pathway is being managed importer by importer, that is, a unique inspection history is being kept for each importer. The importer starts in "census mode", meaning that every consignment is inspected. Once *CN* consecutive consignments pass inspection the importer moves to monitoring mode, in which consignments are inspected with probability *MF* (randomised). If any of these consignments are non-compliant then the importer returns to census mode. The advantage to the importer of being in monitoring mode is convenience: their goods are available as soon as the paperwork is complete.

The CSP-3 variant is more forgiving (Chapter 4. Prevent). This requires two extra parameters, namely:

- Tight census number, *TC*—the number of consignments that must be inspected in temporary census mode after a fail is detected—usually 4; and
- *k*, which is the minimum permitted spacing of two detected contaminated consignments (usually the same as *CN*), in other words, if a subsequent fail is detected within *k* inspections of any other fail (TC of which will be in tight census) then the pathway moves to census mode.

If an importer's consignment fails inspection in monitoring mode, then their next *TC* consignments are subject to mandatory inspection in what is referred to as "tight census mode". If another failure is detected then the importer returns to census mode; if all four consignments pass inspection the importer's next consignments will be inspected at the lower inspection rate until *k* consignments are passed, at which time the importer passes back into monitoring mode.

CSP-1 is easier to communicate to stakeholders, who are thus likely to develop a clearer understanding of the incentives embedded in the inspection rule (Rossiter et al. 2018). Rossiter and Hester (2017) suggested that the CSP-1 algorithm would be preferable from the biosecurity regulator's perspective, particularly where the consequences of BRM leakage are large. Under the scenarios modelled, CSP-1 and CSP-3 (the report didn't test CSP-2) had higher payoffs to the regulator than mandatory inspection, and of those two, CSP-1 had the highest payoff to the regulator (Rossiter and Hester 2017). Usage of the CSP family of plans can be guided by the average outgoing quality (AOQ) curve or the average outgoing quality limit (AOQL), explained in detail in Chapter 4. Prevent.

Each operational parameter can be classified according to its effect on the sampling process and its interpretation in terms of managing biosecurity risk. Regardless of the system, most operational parameters can be classified into one of three types, namely (i) the number of inspected and compliant consignments required to reduce the prescribed inspection level, (ii) the number of inspected and non-compliant consignments required to increase the prescribed inspection level, and (iii) the amount of effort at any given level (number of consignments to be inspected or number of units per consignment).

For example, in CSP-3, each of the four parameters has at least one operational interpretation:

- The clearance number, CN, sets the number of consecutive compliant consignments required to convince the inspector that the pathway can be monitored instead of fully inspected. This can be interpreted as a required level of assurance or an incentive (see Chapter 4. Prevent and Chapter 8. Incentives). That is, the clearance requirement becomes an incentive for stakeholders to ensure that the pathway achieves and maintains a high level of compliance.
- The monitoring fraction, MF, sets the amount of effort imposed on the pathway during monitoring and can be interpreted as a required level of assurance or an incentive. That is, the reduction in sampling effort is a reward to stakeholders for high compliance.
- The minimum spacing of two detected contaminated consignments, k, provides some sensitivity to what constitutes a change in pathway status as opposed to an isolated incident. Having seen one contaminated consignment, we are effectively on alert. How many clear consignments do we need to see before we can relax again? It is also an incentive for stakeholders to reduce the contamination rate to avoid going back to census mode.
- The duration of temporary 100% inspection, TC, sets a preference for how long the pathway is scrutinised after detecting non-compliance, providing further sensitivity when discerning between isolated incidents and important changes in pathway status.

Such schemes have been in use for some time in both Australia (the Department of Agriculture, Fisheries and Forestry's Compliance-Based Intervention Scheme and the Department of Agriculture's Compliance-Based Inspection Scheme, CBIS; DAWE 2021) and New Zealand (risk-based inspection of low-risk fresh produce such as green beans imported from Australia).

Adaptive sampling plans such as CSP involve electing to not inspect some consignments. Consequently, they should only be used on pathways in which some leakage can be tolerated. The principal value of these plans is that they naturally tend to allocate more inspection effort to domains of the pathway that have higher *interception rates* without requiring predictive statistical modelling or machine learning (see Chapter 13. Profiling and Automation). The downside is that adaptive sampling plans will always experience a delay in response to changes in the pathway contamination rate, as these plans are based on past inspection results. In comparison, predictive modelling allows for the possibility of anticipation when pathway contamination is associated with external predictors (season, region of origin of the commodity, etc) and we can monitor or forecast changes in the predictors' values.

TOLERATING LEAKAGE

Although ideally the effect of border intervention would be to totally prohibit entry by contaminated items, the reality is that border intervention can only intercept a portion of contamination (e.g. reduce the approach rate). This would be true even if every consignment and unit were inspected, as some infections are sub-clinical (leading to imperfect detections) and some pathways cannot be regulated (e.g. wind and tide).

One of the most stringent border intervention operations is that imposed by Chevron Corporation for entry into Barrow Island, an environmental heritage area off the coast of Western Australia. Chevron has close to total control over the pathway, wrapping equipment before shipping and undertaking careful inspection of all passengers. Nonetheless, the careful post-border surveillance exercise ongoing at that location shows that leakage still occurs. It is impossible to imagine that a government regulator would have the resources, or indeed the license, to impose such a stringent system (Scott et al. 2017).

Because there is no chance that a border intervention will entirely eliminate risk, the goals of a border intervention should instead be: (1) verify the overall compliance of the pathway to biosecurity regulation, and (2) where necessary and as part of an end-to-end biosecurity framework (see Chapter 2. Biosecurity Systems and International Regulations), reduce the approach rate sufficiently that it is impossible for pests to establish a minimum viable population (MVP). Note that this prescription is not the same as keeping the pathway contamination rate below a given level. For example, half an MVP of pest A and half of an MVP of pest B will not result in an establishment.

This view of border intervention has important implications. In risk-based reduced intervention regimes (e.g. adaptive sampling schemes), the probability of contamination leakage will increase— at least on the part of the pathway that experiences effort reduction. Unfortunately, this element of risk-based intervention is enough to deter some regulators. As noted above, even the most stringent border intervention program cannot guarantee zero leakage. A risk-based approach may increase the leakage marginally, but at a substantial cost saving that may provide the opportunity to decrease risks elsewhere (see Chapter 9. Resource Allocation).

Another factor in tolerating leakage is the need to gather information about portions of the operation that one does not consider risky. Given a risk profile, it is tempting to assume that: (1) the risk profile is known exactly, and (2) the risk profile will never change, and so it is safe and reasonable to devote available resources to reducing known risks. However, risks are never known exactly in operational settings, and risks are not necessarily constant, so it is critically important that the regulator develops some mechanism for assessing the quality and timeliness of the information that underpins risk profiles (see Chapter 3. Anticipate). This means that a small amount of inspection needs to be undertaken on all pathways to guard against changes in prevalence, possible fraud, or in case the original assessment was wrong (see also Chapter 13. Profiling and Automation).

More sophisticated approaches can be taken to allocating effort at the border. Border inspections are not only applied to stop pests at a point of entry but are also used to monitor pathway risk and make decisions (e.g. to shut down a pathway). From this perspective, we also want to choose a sample size that is sufficient to detect important changes in prevalence and quarantine pests. Recognising the trade-off between the need to detect as many contaminated items as possible and the need to acquire knowledge about contamination rates, Robinson, Burgman, and Cannon (2011) developed an algorithm to guide the allocation of inspection resources based on classifying pathways in high- and low-risk categories, using past data. Pathways in the high-risk category are fully inspected, while pathways in the low-risk category are monitored at a rate that would detect with high probability their chance of switching to the high-risk category. Robinson et al. (2011) showed that focusing inspection resources on riskier pathways improved *inspection efficiency*, while modest investments in monitoring low-risk pathways were a valuable source of intelligence.

The sample size can also be determined by optimising an objective function, such as minimising total leakage across pathways or minimising cost. Chen, Epanchin-Niell, and Haight (2018) allocated sampling effort among pathways to minimise the total leakage of infested units coming into the United States. Pathways with high prevalence ended up with a higher sample size than pathways with low prevalence. Since the likelihood of establishment and damages caused by an incursion are likely to vary by pest, this type of analysis requires a thorough understanding of the pathway and biosecurity system involved. There is also the danger of model misspecification or calibration errors which can be damaging to biosecurity, especially if the risk profile is volatile (Powell 2015).

IN A NUTSHELL

- Border inspection plays a dual role of monitoring and reducing biosecurity risk at the border. Given the large volumes of goods arriving at the border, border inspection must rely on sampling rather than a full census.
- Zero risk is unattainable and undesirable, the regulator must decide on the level of confidence required to detect certain levels of contamination, and relate this to an appropriate sample size using a statistical inference framework, usually design-based inference.
- The "600-unit sample" rule is used most often to calculate sample size, but determining optimal sample size, requires a more in-depth analysis of leakage rates and biological and economic information.
- When certain data are available, the more sophisticated techniques of model-based inference and Bayesian inference might be more appropriate ways of understanding results of inspections.
- Adaptive inspection schemes provide methods to choose to inspect or not inspect consignments based on the recent inspection history of the pathway.

REFERENCES

Barron, M. C. 2006. "Effects of Aggregation on the Probability of Detecting Infestations in Fresh Produce Consignments." *New Zealand Plant Protection* 59:103–108. https://doi.org/10.30843/nzpp.2006.59.4435.

Bayes, T. 1763. "An Essay Towards Solving a Problem in the Doctrine of Chances." *Philosophical Transactions of the Royal Society of London* 53: 370–418.

Beale, R., J. Fairbrother, A. Inglis, and D. Trebeck. 2008. One Biosecurity: A Working Partnership. Quarantine and Biosecurity Review Panel, Department of Agriculture, Fisheries and Forestry (Australia).

Chambers, R. L., and R. G. Clark. 2012. *An Introduction to Model-Based Survey Sampling With Applications.* Oxford, UK: Oxford University Press.

Chen, C., R. S. Epanchin-Niell, and R. G. Haight. 2018. "Optimal Inspection of Imports to Prevent Invasive Pest Introduction." *Risk Analysis* 38 (3):603–619. https://doi.org/10.1111/risa.12880.

Cochran, W. G. 1977. *Sampling Techniques.* 3rd ed. New York: Wiley.

Constable, F., G. Chambers, L. Penrose, A. Daly, J. Mackie, K. Davis, B. Rodoni, and M. Gibbs. 2019. "Viroid-Infected Tomato and Capsicum Seed Shipments to Australia." *Viruses* 11 (2):98. https://doi.org/10.3390/v11020098.

DAWE. 2021. "Compliance-Based Intervention Scheme (CBIS)." Last Modified 17 September 2021. https://www.agriculture.gov.au/import/goods/plant-products/risk-return.

Department of Defense. 1996. MIL-STD-1916 Department of Defense Test Method Standard: DoD Preferred Methods for Acceptance of Product.

Department of Defense. 1999. Companion Document to MIL-STD-1916.

Dumelle, M., M. Higham, J. M. Ver Hoef, A. R. Olsen, and L. Madsen. 2022. "A Comparison of Design-Based and Model-Based Approaches for Finite Population Spatial Sampling and Inference." *Methods in Ecology and Evolution* 13 (9):2018–2029. https://doi.org/10.1111/2041-210X.13919.

Fisher, R. A. 1925. *Statistical Methods for Research Workers.* Edinburgh and London, UK: Oliver and Boyd.

Gregoire, T. G. 1998. "Design-Based and Model-Based Inference in Survey Sampling: Appreciating the Difference." *Canadian Journal of Forest Research* 28 (10):1429–1447. https://doi.org/10.1139/x98-166.

Hughes, G., L. V. Madden, and G. P. Munkvold. 1996. "Cluster Sampling for Disease Incidence Data." *Phytopathology* 86:132–137.

IPPC. 2008. International Standard for Phytosanitary Measures. ISPM No. 31: Methodology for Sampling of Consignments.

IPPC. 2018. International Standard for Phytosanitary Measures. ISPM No. 6: Surveillance.

Lane, S. E., R. Gao, M. Chisholm, and A. P Robinson. 2017. Statistical profiling to predict the biosecurity risk presented by non-compliant international passengers. In *arXiv preprint arXiv:1702.04044.*

Lane, S. E., R. Souza Richards, C. McDonald, and A. P. Robinson. 2018. Sample size calculations for phytosanitary testing of small lots of seed. Centre of Excellence for Biosecurity Risk Analysis. The University of Melbourne.

Laplace, P. S. 1812. *Théorie Analytique Des Probabilités*. Paris: Courcier.

Powell, M. R. 2015. "Risk-Based Sampling: I Don't Want to Weight in Vain." *Risk Analysis* 35 (12):2172–2182. https://doi.org/10.1111/risa.12415.

Quinlan, M., M. Stanaway, and K. L Mengersen. 2015. "Biosecurity Surveillance in Agriculture and Environment: A Review." In *Biosecurity Surveillance: Quantitative Approaches*, edited by Frith Jarrad, Samantha Low-Choy and Kerrie L. Mengersen, 9–42. Wallingford, UK: CABI.

Ransom, L. 2017. "Australia's Experience with Risk-Based Sampling." International Symposium for Risk-Based Sampling, Baltimore, Maryland, 26–30 June, 2017.

Robinson, A., M. A. Burgman, and R. Cannon. 2011. "Allocating Surveillance Resources to Reduce Ecological Invasions: Maximizing Detections and Information About the Threat." *Ecological Applications* 21 (4):1410–1417. https://doi.org/10.1890/10-0195.1.

Rossiter, A., and S. M. Hester. 2017. "Designing Biosecurity Inspection Regimes to Account for Stakeholder Incentives: An Inspection Game Approach." *Economic Record* 93 (301):277–301. https://doi.org/10.1111/1475-4932.12315.

Rossiter, A., A. Leibbrandt, B. Wang, F. Woodhams, and S. Hester. 2018. Testing Compliance-Based Inspection Protocols. Centre of Excellence for Biosecurity Risk Analysis. The University of Melbourne.

Scott, J. K., S. J. McKirdy, J. van der Merwe, R. Green, A. A. Burbidge, G. Pickles, D. C. Hardie, K. Morris, P. G. Kendrick, M. L. Thomas, K. L. Horton, S. M. O'Connor, J. Downs, R. Stoklosa, R. Lagdon, B. Marks, M. Nairn, and K. Mengersen. 2017. "Zero-Tolerance Biosecurity Protects High-Conservation-Value Island Nature Reserve." *Scientific Reports* 7 (1):772. https://doi.org/10.1038/s41598-017-00450-y.

Stephens, K. N. 2001. *Handbook of Applied Acceptance Sampling: Plans, Procedures, and Principles*. Milwaukee, WI: American Society for Quality Press.

Trouvé, R., M. Ducey, R. Souza-Richards, D. Dall, and A. Robinson. 2020. Alternative approaches to developing assurance about the regulatory compliance of consignments of plant products. Centre of Excellence for Biosecurity Risk Analysis. The University of Melbourne.

Valliant, R., A. H. Dorfman, and R. M. Royall. 2000. *Finite Population Sampling and Inference: A Prediction Approach*. New York: Wiley.

Venette, R. C., R. D. Moon, and W. D. Hutchison. 2002. "Strategies and Statistics of Sampling for Rare Individuals." *Annual Review of Entomology* 47 (1):143–174. https://doi.org/10.1146/annurev.ento.47.091201.145147.

Whattam, M., G. Clover, M. Firko, and T. Kalaris. 2014. "The Biosecurity continuum and Trade: Border Operations." In *The Handbook of Plant Biosecurity: Principles and Practices for the Identification, Containment and Control of Organisms That Threaten Agriculture and the Environment Globally*, edited by Gordon Gordh and Simon McKirdy, 149–188. Dordrecht: Springer Netherlands.

Section III

Post-Border

6 Detect
Designing Post-Border Surveillance Schemes

James S. Camac

ABSTRACT

Pre-border and border controls reduce (but do not eliminate) the risk of biosecurity threats entering a country, so regulators must prepare for the eventuality of pests and disease incursions. Post-border surveillance is the primary tool used to detect new outbreaks, delimit and manage existing outbreaks, declare eradication success, and provide evidence to trading partners that a threat is absent to gain or maintain market access. This chapter outlines the three broad types of post-border surveillance (active, general, and passive) and reviews their pros and cons. General and passive surveillance are useful sources of first detections but are of limited use for inferring the likelihood of threat absence, due to issues with unknown surveillance reliability and effort. Active surveillance can be used to inform multiple surveillance objectives, but its cost restricts its use to small geographic areas and high-risk threats. Overall, the amount of surveillance to be implemented depends on the objective of surveillance, the reliability of the surveillance method, and the regulator's tolerance for failing to detect a threat that is present. For all objectives, the surveillance strategy and effort required should represent best value for money (i.e. minimising the costs of surveillance and potential costs associated with the threat).

GLOSSARY

Area freedom Declaration of the absence of a threat from a region (e.g. a production zone) for the purposes of access to international or domestic markets.

Proof of freedom Declaring a threat as absent from a region after eradication measures have been successful.

Active surveillance A deliberate and coordinated surveillance effort designed to detect new or priority pests and diseases. Targeted surveillance is a form of active surveillance that is optimised to detect a particular threat.

General surveillance A semi-coordinated, multi-threat focused, surveillance effort commonly implemented by environmental stakeholders (e.g. park rangers), agricultural stakeholders (e.g. farmers, agronomists, vets, and industry organisations), scientists, and/or citizen science groups.

Passive surveillance The chance detection and reporting of threats by the public. Passive surveillance is the most fortuitous and accidental type of surveillance, with little structured or coordinated surveillance effort or reporting framework.

Sensitivity The probability a single surveillance unit (e.g. survey, trap, or test) will detect a threat assuming it is present. Imperfect sensitivity (<1) increases the likelihood of false negatives (i.e. declaring a threat absent when it is in fact present).

Surveillance effort A measure of the amount of effort undertaken to detect a threat, which could be in terms of the number of traps, survey hours, site revisits, or diagnostic tests.

DOI: 10.1201/9781003253204-9

Tracing An intelligence-gathering exercise that aims to identify an incursion's plausible means of introduction and secondary spread.

Natural detection point The point where the pest or disease becomes self-evident (i.e. 100% probability of detection), even without expenditure on active surveillance or additional passive surveillance.

INTRODUCTION

Pre-border and border controls attempt to reduce the risk of exotic species and diseases entering and establishing in a jurisdiction, but these controls do not eliminate risk entirely. A jurisdiction's biosecurity system reduces risk to an appropriate level of protection (ALOP) that balances trade-offs so that the risk of threat entry and establishment is acceptable by the public, industry, and trading partners (see Chapter 2. Biosecurity Systems and International Regulations). The consequence of not reducing risk to zero is that from time to time, even a well-managed biosecurity system will have post-border incursions that establish and spread. In an era of increasing globalisation (Hulme 2009, Seebens et al. 2017), regulators must plan for this eventuality.

Post-border surveillance is the primary tool regulators use to detect new outbreaks before they spread so widely that they become infeasible to eradicate and/or cause significant economic, social, and environmental impacts. It is also used to infer confidence that a threat is absent from a region of interest, which is important for both maintaining and regaining market access for some commodities. Post-border surveillance also plays a role in informing on-the-ground biosecurity response strategies.

This chapter describes the various objectives of post-border surveillance (namely area freedom, early detection, delimitation, and monitoring) and the surveillance methods that are most appropriate to achieving each. These methods (active, general, and passive surveillance) can all play an important role in detecting biosecurity threats. The practical considerations in selecting a method, including where to undertake surveillance, how to infer absence from surveillance data, and how much to spend on surveillance, are also discussed.

OBJECTIVES OF POST-BORDER SURVEILLANCE

Post-border surveillance provides the necessary evidence to inform four objectives: area freedom, early detection, incursion delimitation, and progress towards post-incursion management objectives (e.g. eradication or containment).

SURVEILLANCE FOR MARKET ACCESS (AREA FREEDOM)

To mitigate exposure to high-impact biosecurity threats, many countries (and some jurisdictions within countries) impose restrictions on the movement and trade of plants, animals, and associated goods. To export these items across international borders and to secure or maintain access to premium export markets, governments and industries are required to provide evidence to trading partners that their region or production zones are free from agricultural threats. This evidence of absence, referred to as area freedom, typically involves various forms of post-border surveillance complemented by other sources of information, such as scientific publications, research data, field observations, and other non-survey data (IPPC 2017). The quality and quantity of such evidence is ultimately governed by the requirements of the trading partner, the potential impact of a threat, and the international standards imposed by the World Trade Organisation (see Chapter 2. Biosecurity Systems and International Regulations). The globalisation of human movement and trade has resulted in countries becoming more exposed to new pests and diseases (Hulme 2009;

Seebens et al. 2017, 2021), so trading partners are increasingly requiring more evidence to support claims of area freedom. Data derived from post-border surveillance programs, particularly ongoing active surveillance programs, are often seen as the gold standard of evidence for area freedom.

SURVEILLANCE FOR EARLY DETECTION

Early detection is one of the main lines of defence against the widespread establishment and spread of exotic threats. Early detection surveillance is a form of ongoing monitoring that often complements surveillance for area freedom. However, unlike area freedom, the implementation of surveillance for early detection is not governed by the requirements of trading partners or international standards. Surveillance for early detection is underpinned by the central theory that incursions are an inevitability and that to minimise impacts and maximise the feasibility and cost-effectiveness of containment and eradication, regulators must detect new threats when populations are small and geographically restricted (Ahmed et al. 2022; Leung et al. 2002). Because this form of surveillance is not a mandate set by a trading partner or international body, governments and industry contend with the issue of whether to invest resources (e.g. infrastructure, people, and diagnostics) in the early detection of a potential future outbreak or to only deploy resources when an incursion is detected. This trade-off between prevention and reaction is governed by three factors: (1) the likelihood of an incursion; (2) its potential environmental, economic, and/or social impacts; and (3) the feasibility of containment and eradication (see Chapter 7. Prepare, Respond, and Recover).

SURVEILLANCE FOR DELIMITING THE EXTENT OF AN OUTBREAK

When an incursion occurs, determining the extent of the outbreak is critical. Delimiting an outbreak quickly and accurately can reduce the likelihood of further spread and maximise eradication success through targeted control measures. Initial detections can come from area freedom or early detection surveillance programs, industry stakeholders (e.g. farmers and agronomists), or the public. For example, outbreaks of exotic fruit fly (e.g. the oriental fruit fly *Bactrocera dorsalis*) can be detected by sophisticated pheromone lure traps, industry members, and citizens who notice their fruit tree has become infested.

Delimitation surveillance is a reactive but necessary process that occurs once an outbreak has been detected. Once a detection is made, biosecurity agencies undertake intelligence-gathering activities, commonly referred to as tracing, to identify plausible means of introduction and secondary spread (Leung, Cacho, and Spring 2010; Potts et al. 2013). Typically, these activities involve detailed discussions with landowners in the immediate vicinity of the initial detection (which may or may not be the point of introduction), expanding to other locations based on this information. This local intelligence gathering is used to construct a network of likely spread movements among properties, which can then be used in models that simulate the likely extent of post-border spread (IPPC 2016b). These data and models are ground-truthed by conducting threat-specific surveillance (i.e. targeted or active surveillance) to determine the true extent of the outbreak. Examples of formal tracing schemes include Australia's National Livestock Information System and New Zealand's National Animal Identification and Tracing (MAF 2009).

SURVEILLANCE FOR THREAT MONITORING AND MANAGEMENT SUCCESS (PROOF OF FREEDOM)

Once an outbreak has been detected and its extent has been determined, biosecurity practitioners are faced with deciding how to manage the infestation (see Chapter 7. Prepare, Respond, and Recover). Three management options are possible: (1) attempt eradication; (2) contain (i.e. prevent further spread); or (3) do nothing and shift resources to mitigating and adapting to impacts. If the decision is made to do nothing, there is little need for surveillance, as resources are allocated towards mitigation and adaptation strategies (see Chapter 7. Prepare, Respond, and Recover).

If eradication or containment is the objective, then surveillance becomes a critical tool for evaluating the effectiveness of control measures (see Chapter 10. Monitoring, Evaluation, and Reporting). For containment and eradication purposes, targeted surveillance will typically be positioned within the delimited area of infestation and in surrounding areas (IPPC 2016b). Surveillance within the outbreak area is focused on both monitoring and evaluating the success of control treatments and to provide evidence that the threat has been successfully eradicated. By contrast, surveillance surrounding an outbreak is focused on determining whether the outbreak is contained and not spreading outside the area of treatment. Both forms of surveillance are an essential requirement for declaring a threat has been successfully contained or eradicated (commonly referred to as proof of freedom).

How long such surveillance is maintained depends on the management objective. If the focus is eradication, then it will remain in place until a desired level of statistical confidence is reached such that proof of freedom can be declared (Ramsey, Parkes, and Morrison 2009; Rout 2017; Rout, Salomon, and McCarthy 2009). Alternatively, if eradication is deemed infeasible and containment is the goal, then surveillance may be ongoing, with management focusing on maintaining low numbers and reducing spread.

TYPES OF POST-BORDER SURVEILLANCE

A range of surveillance methods can be used to facilitate the early detection of new outbreaks, monitor and manage existing outbreaks, or provide evidence of eradication success or area freedom. Surveillance methods fall on a continuum that ranges from the deliberate and coordinated use of sophisticated surveillance methods that maximise the detection of a particular threat, to surveillance methods that attempt to benefit from chance detections and reporting made by the public (Hester and Cacho 2017). This continuum is categorised into three main types of surveillance: active surveillance, general surveillance, and passive surveillance (Figure 6.1).

ACTIVE SURVEILLANCE

Active surveillance (also referred to as targeted surveillance) is the deliberate and coordinated effort to detect new or managed pests and diseases. It is typically implemented by biosecurity regulators to meet specific objectives, such as to provide evidence of area freedom to trading partners, to delimit and contain an incursion, or to monitor the performance of an eradication or containment program.

Active surveillance uses sophisticated tools and survey methods that are highly effective in detecting a specific threat. For animal and plant diseases, this often involves planned regular surveys of hosts for signs of illness, coupled with routine sampling for diagnostics such as polymerase chain reaction (PCR) tests and antibody (i.e. serology) tests. For non-disease threats (e.g. invasive vertebrates, invertebrates, and weeds), active surveillance includes repeated site surveys using visual observations and tools such as pheromone lure traps, animal traps, acoustic monitoring, field cameras, seed bank analyses, or detector dogs. New technologies are increasingly used to enhance detection rates, including environmental DNA (eDNA) methods for detecting invasive species in ballast water and freshwater systems, drones and remote sensing, and machine learning for rapid and automated identification of priority threats (see Chapter 13. Profiling and Automation).

Data derived from active surveillance are the most reliable for inferring the presence or absence of a threat. Not only do active surveillance tools and methods have known likelihoods of detection (i.e. sensitivity) and known false positives (i.e. specificity), but they are also implemented by highly trained staff using proven protocols. These detailed protocols describe how and where surveillance should be implemented, what data should be collected (e.g. threat presence/absence and measures of surveillance effort), and how these data can be used to infer threat absence.

Active surveillance is the most resource intensive and costly form of surveillance (Anderson et al. 2017). As such, it is commonly used to delimit and contain outbreaks and to declare eradication

Surveillance continuum

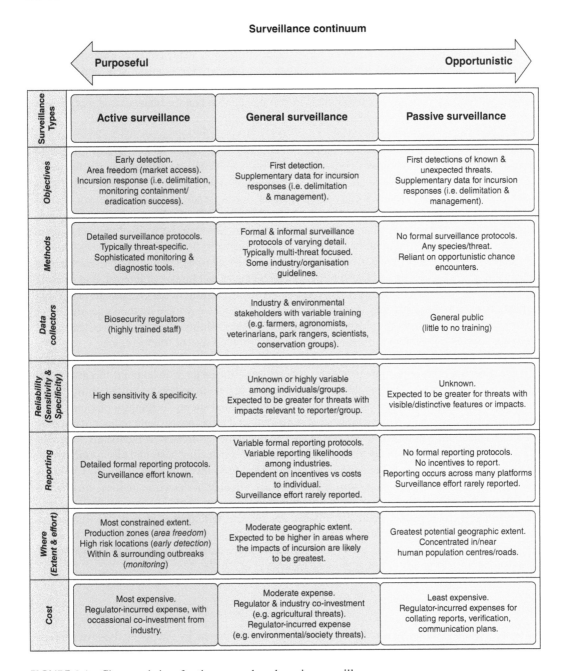

FIGURE 6.1   Characteristics of active, general, and passive surveillance.

success (i.e. proof of freedom). Because of its expense, it is also the most geographically and temporally constrained method, with most effort occurring within and surrounding known outbreaks and only persisting for as long as it takes to meet the required burden of evidence to declare proof of freedom. If active surveillance for a particular threat is expected to provide high value for money compared to other risk-reducing activities, it may be used on an ongoing basis to either facilitate the early detection of incursions or to support claims of area freedom for market access. In these cases, ongoing active surveillance can be positioned near high-risk points of entry (e.g. ports), in areas of high establishment potential, or within high-value production zones.

GENERAL SURVEILLANCE

General surveillance is a semi-coordinated surveillance effort implemented by environmental stake-holders (e.g. park rangers), agricultural stakeholders (e.g. farmers, agronomists, vets, and industry organisations), scientists, or citizen science groups (Hester and Cacho 2017). General surveillance is more opportunistic than active surveillance, in that it capitalises on the interests and motivations of various groups to conduct the surveillance necessary to protect their own assets or the community (Kruger, Ticehurst, and van der Meer Simo 2022). Biosecurity regulators play a lesser role in implementing general surveillance, but regulators sometimes co-invest in such programs if they can improve early detection and bolster existing active surveillance programs (Hester and Cacho 2017).

Because general surveillance is implemented by multiple stakeholders, with costs and logistics being shared among these groups, it can be implemented over a greater geographic extent compared to active surveillance. However, biosecurity regulators face multiple challenges in using and inter-preting data from general surveillance. While reported detections can be verified and acted upon by regulators, estimating the reliability of general surveillance in detecting threats is challenging because the reliability of different groups is unknown and likely to be highly variable. Each group implements general surveillance based on their own objectives and threats of interest, and as such, they might utilise different taxon expertise, training resources, and surveillance methods at varying levels of effort, which may or may not be documented. Without knowing a method's sensitivity or the survey effort undertaken, regulators cannot quantify the reliability of general surveillance, and thus, cannot solely rely on it to estimate the likelihood a threat is absent from a region.

In practice, the sensitivity of general surveillance is assumed to be lower than that of active sur-veillance. Because of its opportunistic nature, general surveillance tends to be multi-threat focused and not optimised to detect any single threat. In contrast, active surveillance is often optimised to detect a specific threat, and it relies on specialist equipment and survey techniques that maximise the likelihood of detection. While estimates of sensitivity and effort are difficult to quantify, it may be assumed that both will be greater in regions where the potential impacts of incursions are great-est. For instance, farmers are likely to have higher expertise in detecting pests and diseases that pose significant threats to their produce, and they are more likely to invest in detection effort in areas most exposed to a threat. The same farmers may be less capable of detecting and less inclined to search for a pest or disease that does not pose a threat to their own assets.

Despite these challenges, general surveillance has been a significant contributor to first detec-tions in various agricultural industries (Hammond et al. 2016), and there is a strong desire by regu-lators and industry to make better use of general surveillance data by integrating those with other methods to infer likelihoods of threat absence (Martin et al. 2017).

PASSIVE SURVEILLANCE

Passive surveillance is the chance detection and reporting of threats by the public. Passive surveil-lance is the most fortuitous and accidental of all types of surveillance (Hester and Cacho 2017), and it differs from active and general surveillance in that there is little coordinated surveillance effort or reporting framework. Detections arise purely by chance, often based on random encounters cou-pled with individual curiosity. Passive surveillance can be described as "the threat comes to you" whereas active surveillance and to a lesser degree, general surveillance, can be described as "you go to the threat". Much like general surveillance, public reportings have been a significant source of first detections for many invasive species (Hester and Cacho 2017).

Reporting of passive surveillance occurs on a variety of formal and informal platforms such as biosecurity hotlines, online citizen science databases (e.g. iNaturalist, the Global Biodiversity Information Facility, and the Atlas of Living Australia), and social media platforms. Individuals making chance detections do so with no underlying objective, and often, with variable capability to identify what they have found, let alone to whom it should be reported. Public detections often go unreported, and when they are reported, can be prone to false positives (i.e. a report being misidentified as a threat of concern).

While governments and regulators have little direct costs associated with implementing passive surveillance, there are many indirect costs associated with using such data, including the cost of verifying and collating records from multiple platforms. As with general surveillance, passive surveillance data have unknown sensitivity and survey effort. Detection likelihoods will also vary across space and taxa. Public recordings tend to be geographically biased towards regions of high human population density (e.g. in and around metropolitan areas) or along roads or walking tracks (Figure 6.2; Dodd et al. 2015). Public detections are also more likely for threats with highly

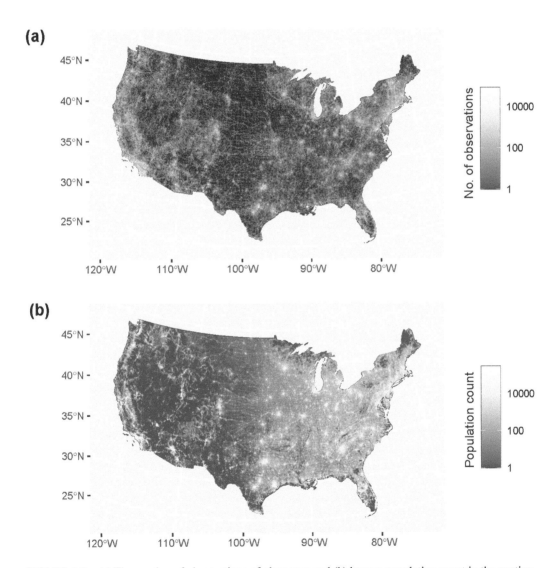

FIGURE 6.2 (a) The number of observations of plant taxa and (b) human population count in the contiguous United States. Note the highest densities of occurrences typically correspond to areas with high human population counts (i.e. close to metropolitan areas and along major road networks). Plant taxa data were extracted from the global biodiversity information facility on 14 December 2021 (GBIF Org 2021) and human population count data for the year 2020 were obtained from Columbia University's Centre for International Earth Science Information Network (CIESIN 2022). Both layers were projected to the USA Albers Equal Area Conic coordinate system (ESRI:102003) and aggregated to a 5-km raster grid.

visible impacts or striking physical attributes (Box 6.1), and reportings generally occur when a threat reaches its natural detection point (e.g. when it has established and reached a population size, extent, or impact that makes encounters with the public more likely; Kompas et al. 2019).

PRACTICAL CONSIDERATIONS IN POST-BORDER SURVEILLANCE

A range of practical considerations must be considered when designing a post-border surveillance system, including:

- Selecting the type(s) of surveillance to use under each objective.
- For early detection, determining where to conduct surveillance.
- For area freedom or proof of freedom, inferring threat absence from surveillance data.

SELECTING SURVEILLANCE TYPE(S) FOR AN OBJECTIVE

A fundamental challenge faced by biosecurity regulators is deciding which surveillance type to prioritise and fund under different surveillance objectives. Each surveillance type (active, general, and passive) has advantages and disadvantages, and each type differs in terms of cost, reliability, and capacity to infer likelihoods of absence (Figure 6.1).

Active surveillance is almost always used as the primary tool to delimit and monitor an outbreak, and it is increasingly required by trading partners for declaring area freedom. When using active surveillance, regulators can easily coordinate and quantify surveillance effort within and surrounding an outbreak or production zone. Reliability (i.e. sensitivity and specificity) in detecting a threat is known, hence data from active surveillance can be used to infer the likelihood a threat is absent and inform when to declare eradication success (i.e. proof of freedom) or area freedom for market access.

In contrast, regulators have limited capacity to coordinate passive and general surveillance. Regulators can conduct awareness campaigns in and around outbreak zones to increase public and stakeholder detection and reporting, but the efficacy of such programs is difficult to quantify. Even with awareness campaigns, the reliabilities of passive and general surveillance are often unknown, because sensitivity and survey effort data are not commonly collected. Passive and general surveillance are therefore rarely used in statistical models to infer likelihoods of absence or eradication success. Instead, detections derived from these forms of surveillance act as supplemental data to identify new outbreaks, support existing evidence of absence, and position active surveillance resources (IPPC 2016a).

Active surveillance is also the gold standard for early detection, but its high cost often restricts its use to threats posing the highest risk to a jurisdiction (i.e. where the benefits of earlier detection outweigh increases in surveillance cost; see Chapter 9. Resource Allocation). Active surveillance for early detection is undertaken when a trading partner requires its use in informing area freedom declarations. For example, active surveillance networks using pheromone lure traps are implemented in the United States, Australia, and New Zealand for both area freedom and early detection of exotic fruit fly threats such as the Mediterranean fruit fly (*Ceratitis capitata*) and the Oriental fruit fly (*B. dorsalis*). These species feed on many hosts, have significant impacts on horticultural yield and quality, and they are difficult to eradicate and can disrupt access to premium export markets. As such, there is an incentive for countries to detect these threats early to minimise impacts.

For most threats, where the costs of active surveillance are not expected to provide good value for money, regulators rely on both general and passive surveillance for first detections. While neither is truly optimised for early detection, the implicit assumption is that both general and passive surveillance will detect threats early enough so that they can be contained or eradicated. However, this assumption requires scrutiny. In Australia, detection of the non-descript Russian wheat aphid (*Diuraphis noxia*; Figure 6.3c) relied on general and passive surveillance systems. The species remained undetected until 2016, by which time it was widespread in cereal cropping regions across the south-east and was deemed infeasible to control or eradicate (Yazdani et al. 2018).

To reduce the chance of incursions being detected at a point where it is infeasible to control or eradicate, regulators must consider:

- How the attributes of the threat (e.g. size and morphology) influence the reliability of passive and general surveillance (Box 6.1).
- Where general and/or passive surveillance effort is greatest (Figure 6.2), and how this compares to where a threat is most likely to establish.
- Whether the costs of active surveillance provide good value for money in terms of earlier detection.

BOX 6.1. SPECIES-SPECIFIC FACTORS INFLUENCE THE PROBABILITY OF DETECTION

An analysis of citizen reports of beetles (Coleoptera) and true bugs (Hemiptera) in the Atlas of Living Australia found that the probability of reporting by a member of the public (and by implication, the probability of detection) was strongly influenced by the physical attributes of a species (Caley, Welvaert, and Barry 2019). Species with large body size, large geographic range, or striking colour patterns or morphological features had a significantly higher annual probability of being reported by a member of the public. The black pine sawyer beetle (*Monochamus galloprovincialis*; Figure 6.3a) and the Colorado potato beetle (*Leptinotarsa decemlineata*; Figure 6.3b), two relatively large (>10 mm) beetles with distinctive morphological features, had average annual detection probabilities of 0.91 and 0.76, respectively. At the other end of the spectrum, the small (<5 mm) and non-descript Russian wheat aphid (*D. noxia*; Figure 6.3c) and the generic-looking mountain pine beetle (*D. ponderosae*; Figure 6.3d) had citizen detection probabilities of 0.03 and 0.02, respectively.

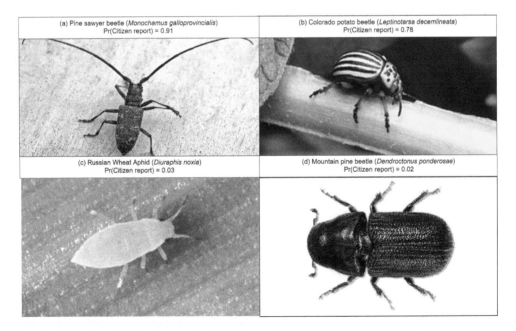

FIGURE 6.3   Probabilities of citizen reporting four threats in Australia. Note that pests with less distinctive features have lower reporting rates (From Caley et al. 2019).

DETERMINING WHERE TO CONDUCT SURVEILLANCE FOR EARLY DETECTION

There is often a clear geographic disconnect between where general surveillance effort is concentrated and where initial establishment events are most likely to occur. Most successful exotic threats have generalist attributes and enter and spread via human movement and trade. Propagule pressure and establishment potential for most threats will therefore be greatest in regions with highest human activity, such as points of entry (i.e. airports and ports) and populated areas where people and goods are most likely to disperse (i.e. cities and urban areas). In contrast, general surveillance for most agricultural and environmental biosecurity threats tends to be concentrated in sparsely populated areas where the potential impacts of incursions are greatest (e.g. production zones and national parks), but where propagule pressure and establishment potential may be low.

While passive surveillance is expected to be concentrated in areas of high human activity or establishment potential (Figure 6.2), its ability to inform regulators of new incursions in a timely manner depends on the attributes of the threat (Box 6.1). A threat could establish in an urban area and remain undetected until it has reached its natural detection point, by which time it may have spread far and wide, making control and eradication infeasible.

Active surveillance can be labour intensive and costly to maintain and is rarely implemented uniformly across geographic space. Rather, it should be concentrated in regions expected to have high entry or establishment potential. The establishment potential of an exotic threat is governed by three spatial factors (Camac, Baumgartner, Hester, et al. 2021, Camac, Baumgartner, et al. 2020, Catford, Jansson, and Nilsson 2009), where all three factors must be met for establishment to occur:

- Can the threat reach the location of interest (i.e. propagule pressure)?
- Are abiotic conditions suitable (e.g. climate suitability)?
- Are biotic conditions suitable (e.g. presence of host or food)?

Based on these factors, two approaches are commonly used to inform early detection surveillance. Pathway models estimate contamination or leakage rates reaching a country's border or points of entry (Camac, Baumgartner, Garms, et al. 2021; also see Chapter 9. Resource Allocation), while species distribution models identify areas of suitable environment that may be conducive for establishment. Both approaches provide critical information for where to prioritise surveillance and are outlined in Chapter 14. Map.

INFERRING THREAT ABSENCE FROM SURVEILLANCE DATA

A fundamental problem with declaring an exotic threat absent from a region based only on a lack of detections is that this assumes the surveillance program has perfect sensitivity. That is, if the threat is present, the surveillance program will always detect it. In practice, post-border surveillance programs never have perfect sensitivity. The complication of imperfect sensitivity means that a lack of detections from a surveillance program can arise from one of two processes: (1) the threat is truly absent; or (2) the threat is present, but the surveillance program failed to detect it (i.e. a false negative). While imperfect specificity leading to false positives can be cross-checked, it is impossible to be certain a threat is truly absent.

Given the issue of not being able to definitively determine the absence of a threat, scientists and biosecurity practitioners use statistical approaches to quantify the likelihood of threat absence. Which approach to use largely depends on the decision context and the data at hand. If the focus is to determine the success of an eradication program, models can be used to quantify the certainty of pest absence as a function of a time series of sightings or detections, detailed information on the number of individuals removed during an eradication program, and/or management effort over time (for a detailed summary of approaches see Rout 2017). By contrast, if the focus is to determine pest absence without an outbreak, such as for the purposes of early detection or area freedom, then negative observations (i.e. the number of non-detections) from surveillance programs can be used to inform likelihoods of absence (Barrett et al. 2010).

Irrespective of the approach used, three fundamental questions must be answered to estimate likelihoods of absence:

- What is the target prevalence or population size of the threat to be detected?
- What is the sensitivity of the surveillance method for the threat?
- What is the tolerance for being wrong?

Determining the Target Prevalence or Population Size of the Threat

In addition to the unit of area one wishes to infer absence for, estimating the likelihood of absence requires a clear definition of the minimum prevalence (i.e. design prevalence) of the threat to be detected or, in the case of non-disease threats, the population size to be detected. In a biosecurity context, prevalence could be the proportion of infected individuals or sites within a pre-defined sampling unit (e.g. a herd or a region).

When determining the minimum prevalence and/or population size to be detected, the answer will be a trade-off between being high enough to detect the threat and low enough that it can still be controlled or eradicated—a maximum tolerable level (Martin et al. 2017; Whittle et al. 2013). A design prevalence of 1% or 0.1% is commonly set for screening consignments at the border (see Chapter 5. Screen). This prevalence rarely translates to the post-border context as the surveillance unit changes from consignments to susceptible individuals or locations. Post-border, prevalence should be set by first considering the number of susceptible units (e.g. individuals or sites) in an area and then determining what number of undetected infected individuals would be tolerable. If a regulator optimised post-border surveillance effort to be 95% confident of detecting a 1% prevalence of a citrus disease in an area containing 1,000,000 susceptible host trees, the post-border surveillance system will be optimally designed to detect an outbreak containing more than 10,000 infected trees. A question the regulator should ask is whether 10,000 possible undetected infections is tolerable.

For pests, minimum population size may be more relevant and tractable than prevalence. Larger populations are generally more easily detectable, but they are also more difficult to control and eradicate. The sensitivity of a surveillance unit is typically measured in terms of the probability of detecting a single individual at a site and/or for a certain period of time. For many threats, detection probabilities will be extremely low, so sensitivity may be re-scaled such that it becomes the probability of detecting at least one individual from a population of N individuals. The size of this population is governed by the probability of detecting the threat and the probability of containing or eradicating it.

Estimating Surveillance Sensitivity

Surveillance programs never have perfect sensitivity, and estimating surveillance sensitivity is difficult because it can vary substantially among species and is influenced by multiple factors, including species traits (see Box 6.1), surveillance effort (e.g. the number of surveys, tests, or traps and survey time (Garrard et al. 2008; Hauser and McCarthy 2009); the local abundance of the species (McCarthy et al. 2013), site conditions, and observer or surveillance attributes (Bailey, Simons, and Pollock 2004).

Despite these difficulties, information about sensitivity can be obtained in many situations, including by conducting experiments designed to estimate sensitivity under a variety of conditions (Hauser et al. 2016), by using (or estimating) detection rates for species with similar traits (Box 6.1; Caley et al. 2019; Garrard et al. 2012), by conducting meta-analyses, or, when no other data are available, by using expert judgement (see Chapter 12. Elicit).

What is the Tolerance for Being Wrong?

Regardless of the method used to infer threat absence, a regulator will have to define their tolerance for incorrectly concluding the threat is absent when in fact it is present (Figure 6.4). In a biosecurity context, this tolerance is often described in terms of either the likelihood of

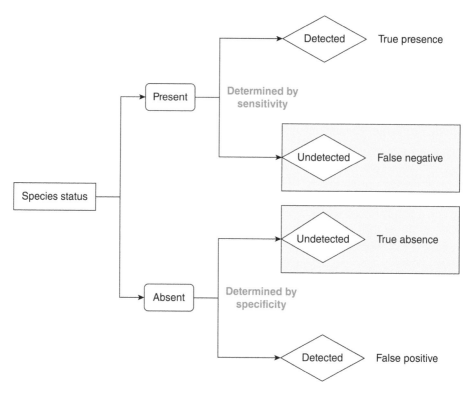

FIGURE 6.4 Scenario tree outlining the two processes leading to absence observations when surveillance sensitivity and specificity are imperfect.

detection failure (i.e. 1 – probability of detection) or likelihood of presence (i.e. 1 – probability of absence) after accounting for zero detections, design prevalence, and surveillance sensitivity and effort.

To declare threat absence, the likelihoods of detection failure or threat presence are set to some tolerable level and are typically governed by rules of thumb coupled with regulatory requirements and a country's ALOP (see Chapter 2. Biosecurity Systems and International Regulations). Commonly, this tolerance is set to 0.05%—we accept that we will incorrectly declare a threat as absent (when in fact it is present) 5% of the time (European Food Safety Authority et al. 2020). However, this tolerance level is somewhat arbitrary and does not consider the cost of surveillance or the potential economic, social, and environmental costs of failing to detect a threat. If the impacts of failing to detect a threat are high and the cost of surveillance is low, then it may be prudent for regulators to lower the tolerance level to reduce the chance of incorrectly declaring absence and maximise the likelihood of avoiding those damages. If the impacts of failed detections are low and the cost of surveillance is high, then a regulator may be more tolerant to failed detections and set a higher tolerance level (Camac, Dodd, et al. 2020;Hauser and McCarthy 2009; Kompas, Chu, and Nguyen 2016).

Multiple model types can be applied to infer the absence of a biosecurity threat, and which one to use will depend on how surveillance sensitivity and effort are measured and whether spatial information on the likelihood of threat presence is available. Here, we outline two of the simplest models used to infer threat absence: the binomial model and Bayes theorem (Box 6.2). For alternative models and a more comprehensive discussion of their assumptions and limitations, see Chapter 5. Screen and Hester, Hauser, and Kean (2017).

BOX 6.2. INFERRING THREAT ABSENCE WITH THE
BINOMIAL MODEL AND BAYES THEOREM

Binomial Model

The binomial model is the most commonly used approach to estimate the likelihood of threat absence. The model is based on the confidence of rejecting the null hypothesis that the threat is present, where confidence is defined as the probability a surveillance program detects the threat. This is estimated as a function of the sensitivity (S) of a surveillance unit (e.g. diagnostic test, trap, or survey) at threat prevalence (p) and a given amount of surveillance effort (N_{effort}; e.g. the number of tests or traps). The probability of detection is expressed as:

$$\Pr(Detection) = 1 - (1 - p \times S)^{N_{effort}} \qquad \text{(Eq. 6.1)}$$

Absence is inferred if no detection is recorded and the confidence of detection, $\Pr(Detection)$, is at or above a pre-defined level. If tolerance for being wrong is 0.05, then confidence would need to be 0.95 or higher to reject the null hypothesis that the threat is present (i.e. declare the threat absent). This model can be further expanded to account for other technical considerations, such as clustered sampling (see Chapter 5. Screen). While this approach provides a measure of confidence in threat absence, it does so solely as a function of surveillance sensitivity at a pre-defined prevalence and the amount of surveillance effort. The method ignores potential differences in the likelihood of pest establishment across space or time: confidence of absence may be overestimated in places where establishment likelihood is high and underestimated in areas where establishment likelihood is low (see Chapter 14. Map).

Bayes Theorem

Compared to the binomial model, which quantifies the conditional probability of detection assuming the threat is present, Bayes theorem uses a more logical measure of pest absence—the probability of pest absence (Barrett et al. 2010; McArdle 1990). Another advantage of the Bayesian approach is that it can directly incorporate additional sources of information, using a model parameter known as the prior (McCarthy 2007). In the context of estimating likelihoods of threat absence, the prior describes the belief a threat is present at a location, $\Pr(Presence)$. The probability of threat absence is given by:

$$\Pr(Absence) = 1 - \frac{(1 - \Pr(Detection)) \times \Pr(Presence)}{(1 - \Pr(Presence)) + (1 - \Pr(Detection)) \times \Pr(Presence)} \qquad \text{(Eq. 6.2)}$$

$$\text{where } \Pr(Detection) = 1 - (1 - Pr(S))^{N_{effort}} \qquad \text{(Eq. 6.3)}$$

Because the Bayesian approach can explicitly account for information on differential risk across space and/or time, it is less susceptible to over- or under-estimation of the likelihood of absence. The Bayesian approach can also identify regions that require greater surveillance effort based on estimated likelihoods. The prior probability of threat presence can be informed using a variety of data sources, such as expert elicitation (see Chapter 12. Elicit), a map of establishment likelihood (see Chapter 14. Map), or data from past surveillance programs or incursions. If no prior knowledge is available, the prior can be set at 0.5 (i.e. a 50% chance of the threat being present) and the posterior probability of absence will be driven solely by surveillance records (Rout 2017).

VALUE FOR MONEY IN SURVEILLANCE PLANNING

As surveillance effort increases, so too does the confidence that a threat is absent. Higher confidence implies lower chances of incorrectly claiming absence and, thus, lower chances of incurring additional surveillance and market costs (e.g. time out of market and re-eradication costs). However, this confidence comes at the cost of investing in surveillance methods with higher sensitivity and/or investing in more surveillance effort (e.g. additional people, time, or money). The optimal amount of surveillance is the one that minimises surveillance costs and the expected costs of incorrectly claiming absence (Box 6.3).

Irrespective of the objective, the surveillance strategy and effort required should be determined by what represents the best value for money (see Chapter 9. Resource Allocation). When surveillance occurs for early detection, delimiting the extent of an outbreak, or for threat monitoring and management, surveillance requirements are set by the regulator. When the objective of surveillance is market access, trading partners typically set surveillance requirements. However, the cost of implementing those requirements may outweigh the benefit of market access, so regulators should still minimise net expected costs.

Practitioners should also understand that a given budget could be used to survey any number of pests simultaneously, so they should consider whether an investment into surveillance for one pest gives better value than surveillance for other pests, rather than considering each pest in isolation. The "value for money" premise holds regardless of whether budget allocations are being made for a single threat, across a number of threats, or indeed at different stages of the biosecurity continuum (see Chapter 9. Resource Allocation). Allocating a surveillance budget across pests and diseases will depend on the difficulty (cost) of detection and eradication of each pest, and the avoided damages from preventing or removing the pest. Deciding where best to locate surveillance, and over which time period, has important budgetary implications. Finally, surveillance activities do not occur in isolation, they are one of a suite of activities undertaken to manage pest and disease risks (e.g. research, control, treatment, and community engagement) and thus should not be considered in isolation (see Chapter 7. Prepare, Respond, and Recover).

BOX 6.3. DETERMINING OPTIMAL SURVEILLANCE EFFORT

The optimal surveillance effort (n) of a threat can be estimated by minimising the total costs of surveillance and the costs of failing to detect the threat, using the following equation:

$$Total\ costs(n) = Cost_{Surveillance}(n) + \Pr(Wrong \mid n) \times Cost_{Wrong} \qquad \text{(Eq. 6.4)}$$

The first term ($Cost_{Surveillance}$) captures the costs associated with implementing n units of surveillance effort (e.g. infrastructure, diagnostics, and logistics). The probability of being wrong given with n surveillance units, $\Pr(Wrong \mid n)$, is either the probability of failing to detect the threat $(1 - \Pr(Detection))$ derived from the binomial model or the estimated probability of presence $(1 - \Pr(Absence))$ derived from the Bayesian approach (Box 6.2). The cost of being wrong, $Cost_{Wrong}$, is the cost associated with incorrectly declaring a threat is absent, which may include immediate costs (e.g. containment and eradication protocols) and damages that accrue over time (e.g. agricultural yield losses, time out of market, and the cost and time required to declare eradication success). Optimal surveillance effort can be identified by calculating total cost for a range of surveillance efforts, with optimal surveillance effort being the point where total cost is minimised.

In the theoretical scenario plotted in Figure 6.5, we assume that:

- If the threat is wrongly thought to be absent and allowed to spread, the damages would be worth $10 million ($Cost_{Wrong}$).
- The threat is to be detected at a prevalence of 1% using a surveillance method with a unit sensitivity S of 0.2.
- The cost of each surveillance unit (e.g. hardware and logistical costs) is $5,000 and scales linearly with the number n of surveillance units added (blue line).

Using the binomial model defined in Box 6.2, the probability of being wrong given n surveillance units $Pr(Wrong \mid n)$ is $1 - Pr(Detection)$. Total costs are estimated using Eq. 6.4. Under this scenario, the total cost is U-shaped (black line), meaning that increasing surveillance effort decreases total costs, up until a point where high surveillance costs outweigh the benefits of avoiding damages. Note that very high surveillance levels (>1,960 surveillance units) result in greater surveillance expenditure relative to the overall damages to be avoided ($10 million).

In this scenario, the optimal amount of surveillance is 693 units at a cost of about $3.5 million (where total cost is lowest), with a $Pr(Detection)$ of 0.75. The optimal amount of surveillance is lower than the 1,449 samples ($7.3 million) that would traditionally be required to meet the arbitrary confidence threshold of 0.95. These surveillance costs translate to total costs (cost of surveillance + cost of incorrectly declaring absence) of $5.7 million for the optimal scenario and $7.8 million for the traditional confidence threshold scenario, translating into a saving of $2.1 million.

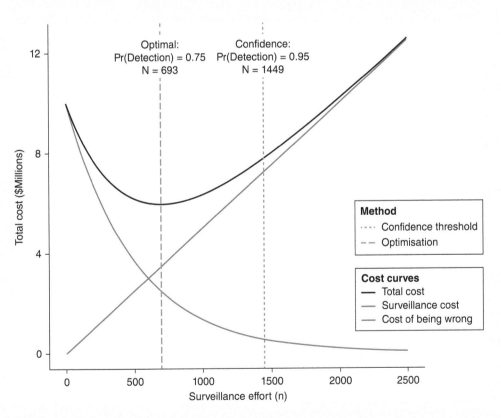

FIGURE 6.5 Theoretical example of the cost of surveillance and the cost of potential damages as surveillance effort (n) increases.

IN A NUTSHELL

- Post-border surveillance can be used to meet four objectives: (1) claiming area freedom, (2) early threat detection, (3) outbreak delimitation, and (4) threat monitoring and management (including declaring proof of freedom).
- Surveillance methods range from the deliberate and coordinated use of sophisticated surveillance tools (i.e. active surveillance), to those that attempt to benefit from detections made by stakeholders (i.e. general surveillance) or the public (i.e. passive surveillance).
- Early detection surveillance should be prioritised in areas where the risk of establishment is greatest.
- Due to variable reliability of the data, significant care should be taken when using data from general and passive surveillance to make claims of threat absence.
- How much surveillance is required depends on the objective of surveillance, the tolerance for being wrong and the magnitude of the cost and benefits (avoided damages) of surveillance.

REFERENCES

Ahmed, D. A., E. J. Hudgins, R. N. Cuthbert, M. Kourantidou, C. Diagne, P. J. Haubrock, B. Leung, C. Liu, B. Leroy, S. Petrovskii, A. Beidas, and F. Courchamp. 2022. "Managing Biological Invasions: The Cost of Inaction." *Biological Invasions*:1–20. https://doi.org/10.1007/s10530-022-02755-0.

Anderson, C., S. Low-Choy, P. Whittle, S. Taylor, C. Gambley, L. Smith, P. Gillespie, H. Löcker, R. Davis, and B. Dominiak. 2017. "Australian Plant Biosecurity Surveillance Systems." *Crop Protection* 100:8–20. https://doi.org/10.1016/j.cropro.2017.05.023.

Bailey, L. L., T. R. Simons, and K. H. Pollock. 2004. "Estimating Site Occupancy and Species Detection Probability Parameters for Terrestrial Salamanders." *Ecological Applications* 14 (3):692–702. https://doi.org/10.1890/03-5012.

Barrett, S., P. Whittle, K. Mengersen, and R. Stoklosa. 2010. "Biosecurity Threats: The Design of Surveillance Systems, Based on Power and Risk." *Environmental and Ecological Statistics* 17 (4):503–519. https://doi.org/10.1007/s10651-009-0113-4.

Caley, P., M. Welvaert, and S. C. Barry. 2019. "Crowd Surveillance: Estimating Citizen Science Reporting Probabilities for Insects of Biosecurity Concern." *Journal of Pest Science* 45:1–8. https://doi.org/10.1007/s10340-019-01115-7.

Camac, J., J. Baumgartner, B. Garms, A. Robinson, and T. Kompas. 2021. Estimating trading partner exposure risk to new pests or diseases. Technical Report for CEBRA project 190606. Centre of Excellence for Biosecurity Risk Analysis.

Camac, J., J. Baumgartner, S. Hester, R. Subasinghe, and S. Collins. 2021. Using edmaps & Zonation to inform multi-pest early-detection surveillance designs. Technical Report for CEBRA project 20121001. Centre of Excellence for Biosecurity Risk Analysis.

Camac, J., J. Baumgartner, A. Robinson, and J. Elith. 2020. Developing pragmatic maps of establishment likelihood for plant pests. Technical Report for CEBRA project 170607. Centre of Excellence for Biosecurity Risk Analysis.

Camac, J., A. Dodd, N. Bloomfield, and A. Robinson. 2020. Sampling to support claims of area freedom: Technical report for the Department of Agriculture. Centre of Excellence for Biosecurity Risk Analysis.

Catford, J. A., R. Jansson, and C. Nilsson. 2009. "Reducing Redundancy in Invasion Ecology by Integrating Hypotheses into a Single Theoretical Framework." *Diversity and Distributions* 15 (1):22–40. https://doi.org/10.1111/j.1472-4642.2008.00521.x.

CIESIN. 2022. "Center for International Earth Science Information Network." accessed 14 December 2021. http://www.ciesin.org/.

Dodd, A. J., M. A. McCarthy, N. Ainsworth, and M. A. Burgman. 2015. "Identifying Hotspots of Alien Plant Naturalisation in Australia: Approaches and Predictions." *Biological Invasions* 18 (3):631–645. https://doi.org/10.1007/s10530-015-1035-8.

European Food Safety Authority, E. Lázaro, S. Parnell, A. V. Civera, J. Schans, M. Schenk, J. C. Abrahantes, G. Zancanaro, and S. Vos. 2020. "General Guidelines for Statistically Sound and Risk-Based Surveys of Plant Pests." *EFSA Supporting Publications* 17 (9):1919E. https://doi.org/10.2903/sp.efsa.2020.EN-1919.

Garrard, G. E., S. A. Bekessy, M. A. McCarthy, and B. A. Wintle. 2008. "When Have We Looked Hard Enough? A Novel Method for Setting Minimum Survey Effort Protocols for Flora Surveys." *Austral Ecology* 33 (8):986–998. https://doi.org/10.1111/j.1442-9993.2008.01869.x.

Garrard, G. E., S. Ramula, M. A. McCarthy, N. S. G. Williams, S. A. Bekessy, and B. A. Wintle. 2012. "A General Model of Detectability Using Species Traits." *Methods in Ecology and Evolution* 4 (1):45–52. https://doi.org/10.1111/j.2041-210x.2012.00257.x.

GBIF Org. 2021. "Occurrence Download." The Global Biodiversity Information Facility. https://www.gbif.org/occurrence/download/0079039-210914110416597.

Hammond, N. E. B., D. Hardie, C. E. Hauser, and S. A. Reid. 2016. "Can General Surveillance Detect High Priority Pests in the Western Australian Grains Industry?" *Crop Protection* 79:8–14. https://doi.org/10.1016/j.cropro.2015.10.004.

Hauser, C. E., and M. A. McCarthy. 2009. "Streamlining 'search and destroy': Cost-Effective Surveillance for Invasive Species Management." *Ecology Letters* 12 (7):683–692. https://doi.org/10.1111/j.1461-0248.2009.01323.x.

Hauser, C. E., J. Weiss, G. Guillera-Arroita, M. A. McCarthy, K. M. Giljohann, and J. L. Moore. 2016. "Designing detection experiments: Three more case studies." 20th Australasian Weeds Conference, Perth, Western Australia, 11-15 September 2016.

Hester, S. M., and O. J. Cacho. 2017. "The Contribution of Passive Surveillance to Invasive Species Management." *Biological Invasions* 19 (3):737–748. https://doi.org/10.1007/s10530-016-1362-4.

Hester, S. M., C. E. Hauser, and J. M. Kean. 2017. "Tools for Designing and Evaluating Post-Border Surveillance Systems." In *Invasive Species: Risk Assessment and Management*, edited by Andrew P. Robinson, Terry Walshe, Mark A. Burgman and Mike Nunn, 17–52. Cambridge: Cambridge University Press.

Hulme, P. E. 2009. "Trade, Transport and Trouble: Managing Invasive Species Pathways in an Era of Globalization." *Journal of Applied Ecology* 46 (1):10–18. https://doi.org/10.1111/j.1365-2664.2008.01600.x.

IPPC. 2016a. International Standard for Phytosanitary Measures. ISPM No. 6: Guidelines for Surveillance.

IPPC. 2016b. International Standard for Phytosanitary Measures. ISPM No. 9: Guidelines for Pest Eradication Programmes.

IPPC. 2017. International Standard for Phytosanitary Measures. ISPM No. 4: Requirements for the Establishment of Pest Free Areas.

Kompas, T., L. Chu, and H. T. M. Nguyen. 2016. "A Practical Optimal Surveillance Policy for Invasive Weeds: An Application to Hawkweed in Australia." *Ecological Economics* 130:156–165. https://doi.org/10.1016/j.ecolecon.2016.07.003.

Kompas, T., L. Chu, P. Van Ha, and D. Spring. 2019. "Budgeting and Portfolio Allocation for Biosecurity Measures." *Australian Journal of Agricultural and Resource Economics* 63 (3):412–438. https://doi.org/10.1111/1467-8489.12305.

Kruger, H., J. Ticehurst, and A. van der Meer Simo. 2022. Guidelines for General Surveillance Programs: Insights and Considerations from Systems Thinking and Nine Case Studies. Canberra: ABARES.

Leung, B., O. J. Cacho, and D. Spring. 2010. "Searching for non-Indigenous Species: Rapidly Delimiting the Invasion Boundary." *Diversity and Distributions* 16 (3):451–460. https://doi.org/10.1111/j.1472-4642.2010.00653.x.

Leung, B., D. M. Lodge, D. Finnoff, J. F. Shogren, M. A. Lewis, and G. Lamberti. 2002. "An Ounce of Prevention or a Pound of Cure: Bioeconomic Risk Analysis of Invasive Species." *Proceedings of the Royal Society of London. Series B: Biological Sciences* 269 (1508):2407–2413. https://doi.org/10.1098/rspb.2002.2179.

MAF. 2009. Review of Selected Cattle Identification and Tracing Systems Worldwide: Lessons for the New Zealand National Animal Identification and Tracing (NAIT) Project. Ministry of Agriculture and Forestry.

Martin, T., A. P. Robinson, T. Walshe, M. A. Burgman, and M Nunn. 2017. "Surveillance for Detection of Pests and Diseases: How Sure can We Be of Their Absence?" In *Invasive Species: Risk Assessment and Management*, edited by Andrew P. Robinson, Terry Walshe, Mark A. Burgman and Mike Nunn, 348–384.

McArdle, B. H. 1990. "When Are Rare Species Not There?" *Oikos* 57 (2):276–277. https://doi.org/10.2307/3565950.

McCarthy, M. A. 2007. *Bayesian Methods for Ecology*. Cambridge: Cambridge University Press.

McCarthy, M. A., J. L. Moore, W. K. Morris, K. M. Parris, G. E. Garrard, P. A. Vesk, L. Rumpff, K. M. Giljohann, J. S. Camac, S. S. Bau, T. Friend, B. Harrison, and B. Yue. 2013. "The Influence of Abundance on Detectability." *Oikos* 122 (5):717–726. https://doi.org/10.1111/j.1600-0706.2012.20781.x.

Potts, J. M., M. J. Cox, P. Barkley, R. Christian, G. Telford, and M. A. Burgman. 2013. "Model-based Search Strategies for Plant Diseases: A Case Study Using Citrus Canker (*Xanthomonas citri*)." *Diversity and Distributions* 19 (5–6):590–602. https://doi.org/10.1111/ddi.12065.

Ramsey, D. S. L., J. Parkes, and S. A. Morrison. 2009. "Quantifying Eradication Success: The Removal of Feral Pigs from Santa Cruz Island, California." *Conservation Biology* 23 (2):449–459. https://doi.org/10.1111/j.1523-1739.2008.01119.x.

Rout, T. M. 2017. "Declaring Eradication of an Invasive Species." In *Invasive Species: Risk Assessment and Management*, edited by Andrew P. Robinson, Terry Walshe, Mark A. Burgman and Mike Nunn, 334–347.

Rout, T. M., Y. Salomon, and M. A. McCarthy. 2009. "Using Sighting Records to Declare Eradication of an Invasive Species." *Journal of Applied Ecology* 46 (1):110–117. https://doi.org/10.1111/j.1365-2664.2008.01586.x.

Seebens, H., S. Bacher, T. M. Blackburn, C. Capinha, W. Dawson, S. Dullinger, P. Genovesi, P. E. Hulme, M. van Kleunen, I. Kühn, J. M. Jeschke, B. Lenzner, A. M. Liebhold, Z. Pattison, J. Pergl, P. Pyšek, M. Winter, and F. Essl. 2021. "Projecting the Continental Accumulation of Alien Species Through to 2050." *Global Change Biology* 27 (5):970–982. https://doi.org/10.1111/gcb.15333.

Seebens, H., T. M. Blackburn, E. E. Dyer, P. Genovesi, P. E. Hulme, J. M. Jeschke, S. Pagad, P. Pyšek, M. Winter, M. Arianoutsou, S. Bacher, B. Blasius, G. Brundu, C. Capinha, L. Celesti-Grapow, W. Dawson, S. Dullinger, N. Fuentes, H. Jäger, J. Kartesz, M. Kenis, H. Kreft, I. Kühn, B. Lenzner, A. Liebhold, A. Mosena, D. Moser, M. Nishino, D. Pearman, J. Pergl, W. Rabitsch, J. Rojas-Sandoval, A. Roques, S. Rorke, S. Rossinelli, H. E. Roy, R. Scalera, S. Schindler, K. Štajerová, B. Tokarska-Guzik, M. van Kleunen, K. Walker, P. Weigelt, T. Yamanaka, and F. Essl. 2017. "No Saturation in the Accumulation of Alien Species Worldwide." *Nature Communications* 8 (1):14435. https://doi.org/10.1038/ncomms14435.

Whittle, P. J. L., M. Burgman, R. Stoklosa, S. Barrett, F. C. Jarrad, J. D. Majer, P. A. J. Martin, and K. Mengersen. 2013. "A Method for Designing Complex Biosecurity Surveillance Systems: Detecting non-Indigenous Species of Invertebrates on Barrow Island." *Diversity and Distributions* 19 (5–6):629–639. https://doi.org/10.1111/ddi.12056.

Yazdani, M., G. Baker, H. DeGraaf, K. Henry, K. Hill, B. Kimber, M. Malipatil, K. Perry, I. Valenzuela, and M. A. Nash. 2018. "First Detection of Russian Wheat Aphid *Diuraphis noxia* Kurdjumov (Hemiptera: Aphididae) in Australia: A Major Threat to Cereal Production." *Austral Entomology* 57 (4):410–417. https://doi.org/10.1111/aen.12292.

7 Prepare, Respond, and Recover
Selecting Immediate and Long-Term Strategies to Manage Invasions

Susan M. Hester and Lucie M. Bland

ABSTRACT

When an invasion of a pest or disease is first discovered, the immediate response is typically to determine the extent of the incursion as quickly as possible and to control detected outbreaks. Ideally, this initial response preserves all longer-term management options, whether they be eradication, containment, impact reduction, or mitigation. Although eradication may be an appealing strategy for decision makers (as there is an expected end-date to expenditure and threat impacts), eradication is not always the most appropriate or cost-efficient strategy. The socio-political, technical, and economic feasibility of different management strategies must be carefully assessed prior to making any long-term decisions. Understanding how to efficiently allocate limited funding to manage outbreaks of biosecurity threats is a huge challenge for biosecurity agencies. In this chapter, we discuss the feasibility of longer-term management strategies and provide a decision-making framework to show how a biosecurity agency might select a cost-efficient strategy to respond to a pest incursion.

GLOSSARY

Eradication The elimination of every single individual or propagule (e.g. seed) of a species from an area in which recolonisation from new invasions is unlikely (Myers, Savoie, and van Randen 1998).

Containment Aims to restrict the spread of a threat by containing it to a defined area. Containment can be full (i.e. all spread beyond the containment area is prevented) or partial (i.e. spread beyond the containment area is slowed).

Impact reduction (also known as maintenance control) Aims to suppress the population level of a threat to below an acceptable threshold. The acceptable threshold relates to the level of impact on the ecosystem or area invaded, which can be expressed in terms of a threat's distribution or density, or a combination of both (Wittenberg and Cock 2001). Impact reduction is typically selected when eradication and containment are no longer feasible.

Mitigation (also known as asset protection) Involves finding the best way to live with the introduced species by managing impacted assets (e.g. native species or ecosystems), rather than managing the introduced species directly. Within an agricultural context, mitigation is often referred to as adaptation.

Detectability The conspicuousness of the invasive species within the invaded landscape (Panetta 2009), which influences the feasibility of management (Cacho et al. 2006).

DOI: 10.1201/9781003253204-10

Delimitation The process of establishing the boundaries and extent of an incursion. Delimitation maps (i.e. probabilistic maps of detection or absence derived from surveillance data) can be created to assist with this process.

Control methods The management methods used to reduce the abundance and/or distribution of a pest, disease, or weed. Control methods may be manual (e.g. use of hands or hand-held tools), mechanical (e.g. machinery or powered tools), chemical (e.g. herbicides, pesticides, or fungicides), or biological (i.e. ongoing control by another species; Panetta et al. 2011).

Search and control The key activities involved in managing invasive species and integral to the calculation of feasibility of eradication and containment (Cacho et al. 2006). Search and control are often undertaken together (i.e. the threat is controlled as it is found).

Surveillance Searching activities that are undertaken to detect threats. Active, general, and passive surveillance are types of surveillance that differ in terms of the level of coordinated and deliberate searching involved (see Chapter 6. Detect).

Feasibility The degree of ease anticipated in attempts to manage an invasive species. The feasibility of management strategies is influenced by socio-political, technical, and economic factors (Wilson, Panetta, and Lindgren 2016).

INTRODUCTION

Although preventing the entry of pests and diseases is the most cost-effective and environmentally sustainable option for dealing with invasive species (Finnoff et al. 2007; Leung et al. 2002), some pests and diseases will still slip through prevention and screening mechanisms and establish. This occurs even in biosecurity systems that are considered highly effective and comprehensive (Scott et al. 2017).

When a biosecurity threat (i.e. exotic pest or disease) is detected, a rapid and well-coordinated emergency eradication response can reduce adverse impacts and limit the need for post-incident recovery and adaptation. If the emergency response fails to eradicate the biosecurity threat, managers are faced with four long-term management alternatives: (1) continue to pursue eradication, or switch strategies to (2) containment, (3) impact reduction, or (4) mitigation. Containment aims to prevent or reduce the likelihood of establishment and reproduction beyond a predefined geographical range (Wittenberg and Cock 2001), whereas impact reduction attempts to reduce a threat's impacts without necessarily restricting its range (e.g. by suppressing population levels; Grice 2009). Impact reduction methods (e.g. biological control) are usually the only economically feasible, long-term management option for widespread pests and diseases (Cacho and Hester 2022). A fourth course of action, mitigation, does nothing to stop the further spread and establishment of the threat, but rather focuses on mitigating the impacts of the threat on affected assets (e.g. native species; Blackburn et al. 2011; Wittenberg and Cock 2001).

Unfortunately, there are many examples of failed management responses, particularly failed eradication programmes (Pluess et al. 2012; Wittenberg and Cock 2001). Many response programmes start as eradication programmes and change by default to containment, impact reduction, or mitigation when the threat proves too difficult to control (Fletcher et al. 2015; Grice 2009). For example, efforts to manage the spread of the emerald ash borer (EAB) in North America changed from eradication to containment due to the extent of its spread, and then to the control of satellite populations when the pest was detected beyond the containment zone (Liebhold and Kean 2019), and finally to impact reduction via biological control (APHIS 2020; Duan et al. 2022). The continued spread of EAB has cost municipalities, property owners, nursery operators, and forest products industries hundreds of millions of dollars per year (Aukema et al. 2011); the predicted impact of the pest in the eastern United States over a 10-year period (2009–2019) was in the order of billions of dollars (Kovacs et al. 2010).

When faced with an incursion, biosecurity agencies must carefully decide which management strategy is most suitable, as embarking on programmes that are unlikely to be successful will waste the agencies' resources and taxpayer money. This chapter discusses available strategies for managing incursions of pests, diseases, and weeds when funding is drawn from the public purse. First, we describe responses that occur immediately following the detection of an incursion, also known as "emergency" responses. If eradication is not achieved at this stage, decision makers are faced with selecting a long-term management option (namely eradication, containment, impact reduction, or mitigation) based on its feasibility. The feasibility of a management strategy is determined by the interplay of multiple socio-political, technical, and economic factors (Bomford and O'Brien 1995; Hulme 2020; Moon, Blackman, and Brewer 2015). Finally, we discuss practical considerations for selecting and implementing post-border management strategies.

EMERGENCY RESPONSE STRATEGIES

The first detection(s) of a pest, disease, or weed may occur as part of a surveillance programme or via passive or general surveillance (see Chapter 6. Detect). Once a detection is made, biosecurity agencies are tasked with limiting the spread of the threat and preserving eradication as a viable option, while more information is gathered to inform the long-term management of the threat. Generally, eradication or containment are appropriate initial responses to an incursion, put in place until the extent of the incursion and the costs and benefits of long-term management are better understood (Hester, Hauser, and Kean 2017). Biosecurity incident management teams often follow a sequence of activities prescribed in pre-agreed response plans, which can include:

- Incident definition and initial investigation.
- Emergency response activities.
- Providing proof of freedom (where eradication is feasible), particularly where market access is important.
- Options for transitioning to long-term management (where eradication is not feasible).

The emergency response phase often relies on well-developed plans of activities that are enacted immediately when a particular pest or disease is detected (e.g. APHIS 2019; MPI 2018). Eradication of notable pests and diseases is an automatic, pre-determined decision in many countries where loss of market access or unacceptable environmental impacts would occur if the pest or disease were to establish—for example, for foot-and-mouth disease in Australia (Animal Health Australia 2014) and New Zealand (MPI 2022), khapra beetle in the United States (APHIS 2022), and avian influenza in Canada (Government of Canada 2022). When pre-agreed emergency response plans are not available, agencies still need to act quickly in response to the detection of unwanted organisms if they are to preserve all long-term management options, particularly eradication.

The effort (and hence investment) involved in an emergency response comprises the surveillance effort required to delimit an invasion (i.e. determine the extent of its spread) plus the search-and-control effort required to remove the threat from the entire infested area and prevent further reproduction (Figure 7.1). A short-term research budget is also necessary to fund initial investigations and to create a map to direct search-and-control efforts (see Chapter 14. Map).

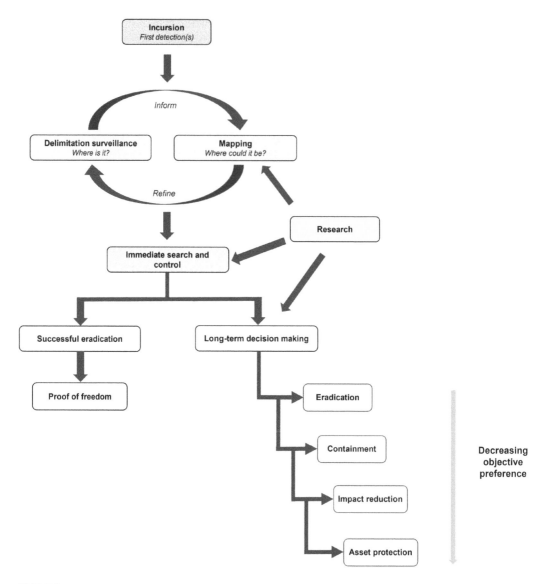

FIGURE 7.1   Short and long-term decision-making processes in response to the detection of a new biosecurity threat.

DELIMITATION SURVEILLANCE AND THREAT MAPPING

As a matter of priority, agencies need to understand the current extent of an incursion. Delimitation should occur as quickly as possible to limit further spread and to limit the effort and cost required to manage the incursion (Panetta and Lawes 2005). Initial delimitation is difficult because (1) the pest or disease may only have been observed in a few locations, (2) the pathway of entry may be unknown, (3) the first detection(s) may not be where the species first arrived or in the epicentre of the outbreak, and (4) the potential dispersal of the species in the new environment may be unknown (Leung, Cacho, and Spring 2010).

   Trace-back and trace-forward techniques, combined with pathways analysis, can be used to gather information on threat introduction and spread (see Chapter 6. Detect). Trace-back activities are used to locate the likely site of introduction, and if this is successful, trace-forward activities are used to locate areas, infrastructure, or organisms that may be infested and need to be surveyed

(Hester et al. 2015). Tracing can more easily be implemented for some threats (e.g. livestock diseases), but it is much more difficult for other threats (e.g. hitchhikers, plant pests). National databases of animal location and movement are useful for tracing. For example, Australia's National Livestock Information System (NLIS) and New Zealand's National Animal Identification and Tracing (NAIT) programme collect data that can be used to delimit incursions (Integrity Systems Company 2022; MPI 2018).

When pest and disease incursions are likely to impact market access, survey designs for delimitation are often guided by international regulations (see Chapter 6. Detect). When delimiting incursions for which there is little or no trade imperative, there are no clear rules about where to search, how much search effort should be applied, or the point at which searching should stop. Rather, these decisions can be conceptualised as a trade-off between the possibility of the invasive species escaping the delimited area versus wasted effort if the search area, time, or intensity are too high (Leung et al. 2010). The speed at which delimitation occurs is also important because as the species continues to spread and increase its population size, the likelihoods of escape and reproduction outside the search area increase, in turn increasing management costs. Delayed delimitation was thought to be responsible for a significant increase in the cost of managing red imported fire ant in Australia (Spring and Kompas 2015), a campaign that has been underway for more than 20 years.

The emergency management response phase will be guided by a delimitation map, which informs the likelihood of a species being present, or a habitat suitability map of abiotic and/or biotic conditions suitable for establishment. More sophisticated maps approximate the establishment potential of the species by accounting for the likely propagule pressure caused by different pathways of entry and spread (see Chapter 14. Map). At this early stage, maps are used to inform where to undertake preliminary delimitation surveys. The surveys undertaken are then used in combination with statistical models to infer likelihoods of absence, which in turn, allow delimitation and refining of the spatial extent of the incursion (see Chapter 6. Detect). If time permits, surveillance data can be complemented with additional research on the biology, ecology, and management feasibility of the threat (Cacho et al. 2018; Epanchin-Niell et al. 2014; Hester et al. 2010; Jentsch et al. 2020).

SEARCH AND CONTROL

Search and control activities carried out in the emergency response phase are integral to the development of delimitation maps and to determine whether eradication in this phase is possible (Hauser et al. 2016). For example, emergency biosecurity measures were implemented on French poultry farms in response to outbreaks of highly pathogenic avian influenza between 2015 and 2017 (Delpont et al. 2021) and in the United Kingdom foot-and-mouth disease outbreaks of 2001 (NAO 2002) and 2007 (DEFRA 2008). The 2007 outbreak of equine influenza in Australia was eradicated in the emergency response phase using a range of outbreak control measures, including movement restrictions, vaccination, quarantining of properties, and issuing of biosecurity guidelines (Schemann et al. 2012). Emergency measures with a view to eradication were put in place following the detection of the red imported fire ant (a pest with significant environmental, social, and domestic economic impacts) in Queensland, Australia, in 2001. Eradication did not occur in the emergency response phase, however, due to the rapid spread of the pest and difficulties delimiting the incursion, but several outbreaks were subsequently eradicated.

If pre-agreed emergency response plans are in place in a jurisdiction, then search-and-control methods will be clearly defined. In general, eradication attempts during the initial response phase are more likely to be successful for pests and diseases that have pre-agreed emergency response plans compared to those who have not, as there is an imperative for a jurisdiction to regain trade access as quickly as possible for high-priority threats. To regain market access, surveillance evidence is used in combination with statistical models to infer the likelihood of species absence (also known as proof of freedom; see Chapter 6. Detect). If this likelihood is above a predefined level of confidence, the threat is deemed eradicated and management activities end here.

For weeds, eradication requires the elimination of every viable seed, which may mean that longer-term management is required for most weed emergency responses. The minimum duration for an eradication programme for weeds will be determined by seed persistence, which may range from several weeks to several decades (Panetta 2015). This contrasts with animal diseases, where eradication can occur quickly when control involves culling animals (Animal Health Australia 2021).

LONG-TERM RESPONSE STRATEGIES

If the emergency response phase determines that immediate eradication of a threat is not feasible or if immediate search and control is not successful, management will transition to a long-term response. At this stage, biosecurity agencies need to decide which long-term management strategy (namely, eradication, containment, mitigation, or asset protection) is most suitable (Figure 7.1). A key challenge for biosecurity agencies is understanding which management strategy (and combination of activities within a chosen strategy) will achieve the best long-term outcomes for society.

Assessing and comparing the benefits and costs of different actions requires a clear definition of the proposed actions, based on information collected during delimitation and search and control. Such data may include:

- Whether the species is known to be invasive elsewhere, and its establishment potential in the local environment (see Chapter 14. Map).
- The likely consequences of establishment.
- Control methods and their costs.
- The likelihood that the species will invade again in future.
- Information on political will and community sentiment with regards to management.

Managers must also anticipate how a given strategy might unfold if selected. This requires determining the activities that will form part of the strategy, the scale at which these activities will be applied (e.g. intensity, area of control, and duration), and predicting the effects of control over time (e.g. through population or distribution modelling). Many of these factors are uncertain at the time of decision making.

FEASIBILITY OF RESPONSE STRATEGIES

Feasibility should guide the choice of long-term strategy (Panetta 2009). Feasibility is determined by the interplay of multiple factors, which can be broadly classified as socio-political, technical, or economic:

- *Socio-political feasibility*: Relevant socio-political considerations include the land tenures over which control will be required, whether the species is regarded by the public as native or non-native (e.g. dingoes in Australia; Ballard and Wilson 2019), societal values around the species being managed (e.g. feral horses in Australia; Nimmo and Miller 2007) and control methods (e.g., live trapping and euthanasia of grey squirrels in Italy; Bertolino and Genovesi 2003), and crucially, whether long-term financial and institutional commitment to a strategy is possible (Panetta and Timmins 2004). Although socio-political support is crucial to implementing an effective management strategy, social values are often overlooked in decision making or assumed to be in favour of proposed government responses. This in turn can lead to failed management programmes or de facto implementation of mitigation ("living with the threat") when proposed strategies are rejected by communities

(Baker et al. 2022). Value elicitation methods can help develop an understanding of social values to integrate into management (see Chapter 12. Elicit).

- *Technical feasibility*: Once socio-political considerations are deemed suitable for management, selecting a strategy becomes a question of technical feasibility. Technical feasibility refers to multiple factors that together determine management success, such as the biology of the invasive species, its detectability, control effectiveness, and logistical considerations around the number, spatial distribution, and accessibility of incursions (Panetta and Timmins 2004). Treatment methods under different strategies may include mechanical, chemical, and biological control, separately or in combination. Surveillance will inform where treatment should occur and may involve trained experts in surveillance, members of the public, dogs, satellites, drones, or other aircraft. At this stage, it may be possible to rule out some strategies as they are unlikely to be technically feasible given available technology, resources, and/or the extent of the incursion.
- *Economic feasibility*: While a proposed strategy may be technically feasible, it may be economically inefficient compared to other strategies. Understanding economic feasibility requires calculating the costs of a management strategy over time and the benefits generated for society (Cacho et al. 2006). Understanding technical feasibility (i.e. the technical factors that influence implementation) is a prerequisite for assessing economic feasibility. Ideally, the decision maker would select the strategy that maximises "value for money" when generating benefits for society (i.e. the most economically efficient strategy; see Chapter 9. Resource Allocation).

Below, we discuss how each of these factors influences the feasibility of eradication, containment, impact reduction, and mitigation.

Eradication

Eradication involves the complete removal of the target pest or disease from an infested area. Eradication can be appealing compared to containment or control, which require an ongoing investment of resources unless the threat can be brought under effective biological control (i.e. ongoing control by another species; Panetta et al. 2011).

Successful eradications have been conducted for most taxonomic groups (Hester et al. 2004; Mack et al. 2000; Rejmánek and Pitcairn 2002). Eradication is more likely to be successful for land vertebrates on islands regardless of size, compared to non-island eradications, with 85% of 1,200 mammal eradications on islands having been successful worldwide (Holmes et al. 2019). Compared with animal eradications, weed eradication programmes are less often successful. When persistent seedbanks occur, eradication requires the elimination of every viable seed, which may take years or even decades beyond the elimination of adult plants (Panetta 2015; Wilson et al. 2016). Further reproduction must be prevented, otherwise the seedbank will be replenished, and the duration of the eradication programme will be extended, or eradication may no longer be the most appropriate course of action. Weed eradication programmes also face challenges in terms of detection. Although invasive animals are typically attracted by lures and baits and in some cases are amenable to control via aircraft, invasive plants must be detected in situ, with regular search and control for as long as it takes to deplete the seedbank (Panetta 2009). For example, seeds of miconia (*Miconia calvescens*), often dispersed over long distances by birds, can persist for 15 years (Hester et al. 2010), while seeds from Scotch broom may persist for 20–25 years in introduced habitats (Hosking, Smith, and Sheppard 1996).

The feasibility of eradication will generally be determined by:

- *Infestation area*: Eradication is generally only possible during the earliest part of the expansion of a threat because species abundance and density are usually relatively low in this stage of invasion (Cacho 2004; Mack and Foster 2009; McNeely et al. 2001) or for isolated populations (Grice 2009). The level of effort required throughout an eradication programme (to search, control, monitor, and apply follow-up control) means that the economic feasibility of eradication declines rapidly with increasing area. A review of 136 eradication programmes across 75 species of invasive alien invertebrates, plants, and plant pathogens (Pluess et al. 2012) analysed the impact of: reaction time; size of the infested area; biological knowledge and preparedness to react; and insularity (whether the campaign was on an island or continent). The study revealed that only the spatial extent of the infestation was significantly related to the eradication outcome: a smaller infestation area was more likely to lead to successful eradication (Pluess et al. 2012). In most cases, well-established populations and large infestation areas are unsuitable for eradication (Harris and Timmins 2009; Rejmánek and Pitcairn 2002; Wittenberg and Cock 2001).

- *Invasive species characteristics*: Pests and diseases with traits that facilitate rapid expansion, or species that are cryptic or harder to detect, may be more difficult to eradicate (Grice 2009). For example, the EAB colonises the upper portions of the canopy of large trees before the main trunk, making early infestations difficult to detect (Herms and McCullough 2014).
- *Environmental characteristics*: Inaccessible landscapes can reduce threat detectability and increase the cost of control actions (or prohibit the use of some controls altogether). For example, orange hawkweed can be difficult for human searchers to distinguish visually from the surrounding vegetation in forested and sensitive alpine areas, particularly when not in flower. Detector dogs and drones can be used to assist with detection, with drones being particularly useful over large areas (Cacho et al. 2018; NSW Department of Planning and Environment 2020).
- *Land management*: Eradication tends to be easier in landscapes that are intensively managed (e.g. agricultural areas where the imperative to eradicate is high). Heavily populated landscapes may pose social challenges to eradication, for example when the aerial application of biological control agents or chemicals is unpopular with urban residents (Brockerhoff et al. 2010).

CONTAINMENT

When eradication fails or is not feasible, containing or delaying the spread of a pest or disease may still prevent substantial ecological damage (Mack and Foster 2009; Simberloff 2003). A containment strategy aims to prevent or slow the establishment and reproduction of a threat beyond a predefined geographical range (Grice et al. 2020; Wittenberg and Cock 2001). In contrast to eradication, containment requires control activities to be applied indefinitely. A containment strategy involves putting most surveillance and control efforts at the invasion front, or "barrier zone", to halt expansion (Grice 2009). One notable containment programme has been the management of the flighted spongy moth complex (*Lymantria* spp.) in the United States, where traps were placed close to the expanding population front to slow the expansion of the pest (Sharov and Liebhold 1998).

The feasibility of containment should be assessed once eradication has been rejected as a goal (i.e. when eradication is deemed infeasible), rather than as the default result of an unsuccessful eradication (Grice 2009). Deciding whether to embark on a containment strategy should be based on clearly defined goals, such as preventing the invasive from spreading beyond a nominated perimeter. Different management activities that can take place within barrier zones should be considered and evaluated in terms of their costs and benefits. Many jurisdictions also establish domestic biosecurity or quarantine zones to restrict the movement of high-risk material between zones. In Australia, phylloxera management zones have been established to limit pest movement within the state of Victoria (DJPR 2022).

Factors that affect the feasibility of eradication also affect the feasibility of containment (specifically, infestation area, species characteristics, environmental conditions, and local land tenure). Although containment can be applied at any management scale (e.g. country, region, or landscape), the effort required to contain the edge of an invasion front increases with invasion size. This is particularly true for threats that are capable of long-distance dispersal (Fletcher et al. 2015; Panetta and Cacho 2014), making containment more likely to be feasible during the early stages of invasion when the boundary of the incursion is relatively small (Grice 2009). Permanent physical barriers may be erected as part of containment programmes, with exclusion fences for rabbits in Australia extending many thousands of kilometres (Crawford 1969). However, most containment programmes relying on physical barriers have not been successful (Grice 2009; McKnight 1969).

IMPACT REDUCTION

Impact reduction aims for a long-term reduction in density or abundance to below a threshold where impacts caused to the environment, community, or economy are considered acceptable (Wittenberg and Cock 2001). Impact reduction typically occurs when the use of traditional control techniques over large areas has become economically inefficient, and thus eradication and containment are no longer considered feasible. In contrast to containment, impact reduction attempts to reduce the impact of an invasive species without necessarily restricting its range, and it is most often applied in areas where the impact of the threat is the greatest (Grice 2009). Like containment, impact reduction requires activity indefinitely. In Australia, national strategies and frameworks guide biosecurity participants in their efforts to continually manage invasive plants, animals, and pests and diseases of national significance (e.g. IPAC 2016a, 2016b). Participants that fail to control noxious weeds on their property can be directed to do so by state and territory legislation.

Establishing biological control agents to permanently control the pest is usually the only economically feasible, long-term impact reduction option. There are many successful examples of biological control programmes, including the control of the prickly pear cactus in Australia by the *Cactoblastis* moth from Argentina (Dodd 1940), control of European rabbits in Australia using various pathogens (Cooke et al. 2013), control of the South American alligator weed in Florida by a flea beetle, and control of the South American cassava mealybug in Africa by a South American parasitic wasp (Mack et al. 2000).

Whether a biological control programme is worthwhile pursuing depends on the potential benefits generated from a reduction in pest abundance and the costs of research, testing, breeding, release, and monitoring of the biocontrol agent (Hester et al. forthcoming). The efficacy of the proposed agent in supressing the pest (i.e. technical feasibility) is fundamental, and information about agent efficacy, costs, and benefits should be understood prior to spending significant time in screening and testing (Hajek et al. 2016). For example, Cacho, Hester, and Tait (2023) and Cacho and Hester (2022, 2023) developed a bioeconomic model to understand whether the European wasp (*Vespula germanica*) could be a candidate for a management programme in south-eastern Australia given the availability of a biocontrol agent (*Sphecophaga vesparum vesparum*), following successful screening and testing of an agent several decades earlier. The authors found that although the economics of controlling European wasp using biological control appear promising due to the benefits of control, proceeding with the chosen agent was not recommended based on current knowledge about its performance.

MITIGATION

When eradication, containment, or impact reduction have failed or are not feasible, a final course of action is to do nothing to directly influence the abundance of the threat, but rather to focus on mitigating impacts on affected biodiversity and ecosystem assets (Blackburn et al. 2011; Wittenberg and Cock 2001). For example, when extensive rat control measures failed to alleviate extinction pressures on the Critically Endangered Seychelles black parrot (*Coracopsis nigra*), a trial program of building and installing rat-proof nesting boxes on the Seychelles islands was successful in improving breeding outcomes (Wittenberg and Cock 2001). Mitigation also involves threat monitoring so that managers can detect and respond to changes in the abundance and/or range of the threat. Managers must also account for new technologies that may render new or previously discarded management strategies feasible, including the development of new vaccines, camera sensors, or the introduction of biological control agents (Wittenberg and Cock 2001).

PRACTICAL CONSIDERATIONS FOR POST-BORDER RESPONSES

A huge challenge for biosecurity agencies is deciding how to efficiently allocate limited resources to activities that will generate the best outcomes for the economy, society, and/or the environment (Bomford and O'Brien 1995; Epanchin-Niell 2017; Fletcher et al. 2015; Hulme 2020; McNeely et al. 2001; Panetta and Timmins 2004). Responding quickly and cost-efficiently to incursions requires planning to assess the feasibility of response strategies and to select the best strategy from a well-defined set of options (Box 7.1).

A cost-efficient approach to selecting long-term biosecurity response strategies starts with clearly defining the response strategies (and different activity combinations within each strategy, where relevant) under consideration (Box 7.1). Based on information collected during initial delimitation and search-and-control activities, managers must outline the specific activities that will form part of each strategy, including: the scale and timing of these activities; associated monetary, human, or other costs; and expected benefits in controlling the threat.

Within the long-term decision-making process, eradication should be assessed first as the pre-ferred option (Box 7.1). The details of one or more eradication scenarios should be assessed, includ-ing details of the area to be managed, timeframes to achieve management goals, and the conditions under which the management decision will be revisited (Wilson et al. 2016). Although eradication may be politically appealing because resources are applied for a finite amount of time, it should only be attempted and persisted with if it remains feasible (e.g. during the emergency response or as a carefully selected and monitored strategy; Mack et al. 2000). Psychological frailties such as the "sunk cost fallacy" (i.e. where a decision maker is reluctant to abandon a strategy because they

BOX 7.1. A COST-EFFICIENT APPROACH TO SELECTING LONG-TERM BIOSECURITY RESPONSE STRATEGIES

Here, we describe how a biosecurity manager can select a cost-efficient biosecurity response strategy for a hypothetical pest based on socio-political, technical, and economic feasibility (Figure 7.2). For the hypothetical pest, a suite of long-term response strategies may encompass:

- Full eradication of the incursion (strategy A).
- Containment involving only measures at the barrier zone (strategy B).
- Containment involving measures at the barrier zone and the establishment of a quar-antine zone (strategy C).
- Impact reduction using a biological control agent (strategy D).
- Impact reduction using manual control (strategy E).
- Mitigation focusing on affected native species (strategy F).

Within a cost-efficient approach, socio-political feasibility is assessed first. In this hypo-thetical example, asset protection is deemed politically unpalatable as an initial strategy given the potentially large adverse impacts of the pest. As a second step, technical feasibility is as-sessed using detailed information on activities to be implemented. At this stage, eradication is deemed technically infeasible given the low detectability of the pest using current approaches. Using an economic analysis, the remaining potential strategies (strategies B, C, D, and E) are compared to each other using information on their relative costs and benefits (i.e. their return on investment; see Chapter 9. Resource Allocation). In our example, strategy C (containment with a barrier zone and quarantine zone) is found to provide the highest value for money and is selected as the initial long-term strategy to manage the threat.

This decision is revisited at regular intervals by managers based on new information con-cerning the spread of the threat, its impacts, and the feasibility of different management strate-gies (Figure 7.2).

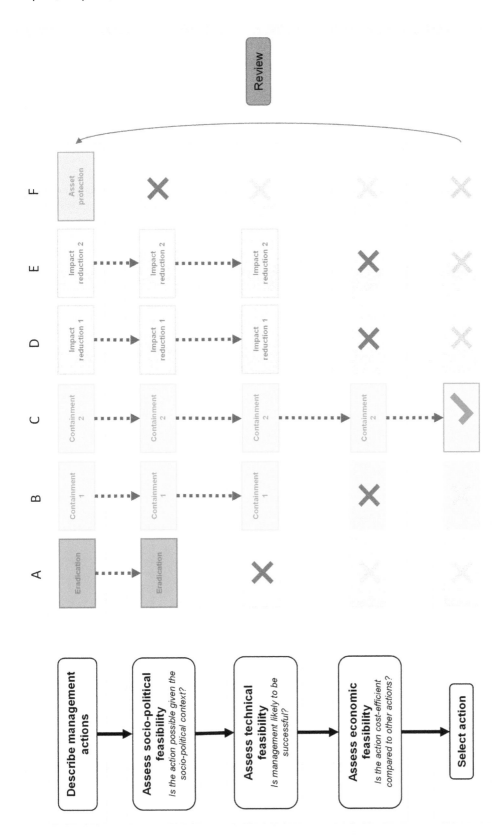

FIGURE 7.2 Decision-making framework to select a long-term cost-efficient strategy to respond to a threat incursion.

have invested heavily in it; Arkes and Blumer 1985) may result in delaying the decision to give up on eradication even when it is clear that abandonment is the preferred option. Understanding if and when to cease eradication (or any other strategy) relies on objectively monitoring and evaluating the outcomes of management over time (see Chapter 10. Monitoring, Evaluation, and Reporting).

There are many examples of eradication programmes that have failed because feasibility was not assessed appropriately, for example when the initial extent of the incursion was not well understood, management costs were underestimated, and when views about the likely success of the programme were overly optimistic (Wilson et al. 2016). Understanding public sentiment is not always straight-forward, and many proposed eradication programmes have been strongly opposed by members of the community (e.g. feral horses in Australia and grey squirrels in Italy; Wittenberg and Cock 2001). Apart from wasting resources, failed or mis-communicated eradication programmes can erode pub-lic trust in biosecurity agencies, which does not bode well for future incursion responses that are likely to require assistance and collaboration from the community (Wittenberg and Cock 2001).

As such, the selection of strategies should not be a case of "set and forget", and switching strat-egies may be required due to events that render the existing strategy infeasible or because new technology now renders new or previously discarded strategies possible (Liebhold and Kean 2019). Eradication or containment may be an appropriate initial response to an incursion, put in place until the extent of the incursion, and costs and benefits of management are understood, allowing a longer-term management decision to be made (Hester et al. 2017). Although eradication, con-tainment, and impact reduction are typically discussed as if they are mutually exclusive, this is not necessarily the case (Hulme 2006). For example, containment could occur at the edges of an incursion at the "barrier zone", while eradication of small, isolated populations could occur within the barrier zone. It may also be possible to shift the barrier zone backwards to eventually eradicate the entire population (Sharov and Liebhold 1998). To assist with these complex decisions, multiple decision-support tools have been developed to help practitioners select and switch response strate-gies based on the characteristics of incursions and proposed management actions (Box 7.2).

BOX 7.2. DECISION-SUPPORT TOOLS TO SELECT BIOSECURITY RESPONSE STRATEGIES

In recent years, various decision-support tools have emerged to assist biosecurity agencies in determining when to use and when to switch strategies based on economic feasibility. For exam-ple, the Switching Point Tool helps determine when the eradication of a weed is no longer the optimal course of action and when containment or impact reduction should be attempted (Cacho 2010; Cacho and Hester 2013). The tool is based on identifying two switching points: the inva-sion size at which it is no longer optimal to attempt eradication but where containment may be an option; and the invasion size at which it becomes optimal to apply just enough control to maintain the invasion at a steady state (containment point) within which impact reduction is feasible. The model calculates the maximum area that should be targeted for eradication based on minimising the combined costs of controlling the invasion plus the damages caused by the invasion.

To understand weed eradication feasibility, the WeedSearch spreadsheet tool was devel-oped by combining population dynamics and search theory (Cacho and Pheloung 2007). WeedSearch allows users to calculate the probability that a weed invasion will be eradicated based on the amount of time spent searching for it (search effort; Cacho and Pheloung 2007). Spatial models of detection and treatment have been developed for specific weeds, such as for the orange hawkweed in Victoria, Australia, to help determine effort allocation across a land-scape by comparing the costs and benefits of earlier detection (Hauser 2009). The Eradograph tool can also be used to assess progress towards eradication of weeds (Burgman et al. 2013; Panetta and Lawes 2007).

In practice, it may be difficult for biosecurity agencies to obtain the data required to compute and select the most economically efficient set of activities across threats. Yet, adopting principles of economic efficiency is essential if biosecurity agencies are to achieve maximum value for money from the use of limited taxpayer funds. This means understanding the value of risk-reduction across a range of potential management targets and using available funds where value is highest (see Chapter 9. Resource Allocation). In addition, the question of "how much funding is required to efficiently address a nation's biosecurity risks" has not been studied, preventing biosecurity agencies from requesting budgets that can adequately meet risk-reduction requirements.

Addressing biosecurity threats is one of many calls on society's resources, and the division of resources among sectors (e.g. cybersecurity, public health, counterterrorism, and disaster risk management; World Economic Forum 2022) has largely been a political decision, rather than an evidence-based decision underpinned by value for money. Responding quickly and efficiently to a biosecurity threat requires a financially sustainable biosecurity budget—a budget that is large enough to cope with the growing number of likely incursions as the volume of travellers and cargo increases across the globe (CSIRO 2020). Biosecurity response activities are typically funded by taxpayers and the domestic industry stakeholders who stand to be most impacted by particular incursions. Those who inadvertently introduce pests, diseases, or weeds (e.g. travellers, importers, or vessel owners) typically have no role in bearing the cost of responding to incursions. Relying on taxpayer money or impacted parties to respond to incursions is not economically efficient or financially sustainable, and this approach neither penalises nor incentivises those who are causing the damage to domestic economies, environments, and communities.

New thinking around an economically efficient and financially sustainable approach to funding domestic biosecurity responses has recently emerged. Stoneham et al. (2021) outlined a biosecurity risk insurance mechanism whereby risk creators (importers and vessel owners) would be required to purchase insurance. Premiums would be calculated by actuaries and based on the level of biosecurity risk associated with the type and origin of the good imported (or the vessel's risk of biofouling). Funds collected through premiums would be pooled by a government insurance agency and used to fund biosecurity system costs and the cost of controlling outbreaks of pests and diseases if and as they occur (Stoneham et al. 2021). Importantly, incursions are likely to be reduced in number and/or severity because the scheme has powerful incentive properties: importers of goods assessed as low-risk will incur lower premiums compared to high-risk importers, creating an incentive for importers to seek out low-risk sources of goods.

A biosecurity risk insurance mechanism could also be applied to domestic biosecurity policies. For example, it could be applied to the industry levies paid by producers to fund responses to incursions and could be calibrated with the level of biosecurity risk mitigation measures implemented on farms. Farmers whose biosecurity risk-mitigation measures are substantial and verifiable would be levied at lower rates than those who do not apply measures. The latter group would have an incentive to implement biosecurity risk-mitigation measures, thus reducing biosecurity risks to the industry. Biosecurity risk insurance could also be more broadly applied to human health responses, such as in response to pandemics. Domestic health responses (which can amount to many millions of dollars; Richards et al. 2022) could be funded from a levy on travellers from high-risk countries. This group would have an incentive to reduce their "riskiness" by seeking out vaccinations and other risk-mitigation measures prior to arrival.

IN A NUTSHELL

- Because management of invasive species involves spending taxpayer's funds, it is imperative that biosecurity agencies consider the best way to use available funds (or indeed the amount of funds required to achieve the management goal that provides the best outcome for society).
- Eradication, while very appealing, may not always be the most appropriate management strategy, and should not be embarked upon unless it is feasible and offers the best outcome for society.

- Response plans need to be evaluated in terms of the technical and socioeconomic feasibility, not off-the-top-of-one's-head, subjective judgement.
- Switching strategies may be required due to events that render the existing strategy infeasible or because new technology now renders new or previously discarded strategies possible. Ongoing monitoring and evaluation of strategies is therefore critical.

REFERENCES

Animal Health Australia. 2014. Disease strategy: Foot-and-mouth disease (Version 3.4). "Australian Veterinary Emergency Plan (AUSVETPLAN)". Canberra, ACT: Agriculture Ministers' Forum.

Animal Health Australia. 2021. "Informing EAD Responses – AUSVETPLAN." accessed 1 November 2022. https://animalhealthaustralia.com.au/ausvetplan/.

APHIS. 2019. "National Plant Health Emergency Management Framework". Animal and Plant Health Inspection Service, U.S. Department of Agriculture.

APHIS. 2020. "Removal of Emerald Ash Borer Domestic Quarantine Regulations". Animal and Plant Health Inspection Service, U.S. Department of Agriculture.

APHIS. 2022. "Khapra Beetle Program Manual". Animal and Plant Health Inspection Service, U.S. Department of Agriculture.

Arkes, H. R., and C. Blumer. 1985. "The Psychology of Sunk Cost." *Organizational Behavior and Human Decision Processes* 35 (1):124–140. https://doi.org/10.1016/0749-5978(85)90049-4.

Aukema, J. E., B. Leung, K. Kovacs, C. Chivers, K. O. Britton, J. Englin, S. J. Frankel, R. G. Haight, T. P. Holmes, A. M. Liebhold, D. G. McCullough, and B. Von Holle. 2011. "Economic Impacts of Non-Native Forest Insects in the Continental United States." *PLOS ONE* 6 (9):e24587. https://doi.org/10.1371/journal.pone.0024587.

Baker, C. M., P. T. Campbell, I. Chades, A. J. Dean, S. M. Hester, M. H. Holden, J. M. McCaw, J. McVernon, R. Moss, F. M. Shearer, and H. P. Possingham. 2022. "From Climate Change to Pandemics: Decision Science Can Help Scientists Have Impact." *Frontiers in Ecology and Evolution* 10. https://doi.org/10.3389/fevo.2022.792749.

Ballard, J. W. O., and L. A. B. Wilson. 2019. "The Australian Dingo: Untamed or Feral?" *Frontiers in Zoology* 16 (1):2. https://doi.org/10.1186/s12983-019-0300-6.

Bertolino, S., and P. Genovesi. 2003. "Spread and Attempted Eradication of the Grey Squirrel (*Sciurus carolinensis*) in Italy, and Consequences for the Red Squirrel (*Sciurus vulgaris*) in Eurasia." *Biological Conservation* 109 (3):351–358. https://doi.org/10.1016/S0006-3207(02)00161-1.

Blackburn, T. M., P. Pyšek, S. Bacher, J. T. Carlton, R. P. Duncan, V. Jarošík, J. R. U. Wilson, and D. M. Richardson. 2011. "A Proposed Unified Framework for Biological Invasions." *Trends in Ecology & Evolution* 26 (7):333–339. https://doi.org/10.1016/j.tree.2011.03.023.

Bomford, M., and P. O'Brien. 1995. "Eradication or Control for Vertebrate Pests?" *Wildlife Society Bulletin (1973-2006)* 23 (2): 249–255.

Brockerhoff, E. G., A. M. Liebhold, B. Richardson, and D. M. Suckling. 2010. "Eradication of Invasive Forest Insects: Concepts, Methods, Costs and Benefits." *New Zealand Journal of Forestry Science* 40 (suppl): S117–S135.

Burgman, M. A., M. A. McCarthy, A. Robinson, S. M. Hester, M. F. McBride, J. Elith, and F. D. Panetta. 2013. "Improving Decisions for Invasive Species Management: Reformulation and Extensions of the Panetta–Lawes Eradication Graph." *Diversity and Distributions* 19 (5-6):603–607. https://doi.org/10.1111/ddi.12055.

Cacho, O. 2004. "When is it Optimal to Eradicate a Weed Invasion?" 14th Australian Weeds Conference: Weed Management: Balancing People, Planet, Profit, Wagga Wagga, New South Wales, Australia, 6-9 September 2004.

Cacho, O. 2010. "A Tool to Support the Decision to Switch between Eradication and Containment of an Invasion: Switching Point Model Version 2.0 User Manual." Australian Centre of Excellence for Risk Analysis.

Cacho, O., and S. M. Hester. 2013. "A Tool to Support the Decision to Switch between Eradication and Containment of an Invasion." Australian Centre of Excellence for Risk Analysis.

Cacho, O. J., and S. M. Hester. 2022. "Modelling Biocontrol of Invasive Insects: An Application to European Wasp (*Vespula germanica*) in Australia." *Ecological Modelling* 467:109939. https://doi.org/10.1016/j.ecolmodel.2022.109939.

Cacho, O. J., and S. M. Hester. 2023. "Modelling Biocontrol of Invasive Insects Using WaspSim: A MATLAB Simulation Model." *SoftwareX* 21. https://doi.org/10.1016/j.softx.2023.101321.

Cacho, O., S. Hester, and P. Tait. 2023. "Re-Evaluating Management of Established Pests Including the European Wasp, *Vespula germanica* Using Biocontrol Agents." https://doi.org/10.1101/2022.11.22.517291.

Cacho, O., D. Panetta, H. Cherry, S. M. Hester, S. Bower, and H. Bower. 2018. "Weed Surveillance Using Search Theory." 21st Australasian Weeds Conference, Sydney, New South Wales, 9–13 September 2018.

Cacho, O., and P. Pheloung. 2007. "WeedSearch: Weed Eradication Feasibility Analysis Software Manual." University of New England.

Cacho, O. J., D. Spring, P. Pheloung, and S. Hester. 2006. "Evaluating the Feasibility of Eradicating an Invasion." *Biological Invasions* 8 (4):903–917. https://doi.org/10.1007/s10530-005-4733-9.

Cooke, B., P. Chudleigh, S. Simpson, and G. Saunders. 2013. "The Economic Benefits of the Biological Control of Rabbits in Australia, 1950–2011." *Australian Economic History Review* 53 (1):91–107. https://doi.org/10.1111/aehr.12000.

Crawford, J S. 1969. "History of the State Vermin Barrier Fences, Formerly Known as Rabbit Proof Fences." Perth, Western Australia: Department of Agriculture and Food.

CSIRO. 2020. "Australia's Biosecurity Future. Unlocking the Next Decade of Resilience." Commonwealth Scientific and Industrial Research Organisation, accessed 19 February 2020. https://www.csiro.au/en/Do-business/Futures/Reports/Health/Biosecurity-Futures.

DEFRA. 2008. "Foot and Mouth Disease 2007: A Review and Lessons Learned." UK Government Department for Environment, Food & Rural Affairs.

Delpont, M., C. Guinat, J.-L. Guérin, E. Le leu, J.-P. Vaillancourt, and M. C. Paul. 2021. "Biosecurity Measures in French Poultry Farms are Associated with Farm Type and Location." *Preventive Veterinary Medicine* 195:105466. https://doi.org/10.1016/j.prevetmed.2021.105466.

DJPR. 2022. "Phylloxera Management Zones." accessed 1 November 2022. https://agriculture.vic.gov.au/biosecurity/protecting-victoria/victorian-viticulture-biosecurity/compliance-movement-conditions/phylloxera-management-zones.

Dodd, A. P. 1940. *The Biological Campaign Against Prickly-Pear*. Brisbane: A. H. Tucker, Government Printer.

Duan, J. J., R. G. Van Driesche, J. Schmude, R. Crandall, C. Rutlege, N. Quinn, B. H. Slager, J. R. Gould, and J. S. Elkinton. 2022. "Significant Suppression of Invasive Emerald Ash Borer by Introduced Parasitoids: Potential for North American Ash Recovery." *Journal of Pest Science* 95 (3):1081–1090. https://doi.org/10.1007/s10340-021-01441-9.

Epanchin-Niell, R. S. 2017. "Economics of Invasive Species Policy and Management." *Biological Invasions* 19 (11):3333–3354. https://doi.org/10.1007/s10530-017-1406-4.

Epanchin-Niell, R. S., E. G. Brockerhoff, J. M. Kean, and J. A. Turner. 2014. "Designing Cost-Efficient Surveillance for Early Detection and Control of Multiple Biological Invaders." *Ecological Applications* 24 (6):1258–1274. https://doi.org/10.1890/13-1331.1.

Finnoff, D., J. F. Shogren, B. Leung, and D. Lodge. 2007. "Take a Risk: Preferring Prevention Over Control of Biological Invaders." *Ecological Economics* 62 (2):216–222. https://doi.org/10.1016/j.ecolecon.2006.03.025.

Fletcher, C. S., D. A. Westcott, H. T. Murphy, A. C. Grice, and J. R. Clarkson. 2015. "Managing Breaches of Containment and Eradication of Invasive Plant Populations." *Journal of Applied Ecology* 52 (1):59–68. https://doi.org/10.1111/1365-2664.12361.

Government of Canada. 2022. "Overview of Avian Influenza Prevention, Preparedness and Response." accessed 1 November 2022. https://inspection.canada.ca/animal-health/terrestrial-animals/diseases/reportable/avian-influenza/prevention-preparedness-and-response/eng/1375982718329/1375982719875.

Grice, A. C. 2009. "Principles of Containment and Control of Invasive Species." In *Invasive Species Management: A Handbook of Principles and Techniques*, edited by Mick N. Clout and Peter A. Williams, 61–76. Oxford; New York: Oxford University Press.

Grice, A. C., H. T. Murphy, J. R. Clarkson, M. H. Friedel, C. S. Fletcher, and D. A. Westcott. 2020. "A Review and Refinement of the Concept of Containment for the Management of Invasive Plants." *Australian Journal of Botany* 68 (8):602–616. https://doi.org/10.1071/BT20092.

Hajek, A. E., B. P. Hurley, M. Kenis, J. R. Garnas, S. J. Bush, M. J. Wingfield, J. C. van Lenteren, and M. J. W. Cock. 2016. "Exotic Biological Control Agents: A Solution or Contribution to Arthropod Invasions?" *Biological Invasions* 18 (4):953–969. https://doi.org/10.1007/s10530-016-1075-8.

Harris, S., and S. M. Timmins. 2009. "Estimating the Benefits of Early Control of All Newly Naturalised Plants." New Zealand Department of Conservation.

Hauser, C. 2009. "Where and How Much? A Spreadsheet that Allocates Surveillance Effort for a Weed." *Plant Protection Quarterly* 24 (3): 94–97.

Hauser, C. E., K. M. Giljohann, M. Rigby, K. Herbert, I. Curran, C. Pascoe, N. S. G. Williams, R. D. Cousens, and J. L. Moore. 2016. "Practicable Methods for Delimiting a Plant Invasion." *Diversity and Distributions* 22 (2):136–147. https://doi.org/10.1111/ddi.12388.

Herms, D. A., and D. G. McCullough. 2014. "Emerald Ash Borer Invasion of North America: History, Biology, Ecology, Impacts, and Management." *Annual Review of Entomology* 59 (1):13–30. https://doi.org/10.1146/annurev-ento-011613-162051.

Hester, S. M., S. J. Brooks, O. J. Cacho, and F. D. Panetta. 2010. "Applying a Simulation Model to the Management of an Infestation of *Miconia calvescens* in the Wet Tropics of Australia." *Weed Research* 50 (3):269–279. https://doi.org/10.1111/j.1365-3180.2010.00778.x.

Hester, S. M., C. E. Hauser, and J. M. Kean. 2017. "Tools for Designing and Evaluating Post-Border Surveillance Systems." In *Invasive Species: Risk Assessment and Management*, edited by Andrew P. Robinson, Terry Walshe, Mark A. Burgman and Mike Nunn, 17–52. Cambridge: Cambridge University Press.

Hester, S. M., D. I. S. Odom, O. J. Cacho, and J. A. Sinden. 2004. "Eradication of Exotic Weeds in Australia: Comparing Effort and Expenditure: Working Paper Series in Agricultural and Resource Economics." University of New England.

Hester, S. M., E Sergeant, A. P. Robinson, and G Schultz. 2015. "Animal, Vegetable, or …? A Case Study in Using Animal-Health Monitoring Design Tools to Solve a Plant-Health Surveillance Problem." In *Biosecurity Surveillance: Quantitative Approaches*, edited by Frith Jarrad, Samantha Low-Choy and Kerrie Mengersen, 313–333. Wallingford, Oxfordshire: CABI International.

Hester, S. M., P. Tait, R. Kwong, G. Lefoe, and D. Kriticos, and O. J. Cacho. forthcoming. "Biological Control of the Invasive Wasp *Vespula germanica* Biocontrol in Australia: Assessing Socio-Economic Feasibility."

Holmes, N. D., D. R. Spatz, S. Oppel, B. Tershy, D. A. Croll, B. Keitt, P. Genovesi, I. J. Burfield, D. J. Will, A. L. Bond, A. Wegmann, A. Aguirre-Muñoz, A. F. Raine, C. R. Knapp, C.-H. Hung, D. Wingate, E. Hagen, F. Méndez-Sánchez, G. Rocamora, H.-W. Yuan, J. Fric, J. Millett, J. Russell, J. Liske-Clark, E. Vidal, H. Jourdan, K. Campbell, K. Springer, K. Swinnerton, L. Gibbons-Decherong, O. Langrand, M. de L. Brooke, M. McMinn, N. Bunbury, N. Oliveira, P. Sposimo, P. Geraldes, P. McClelland, P. Hodum, P. G. Ryan, R. Borroto-Páez, R. Pierce, R. Griffiths, R. N. Fisher, R. Wanless, S. A. Pasachnik, S. Cranwell, T. Micol, and S. H. M. Butchart. 2019. "Globally Important Islands where Eradicating Invasive Mammals will Benefit Highly Threatened Vertebrates." *PLOS ONE* 14 (3):e0212128. https://doi.org/10.1371/journal.pone.0212128.

Hosking, J. R., J. M. B. Smith, and A. W. Sheppard. 1996. "The Biology of Australian Weeds. 28. *Cytisus scoparius* (L.) Link Subsp. scoparius." *Plant Protection Quarterly* 11 (3): 102–108.

Hulme, P. E. 2006. "Beyond Control: Wider Implications for the Management of Biological Invasions." *Journal of Applied Ecology* 43 (5):835–847. https://doi.org/10.1111/j.1365-2664.2006.01227.x.

Hulme, P. E. 2020. "Plant Invasions in New Zealand: Global Lessons in Prevention, Eradication and Control." *Biological Invasions* 22 (5):1539–1562. https://doi.org/10.1007/s10530-020-02224-6.

Integrity Systems Company. 2022. "NLIS Standards." accessed 1 November 2022. https://www.integritysystems.com.au/identification-traceability/nlis-standards/.

IPAC. 2016a. "Australian Pest Animal Strategy 2017 to 2027." Canberra: Australian Government Department of Agriculture and Water Resources.

IPAC. 2016b. "Australian Weeds Strategy 2017 to 2027." Canberra: Australian Government Department of Agriculture and Water Resources Invasive Plants and Animals Committee.

Jentsch, P. C., C. T. Bauch, D. Yemshanov, and M. Anand. 2020. "Go Big or Go Home: A Model-Based Assessment of General Strategies to Slow the Spread of Forest Pests via Infested Firewood." *PLOS ONE* 15 (9):e0238979. https://doi.org/10.1371/journal.pone.0238979.

Kovacs, K. F., R. G. Haight, D. G. McCullough, R. J. Mercader, N. W. Siegert, and A. M. Liebhold. 2010. "Cost of Potential Emerald Ash Borer Damage in U.S. Communities, 2009–2019." *Ecological Economics* 69 (3):569–578. https://doi.org/10.1016/j.ecolecon.2009.09.004.

Leung, B., O. J. Cacho, and D. Spring. 2010. "Searching for Non-Indigenous Species: Rapidly Delimiting the Invasion Boundary." *Diversity and Distributions* 16 (3):451–460. https://doi.org/10.1111/j.1472-4642.2010.00653.x.

Leung, B., D. M. Lodge, D. Finnoff, J. F. Shogren, M. A. Lewis, and G. Lamberti. 2002. "An Ounce of Prevention or a Pound of Cure: Bioeconomic Risk Analysis of Invasive Species." *Proceedings of the Royal Society of London. Series B: Biological Sciences* 269 (1508):2407–2413. https://doi.org/10.1098/rspb.2002.2179.

Liebhold, A. M., and J. M. Kean. 2019. "Eradication and Containment of Non-Native Forest Insects: Successes and Failures." *Journal of Pest Science* 92 (1):83–91. https://doi.org/10.1007/s10340-018-1056-z.

Mack, R. N., and S. K. Foster. 2009. "Eradicating Plant Invaders: Combining Ecologically-Based Tactics and Broad-Sense Strategy." In *Management of Invasive Weeds*, edited by Inderjit, 35–60. Dordrecht: Springer Netherlands.

Mack, R. N., D. Simberloff, W. M. Lonsdale, H. Evans, M. Clout, and F. A. Bazzaz. 2000. "Biotic Invasions: Causes, Epidemiology, Global Consequences, and Control." *Ecological Applications* 10 (3):689–710. https://doi.org/10.1890/1051-0761(2000)010[0689:BICEGC]2.0.CO;2.

McKnight, T. L. 1969. "Barrier Fencing for Vermin Control in Australia." *Geographical Review* 59 (3): 330–347. https://doi.org/10.2307/213480.

McNeely, J. A., H. A. Mooney, L. E. Neville, P. J. Schei, and J. K. Waage. 2001. "Global Strategy on Invasive Alien Species." Gland, Switzerland: IUCN.

Moon, K., D. A. Blackman, and T. D. Brewer. 2015. "Understanding and Integrating Knowledge to Improve Invasive Species Management." *Biological Invasions* 17 (9):2675–2689. https://doi.org/10.1007/s10530-015-0904-5.

MPI. 2018. "The New Zealand Government Biosecurity Response Guide." Wellington, New Zealand: Ministry for Primary Industries.

MPI. 2022. "Response to Foot-and-Mouth Disease." Last Modified 16 August 2022, accessed 1 November 2022. https://www.mpi.govt.nz/biosecurity/plans-for-responding-to-serious-disease-outbreaks/foot-and-mouth-disease/response-to-foot-and-mouth-disease/.

Myers, J. H., A. Savoie, and E. van Randen. 1998. "Eradication and Pest Management." *Annual Review of Entomology* 43 (1):471–491. https://doi.org/10.1146/annurev.ento.43.1.471.

NAO. 2002. "The 2001 Outbreak of Foot and Mouth Disease: Report by the Comptroller and Auditor General." London: National Audit Office.

Nimmo, D. G., and K. K. Miller. 2007. "Ecological and Human Dimensions of Management of Feral Horses in Australia: A Review." *Wildlife Research* 34 (5):408–417. https://doi.org/10.1071/WR06102.

NSW Department of Planning and Environment. 2020. "Orange Hawkweed." Last Modified 2 October 2020, accessed 1 November 2022. https://www.environment.nsw.gov.au/topics/animals-and-plants/pest-animals-and-weeds/weeds/new-and-emerging-weeds/orange-hawkweed.

Panetta, F. D. 2009. "Weed Eradication—An Economic Perspective." *Invasive Plant Science and Management* 2 (4):360–368. https://doi.org/10.1614/IPSM-09-003.1.

Panetta, F. D. 2015. "Weed Eradication Feasibility: Lessons of the 21st Century." *Weed Research* 55 (3): 226–238. https://doi.org/10.1111/wre.12136.

Panetta, F. D., and O. J. Cacho. 2014. "Designing Weed Containment Strategies: An Approach Based on Feasibilities of Eradication and Containment." *Diversity and Distributions* 20 (5):555–566. https://doi.org/10.1111/ddi.12170.

Panetta, F. D., O. J. Cacho, S. M. Hester, and N. M. Sims-Chilton. 2011. "Estimating the Duration and Cost of Weed Eradication Programs." Island Invasives: Eradication and Management. Proceedings of the International Conference on Island Invasives Gland, Switzerland: IUCN and Auckland, New Zealand, Auckland, 8-12 February 2010. https://www-chicagomanualofstyle-org.eu1.proxy.openathens.net/16/ch15/ch15_sec043.html

Panetta, F. D., and R. Lawes. 2005. "Evaluation of Weed Eradication Programs: The Delimitation of Extent." *Diversity and Distributions* 11 (5):435–442. https://doi.org/10.1111/j.1366-9516.2005.00179.x.

Panetta, F. D., and R. Lawes. 2007. "Evaluation of the Australian Branched Broomrape (*Orobanche ramosa*) Eradication Program." *Weed Science* 55 (6):644–651. https://doi.org/10.1614/WS-07-058.1.

Panetta, F. D., and S. M Timmins. 2004. "Evaluating the Feasibility of Eradication for Terrestrial Weed Incursions." *Plant Protection Quarterly* 19 (1): 5–11.

Pluess, T., R. Cannon, V. Jarošík, J. Pergl, P. Pyšek, and S. Bacher. 2012. "When are Eradication Campaigns Successful? A Test of Common Assumptions." *Biological Invasions* 14 (7):1365–1378. https://doi.org/10.1007/s10530-011-0160-2.

Rejmánek, M., and M. J. Pitcairn. 2002. "When is Eradication of Exotic Pest Plants a Realistic Goal?" Proceedings of the International Conference on Eradication of Island Invasive Species, Auckland.

Richards, F., P. Kodjamanova, X. Chen, N. Li, P. Atanasov, L. Bennetts, B. J. Patterson, B. Yektashenas, M. Mesa-Frias, K. Tronczynski, N. Buyukkaramikli, and C. El Khoury. 2022. "Economic Burden of COVID-19: A Systematic Review." *ClinicoEconomics and Outcomes Research* 14:293–307. https://doi.org/10.2147/CEOR.S338225.

Schemann, K., S. M. Firestone, M. R. Taylor, J. A. L. M. L. Toribio, M. P. Ward, and N. K. Dhand. 2012. "Horse Owners'/Managers' Perceptions about Effectiveness of Biosecurity Measures Based on Their Experiences During the 2007 Equine Influenza Outbreak in Australia." *Preventive Veterinary Medicine* 106 (2):97–107. https://doi.org/10.1016/j.prevetmed.2012.01.013.

Scott, J. K., S. J. McKirdy, J. van der Merwe, R. Green, A. A. Burbidge, G. Pickles, D. C. Hardie, K. Morris, P. G. Kendrick, M. L. Thomas, K. L. Horton, S. M. O'Connor, J. Downs, R. Stoklosa, R. Lagdon, B. Marks, M. Nairn, and K. Mengersen. 2017. "Zero-Tolerance Biosecurity Protects High-Conservation-Value Island Nature Reserve." *Scientific Reports* 7 (1):772. https://doi.org/10.1038/s41598-017-00450-y.

Sharov, A. A., and A. M. Liebhold. 1998. "Bioeconomics of Managing the Spread of Exotic Pest Species with Barrier Zones." *Ecological Applications* 8 (3):833–845. https://doi.org/10.2307/2641270.

Simberloff, D. 2003. "Eradication—Preventing Invasions at the Outset." *Weed Science* 51 (2):247–253. https://doi.org/10.1614/0043-1745(2003)051[0247:EPIATO]2.0.CO;2.

Spring, D., and T. Kompas. 2015. "Managing Risk and Increasing the Robustness of Invasive Species Eradication Programs." *Asia & the Pacific Policy Studies* 2 (3):485–493. https://doi.org/10.1002/app5.105.

Stoneham, G., S. M. Hester, J. S.-H. Li, R. Zhou, and A. Chaudhry. 2021. "The Boundary of the Market for Biosecurity Risk." *Risk Analysis* 41 (8):1447–1462. https://doi.org/10.1111/risa.13620.

Wilson, John R., F. D. Panetta, and C. Lindgren. 2016. *Detecting and Responding to Alien Plant Incursions.* Cambridge: Cambridge University Press.

Wittenberg, R., and M. J. W. Cock, eds. 2001. *Invasive Alien Species: A Toolkit of Best Prevention and Management Practices.* Wallingford, Oxon, UK: CAB International.

World Economic Forum. 2022. "Global Risks Report 2022." accessed 1 November 2022. https://www.weforum.org/reports/global-risks-report-2022/.

Section IV

Whole of System

8 Incentives

Incorporating Incentives into Biosecurity Policies and Regulations

Gary Stoneham, Susan M. Hester, and Arthur Campbell

ABSTRACT

Biosecurity systems and the activities undertaken within them are public goods—no one can be excluded from accessing them or forced to pay for them. As a result, markets that would otherwise indicate the optimal amount of risk mitigation do not evolve naturally. Because of this market failure, governments across the globe intervene in the movement of goods and people, typically with science-based regulations that rarely account for stakeholder behaviours. Biosecurity regulations that do not carefully consider stakeholder responses can deliver unintended and counterproductive policy consequences, potentially undermining biosecurity outcomes. Potential solutions to generate the efficient level of biosecurity effort would be to modify the markets in which biosecurity effort and activities are determined or to create markets where they are missing. This chapter considers the missing biosecurity market problem and efficient solutions, including developing an "incentive-approach" to biosecurity. It also considers laboratory-based economic experiments to assess and refine policy changes before they are introduced into the economy and the role of technologies in the design and implementation of incentive-compatible policies.

GLOSSARY

Incentives Inducements for individuals to take actions that they would otherwise not consider. Incentive-compatible policies align the actions of self-interested individuals with a broader policy objective.

Market Places where buyers and sellers meet to trade goods and services. It is where the price and amount traded are determined; where the basic questions of what should be produced, how it should be produced and for whom, are solved.

Market failure Situations where transactions do not allocate goods and services efficiently. A range of factors can lead to market failure, including public goods, externalities, missing or weak property rights, lack of competition, and transaction complexities.

Public goods Types of goods where use by one person does not prevent access or reduce availability to other people. Those consuming public goods cannot be stopped from accessing them or failing to pay for them. Examples of public goods include ecosystem services, national security, or street lighting.

Market design A method for creating rules and processes to organise transactions to achieve a defined outcome.

Game theory Allows the study of strategic behaviour between two or more agents when they have more than one strategy from which to choose and their choices affect the returns (i.e. payoffs) of another agent in the interaction.

DOI: 10.1201/9781003253204-12

Experimental economics The testing and refining of incentive and information structures on human behaviour via economic experiments undertaken in a controlled environment (e.g. laboratory or field).

Actuarial pricing The principles applied to pricing of risk.

Principal-agent problems Occur where a task is delegated by a principal (who accrues the direct benefits of the task) to an agent (who bears the direct time and effort costs).

Alignment problems The differences that occur between how a principal and agent in an organisation would prefer a particular task to be done. Where alignment problems exist, there are potential gains to both parties from improving the incentives of the agent for undertaking the task.

Incentive-compatibility constraint A policy is "incentive-compatible" when each participant in their own interest makes decisions that are aligned with the objectives of the policy.

INTRODUCTION

In open, decentralised economies, decisions about goods and services produced and consumed are generally made by private individuals who implement their production and consumption choices through markets (i.e. places where buyers and sellers meet to exchange goods and services). For many goods and services, buyers and sellers transact through markets that harness competition between (1) consumers to determine who values the goods or services the most and (2) sellers to reveal who can supply items at the lowest cost.

Efficient markets identify the set of transactions that maximise value through a matching process between buyers and sellers. The prices revealed through markets identify the "best" use of resources (from a value-creation perspective), enable buyers to exercise their preferences between alternative goods, and encourage producers to supply goods and services at lower cost. In this way, markets create incentives (i.e. inducements) that encourage participants to take actions that maximise the value created from transactions.

While markets naturally evolve and work well for many goods and services (e.g. commodities, financial services, and travel), markets are missing in some domains of the economy, along with their embedded incentives. Markets are missing for many types of research, where new discoveries and ideas become freely available to all. Markets are also missing for many environmental goods and services, such as climate change mitigation, biodiversity conservation, and air and water filtration (Alvarado-Quesada, Hein, and Weikard 2014; Stoneham et al. 2003). In the latter example, landholders are rewarded for producing agricultural commodities but not ecosystem services (Teytelboym 2019) even though environmental services are highly valued.

There is no formal market for biosecurity activities. Investment in biosecurity effort is determined in markets for imported goods and commodities, but arguably these markets do not encourage the optimal level of biosecurity effort. While some costs of importing goods (e.g. shipping, product spoilage, damage, loss, and theft) are the responsibility of the importer, the costs of preventing and responding to pest and disease incursions are transferred to the citizens and the natural environment of the importing country. These costs do not apply to each import transaction but emerge as expected costs derived from the probability of pest or disease entry and the economic impact of these pests and diseases should they establish and spread.

Biosecurity is therefore a class of risk, but unlike other types of risk (e.g. loss or injury), it is not technically feasible (or it is prohibitively costly) to assign the costs of an outbreak to a specific importer. In this economic environment, importers under-invest in activities such as testing and screening imported goods for harmful pests and diseases, treating consignments, declaring relevant information, compensating for damages caused, and purchasing insurance against expected losses.

These and other worthwhile activities could be undertaken to reduce biosecurity risk, but they are not rewarded through the incentives embedded in existing markets. As such, missing incentives for biosecurity activities within existing markets do not encourage the efficient level of effort needed by importers to reduce biosecurity threats. Nor do they encourage consumers to select appropriate types and quantities of imported goods. In contrast, import markets create incentives for other desirable attributes of imports (e.g. quality or origin) through price premiums paid by consumers.

The absence of appropriate incentives means that some of the value created from transactions in imported goods is illusory. If flowers imported from a high-risk country cause the introduction of a serious pest, the costs imposed on the destination country could more than exceed the benefits derived by consumers. Where this is the case, such transactions reduce rather than increase gross domestic product, making citizens in the importing country "poorer". In contrast, goods that have relatively low biosecurity costs for the importing country increase national income and well-being, making citizens in both importing and exporting countries better off.

The current strategy adopted by many countries is for the government to identify the type and level of additional effort needed to reduce and respond to biosecurity risk (e.g. through regulations, inspections, surveillance, control, and research). Biosecurity activities are typically funded by taxpayers and implemented by government employees (or contracted agents), with the right level of investment defined as the "appropriate level of protection" (ALOP; see Chapter 2. Biosecurity Systems and International Regulations). Under this approach, it is the biosecurity agency that defines threat mitigation actions required for imported goods. In some instances, this could include offshore fumigation to remove disease risk, testing and reporting of symptoms, and in some cases prohibition of high-risk imports. These and other biosecurity interventions are implemented as regulations that are sanctioned under the World Trade Organisation.

The key problem with this approach is that government officers (or their agents) do not have access to all the information needed to allocate funds in a way that generates the highest returns. For example, governments do not have specific information on the costs of alternative imports that would lower biosecurity risks or the prices that consumers would be prepared to pay for these goods. This and other information needed to identify the efficient (i.e. "right") level of biosecurity effort is unevenly distributed among actors in the economy (i.e. asymmetric information) and some has commercial value such that those who hold relevant information are unwilling to reveal it (i.e. private or hidden information). The ALOP target adopted by many countries is formed in the absence of this information, such that it does not lead to the efficient level of biosecurity effort (i.e. biosecurity investment).

In this chapter, we explore two options to create incentives to improve the behaviour of biosecurity stakeholders and generate an efficient level of biosecurity effort. First, we outline how designing a biosecurity market using an actuarial levy (equivalent to an insurance premium) charged to importers could create a structural link between risk exposure and biosecurity funds. Second, we review how policies and regulations can be designed so that they embed appropriate incentives for stakeholder compliance, and we note the important role that technological advances can play in this process. Finally, we review the opportunities offered by experimental economics to assess and refine policy changes before they are introduced into the economy.

DESIGNING MARKETS FOR BIOSECURITY RISK

A potential solution to generate the efficient level of biosecurity effort would be to modify the markets in which biosecurity effort and activities are determined (e.g. markets for traded goods) or to create markets where they are missing. These approaches would create the incentives needed for producers and importers to invest in biosecurity activities and for consumers to select goods that mitigate biosecurity threats.

Markets for many goods and services evolve organically from the same selection process that is observed in nature (McMillan 2002). The rules and processes that define markets are refined over countless transactions and persist when they are perceived to be efficient and fair. An efficient market is one that maximises value created for buyers and sellers who engage in transactions, while fairness refers to the way markets distribute the value created from transactions. Markets do not promise to distribute value equally between buyers and sellers, rather, they distribute value in predictable and acceptable ways. Many variations on these rules and processes emerge in response to a range of complexities that are relevant to specific goods and services. As in nature, markets evolve in a competitive environment but must also allow collaboration, coordination, and cooperation where these and other forms of behaviour increase the value created from transactions.

Markets evolve in many domains of the economy, but markets may be missing or inefficient where the goods and services being transacted are public goods, property rights are weak or unenforceable, competition cannot be harnessed because of too few participants, or where complexities are present (e.g. policy, transaction, strategic, or timing complexities). Relatively recent advances in market design are now available to address some (but not all) of these issues to improve the efficiency of markets and create them where they are needed. These ideas and techniques, summarised by Roth (2002), have had a profound impact on the institutional landscape of the economy. Notable examples include the creation of matching markets for kidneys (Roth, Sönmez, and Ünver 2004), school and university admissions (Abdulkadiroğlu and Sönmez 2003; Roth and Sotomayor 1989), bespoke auctions to allocate and price mobile phone spectrum (Cramton 1995), network access (McDaniel and Neuhoff 2004), natural resources (Cramton and Soros 2007), environmental goods and services (Stoneham et al. 2003), and energy markets (Wilson 2002). These and other applications (see Roth 2018) expand the boundary of markets in the economy and create incentives to change the behaviour of self-interested individuals to align with broader national goals.

Market design has important implications for biosecurity, as biosecurity markets (and embedded incentives) could be created to reward importers and other stakeholders for their effort (Box 8.1). Applying this approach depends on a government's ability to design and implement a mechanism that efficiently prices biosecurity risk, such that importers face incentives to invest in the right type of biosecurity effort and consumers can purchase an appropriate portfolio of imported goods.

BOX 8.1. THE THREE STAGES OF MARKET DESIGN

Market design can be framed as a three-stage process.

1. DIAGNOSIS

The absence of a market suggests that some impediment (referred to as complexity) diminishes the value that could be created from transactions (Arrow 1969). The first diagnostic phase of market design establishes the reasons why a market is inefficient or missing. While some complexities cannot be resolved at a fundamental level (e.g. public goods), other complexities including policy, transaction, strategic, and timing complexities can be identified and resolved.

Complexities include:

- Information problems (e.g. hidden information).
- Matching problems (e.g. identifying potential counterparties).
- Timing problems (e.g. where buyers and sellers do not arrive at the same time in a market).
- Coordination problems (e.g. where market participants must interact with others to discover the most efficient transactions).
- Strategic behaviour (e.g. when there are too few buyers or sellers, and the market is "thin").
- Synergy problems (i.e. the value of one item depends on other items).

2. Economic Theory

The foundation concepts for market design are framed by game theory. The task of the economist is to identify the specific rules and processes needed to overcome complexities that cause missing or inefficient markets in the first place (Hurwicz and Reiter 2006).

3. Experimental Economics

Experimental economics and test bed techniques (see Plott and Smith 1978) are available to test and refine markets to ensure they are efficient and achieve intended outcomes. In an economic experiment, human subjects are typically required to transact an abstract item under different sets of rules and processes (e.g. a designed market) to uncover the economic properties of the market, how the market distributes value between buyers and sellers, and any implementation issues. These experiments allow the rules and process of markets to be refined.

Although widely used in other domains of the economy, risk markets have not yet been created to manage biosecurity risk. In a recent paper, Stoneham et al. (2021) argued that the reason such markets have not evolved as they have for other classes of risk is that risk creators (i.e. importers of goods and inbound vessels and passengers) are not exposed to the financial losses of their actions. In the case of biosecurity, it is not practical (and in many instances not technically possible) to attribute financial losses arising from the introduction of pests and diseases to specific importers, vessels, and/or passengers. Under these circumstances risk creators pass on the financial consequences of their actions to taxpayers in the importing country such that there is no incentive for importers and other risk creators to take out insurance. It is this externality that causes market failure—not the characteristics of biosecurity risk.

Where it is possible to resolve the externality problem (such as through compulsory purchase of insurance) there are important advantages from incorporating insurance principles (i.e. actuarial pricing) into national biosecurity systems through a compulsory, actuarially determined biosecurity levy. Key among the advantages is that the price of biosecurity risk creates incentives for importers to find lower risk imports to secure lower premiums and for consumers to consider the cost of biosecurity risk when making purchases. These behavioural changes are of interest because they harness hidden information held by importers (about the source of lower-risk imports) and consumers (about their preferences) that is not available to biosecurity managers. When the type, quantity, and origin of imported goods are incorporated into risk-based insurance premiums, market transactions lead to the right type and level of biosecurity effort. There are also financial sustainability advantages from creating a dedicated pool of funds (through an actuarial levy) available to fund government biosecurity effort. This mechanism creates a structural link between risk exposure and funds available to respond to biosecurity incursions (see Chapter 7. Prepare, Respond, and Recover).

Incorporating insurance principles into biosecurity systems has profound implications for governments' ability to create incentives. A significant part of the missing incentive problem might be addressed by creating a mechanism that reveals the price of biosecurity risk. The architecture of this mechanism would draw on the same principles used to price other risks, such as those arising from fire and other natural disasters, threats to life and health, income variability, cyber security, and terrorism threats. However, the structure of the mechanism would need to be adapted to reflect the market complexities relevant to biosecurity. Where this is not possible, government intervention will be needed to design and create incentives and regulatory processes needed to achieve the appropriate level of biosecurity effort. Strategic investment in monitoring technologies will assist with this task (Box 8.2).

CREATING STANDALONE INCENTIVES

PRINCIPAL-AGENT AND ALIGNMENT PROBLEMS

Where markets cannot be created, a second strategy is to design and create the incentive structures currently missing due to the missing market problem. This approach relies on widely understood economic principles developed to solve principal-agent problems (Jensen and Meckling 1976). These problems occur whenever control of a task is delegated to an agent by the principal. In the biosecurity context, the government (the principal) seeks to align the actions of self-interested agents (e.g. import businesses, passengers, farmers, and consumers) with the national interest. In this context, governments can encourage self-interested agents to change their behaviour by offering incentives (i.e. inducements) and/or sanctions through laws and regulations.

Principal-agent problems occur in all aspects of life where a task is delegated by a principal (who accrues the direct benefits) to an agent (who bears the direct time and effort costs). In many instances, this creates an alignment problem between how the principal and the agent would prefer the task to be done. The agent, by virtue of receiving few of the direct benefits of the task and bearing many of the direct costs in terms of time and effort, will always care less than the principal about generating higher benefits through exerting greater effort or time.

For example, a shop owner may employ a salesperson to assist customers in choosing the best product to meet their needs. The effort of the salesperson to assist the customer in finding the right product is important in convincing them to make a purchase, but the direct beneficiary of the sale is the store owner. As a result, the salesperson may not exert as much effort in generating sales as the store owner may wish. The store owner may recognise this issue and endeavour to incorporate incentives into the salesperson's compensation (e.g. sales quotas and associated bonuses) to improve their effort.

The presence of an alignment problem means there are potential gains to the principal and agent from improving the incentives of the agent for undertaking the task. In practice, a wide variety of formal and informal agreements between the principal and agent are used to improve how an agent undertakes a delegated task. These often involve interventions aimed at better aligning the goals of all parties. Examples include compensation schemes that reward employees for their performance through bonuses, equity in the firm and stock options; contracts between a provider of a good or service and the purchaser that reward the provider for better quality or timely provision; and insurance contracts with clauses that reward the insured for safe behaviour through no-claims discounts (as for car insurance). One of the oldest and most common alignment interventions are share-cropping arrangements between landlords and tenant farmers, where farmers pay landlords shares of the harvest. In this way, landlords and farmers both share the risks and benefits of the harvest.

A number of factors limit the degree to which an agreement can realise potential gains, for example, due to:

- *Strategic behaviour of principals and agents*: A salesperson may not engage with customers to provide advice or, when asked, may provide poor advice or unfriendly customer service.
- *Imperfect observation of behaviours*. The shop owner may not be able to observe the quality of the customer service the salesperson provides.
- *Verification and enforcement difficulties*: A contract specifying that the salesperson should receive a bonus if they provide good customer service may not be readily enforced by an outside party who cannot verify the quality of the customer service provided.
- *Acceptability of the agreement to both parties*: The shop owner and salesperson must both agree to the terms of an employment contract.

There are broadly two classes of constraints that limit what any agreement can achieve: participation constraints and incentive-compatibility constraints. In most settings, participation in any agreement is voluntary (i.e. neither party may be coerced into an agreement). Participation constraints require that both the principal and the agent anticipate sufficient benefits from entering into the agreement than what they expect not entering into the agreement. In cases where it is possible to walk away during an agreement at any interim stage, both parties need to anticipate sufficient ongoing benefits from the agreement. For example, when a worker accepts an employment contract, they anticipate being better off accepting the contract than remaining unemployed or searching for alternative work.

All actions chosen by the agent and the principal in an agreement must therefore be compatible with each party pursuing their self-interest. This means that the structure of an agreement needs to embody incentives for each party to follow their required actions. The rewards from following the agreement outweigh the punishments from deviating from the agreement. The set of constraints that require that the actions of each party be consistent with each pursuing their own self-interest are known as incentive-compatibility constraints. Incentive design principles allow economists to create incentives that result in outcomes that may otherwise not evolve in poorly functioning markets.

Incentives in Border Inspection

We illustrate these ideas through an inspection game between an importer and a biosecurity agency. In our stylised setting, the biosecurity agency designs a set of regulations to mitigate risk from the import of a product. In the inspection game, the importer chooses whether to comply or not comply with the regulations, and the biosecurity agency chooses the frequency at which it inspects imported consignments and the penalty for non-compliance.

Table 8.1 shows the net benefits to the importer, depending on whether they comply with the regulations and whether the biosecurity agency inspects the consignment:

- An importer who chooses to comply receives a net benefit of $100 − Cost, where $100 is the sales benefit of importing the product and Cost is the cost of compliance to satisfy biosecurity regulations. This net benefit does not vary with whether the consignment is inspected or not.
- If the importer chooses not to comply and is not inspected, the costs of compliance are avoided, and the importer receives a benefit of $100.
- If the importer is inspected, inspection may identify non-compliance with a probability of Detect and the importer incurs a Penalty (the import may be destroyed and/or the importer may pay a fine). The net benefit to the importer is therefore $100 − Detect × Penalty.

The incentive design problem for the biosecurity agency is to set the frequency of inspections (Frequency) and penalties (Penalty) in such a way that an importer will choose to comply with biosecurity regulations. If the importer chooses to comply with regulations, the importer anticipates a

TABLE 8.1

The Net Benefit to an Importer Depending on Their Probability of Compliance with Biosecurity Regulations and the Frequency of Inspection by a Biosecurity Agency.

Biosecurity agency	Importer	
	Comply	Not comply
Inspect	$100 − Cost	$100 − Detect × Penalty
Not inspect	$100 − Cost	$100

benefit of $100 – *Cost* irrespective of whether an import is inspected for compliance. If the importer does not comply with biosecurity regulations, the benefit will depend on whether biosecurity regulators inspect the import and detect non-compliance. In this case, the expected benefit for the importer is $100 – *Frequency* × *Detect* × *Penalty*.

For importers to comply with biosecurity regulations, the benefit from compliance must outweigh the expected benefits of non-compliance. The incentive-compatibility constraint for the biosecurity agency is therefore:

$$\$100 - Cost > 100 - Frequency \times Detect \times Penalty$$

The importer will comply provided that:

$$Frequency \times Detect \times Penalty > Cost$$

For a given level of detection (*Detect*) and compliance cost (*Cost*), the biosecurity system must set a frequency of inspection and penalty high enough so that this condition is met. A range of frequencies and levels of penalty can induce compliance. For example, if an inspection *Frequency* of 50% and a *Penalty* of $1,000 satisfy this condition, so will a *Frequency* of 100% and a *Penalty* of $500.

An importer, anticipating that it will need to comply with a biosecurity regulation, will weigh up choosing to import a product and complying with the regulation or taking an alternative course of action. If the alternative action (e.g. sourcing a product domestically) generates a net benefit of $40, then the importer will import a product provided that its anticipated benefits from importing are better than the alternative. This is the participation constraint for the importer given by:

$$\$100 - Cost > \$40$$

This is satisfied in this example provided that:

$$Cost < \$60$$

The design of biosecurity regulations balances the benefits of trade with potential costs and risks of bringing in products that may cause harm to local industries, people, and the environment. These constraints place limits on how regulations may be designed if they are going to achieve participation and compliance by importers. These constraints may impact the design of biosecurity regulations in a number of ways:

- If the frequency of inspection is low, then the biosecurity system will need sufficiently high penalties for (detected) non-compliance. If inspections are expensive and can only be conducted at low frequencies, penalties must be high to ensure compliance. In practice, there is a limit to the size of the penalty that can be credibly imposed on a party (e.g. the total value of a company), which may limit how infrequent inspections can be.
- At the other extreme, if the highest frequency of inspection is 100%, the lowest penalty to ensure compliance is $\frac{Cost}{Detect}$. If breaches of the biosecurity regulation are not readily verifiable and the probability of detection is low, this will increase the size of the penalty required to ensure compliance.
- Biosecurity regulations that are readily verifiable may also impose significant costs on some importers, so much so that they may result in an importer choosing not to import a product (the high costs of complying results in the participation constraint not being satisfied). Alternative regulations that may be complied with at lower costs but are less readily detected may be preferable.

BOX 8.2. THE ROLE OF TECHNOLOGY IN
MARKET AND INCENTIVE DESIGN

Many biosecurity decisions are made in the absence of accurate and reliable information. Strategic investment in technologies that improve information about the level, distribution, and spread of pests and diseases, the origin of biosecurity threats, the provenance of goods in the supply chain, transmission pathways, and treatment options is likely to have a profound impact on the efficiency and efficacy of biosecurity systems. For example, rapid development of vaccines and accurate tests have highlighted the importance of technology in managing COVID-19. Other technologies could be important in designing and implementing incentives, regulations, and exchange mechanisms (i.e. markets) to align the actions of individuals and organisations with national biosecurity objectives. Advances in computing, data analytics, machine learning, artificial intelligence, genomics, biological engineering, diagnostic capabilities, and epidemiology (Chui et al. 2020) are likely to have important implications for the type and effectiveness of mechanisms used by biosecurity agencies to manage biosecurity threats.

EXPERIMENTAL ECONOMICS AND INCENTIVE DESIGN

An important innovation in the last 30 years has been the development and use of laboratory experiments for testing human behaviour. Experimental economics techniques (Nobel Prize Committee 2002; Plott and Smith 1978) can now be used to augment the incentive design process. Economic experiments provide a controlled environment to examine the effects of particular incentive and information structures on behaviour. Experiments can be used as a test bed for new ideas and to assess policies before wider implementation. For example, Australia's Department of Agriculture, Fisheries and Forestry (DAFF) and the Centre of Excellence for Biosecurity Risk Analysis (CEBRA) used economic experiments to test incentive-compatible protocols for pre-border regulation (Box 8.3).

An important set of assumptions used in the design of any incentive structure concerns how people will respond to a particular set of incentives. Commonly, it is assumed that individuals:

• Are highly sophisticated in how they process information and make choices.
• Have narrowly defined preferences that value outcomes that benefit themselves but place little value on how an outcome affects others.
• Are sophisticated in not only their own strategic reasoning but are also confident other people are similarly sophisticated in theirs.

These are strong assumptions because nobody exhibits behaviour that is perfectly consistent with these assumptions in all situations. The utility of these assumptions is that they are a robust predictor of human behaviour that provides a benchmark against which to measure actual behaviour. Economic experiments can be used to test whether actual behaviour differs systematically from what is assumed in the design of an incentive structure. Where behaviour deviates systematically from expectations, varying some aspect of the environment to move behaviour back to the direction of the benchmark can reveal which assumption is not being met.

Importantly, economic experiments can examine how incentive and information structures interact to change the decision making of experimental subjects. Imposing regulatory changes without carefully considering stakeholder responses can introduce counterproductive incentive structures and deliver unintended policy consequences that can potentially undermine biosecurity outcomes. Laboratory-based economic experiments offer regulatory agencies significant benefits as a safe, low-cost environment to assess and refine policy changes before they are introduced into the economy.

BOX 8.3. USING ECONOMIC EXPERIMENTS TO TEST
INCENTIVE-COMPATIBLE REGULATIONS

Australia's DAFF and CEBRA used economic experiments to design and test incentive-compatible protocols for pre-border regulation (Rossiter et al. 2018). The aim of the project was to assess the incentives inherent in compliance-based inspection protocols and to select key parameter values that would encourage importers to reduce the likelihood of biosecurity risk material entering Australia (Rossiter et al. 2016).

The Compliance-Based Intervention Scheme (CBIS) was introduced by the Department in 2013 to a number of low-risk plant-based products (Robinson et al. 2012; Rossiter and Hester 2017). The CBIS was based on a continuous sampling plan (CSP) algorithm originally developed in the quality control literature (Dodge 1943; Dodge and Torrey 1951). In the biosecurity context, a CSP determines whether to inspect a consignment based on the recent inspection history of the pathway and parameters set by the pathway manager. The CSP rewards consistently compliant importers with reduced inspection rates.

The focus of the economic experiments was to assess the behavioural responses of importers to key parameters in two CSP algorithms (CSP-1 and CSP-3). Rossiter and Hester (2017) suggested that the less-forgiving CSP-1 algorithm might be preferable to CSP-3 when the consequences of biosecurity risk material passing through the border are high. The CSP-1 algorithm is also simpler to implement and more easily communicated to stakeholders (Rossiter et al. 2018).

Experiments were conducted with Australian university students, using well-established procedures for laboratory experiments in economics (Friedman 2004; Guala 2005). Experimental subjects assumed the role of importers and were required to make choices about their preferred supplier over time. Importers could choose four potential suppliers who were identical in all respects except for their transport cost, purchase cost, and the likelihood of biosecurity risk material being present in a consignment. A computer played the role of the regulator.

The experiment examined particular aspects of CBIS rules likely to be more difficult to assess in the field, namely:

- Different inspection rules from the CSP family.
- The level of information provided to stakeholders about the inspection rule.
- Feedback on an importer's performance under the inspection rule.
- Costs and penalties of being inspected and failing inspection.
- Allowing rule choice from a limited set of options.
- Importers' understanding of the rule.

Key findings from the experiments were that the CSP-1 and CSP-3 algorithms resulted in similar importer behaviours. Providing more information about the inspection parameters and the consequences of failing inspection led to better importer choices from the regulator's perspective. In addition, providing targeted feedback to importers supported behaviour consistent with improved compliance (Rossiter et al. 2018). These experiments led to changes in how border inspection rules are implemented in Australia, specifically around the level of disclosure about rules and how feedback on regulatory performance is provided to importers of plant-based products.

IN A NUTSHELL

- The public good nature of biosecurity means that efficient markets for biosecurity risk-mitigation activities do not evolve naturally, and associated incentives are missing.
- Governments impose biosecurity regulations because, without them, the behaviour of importers would not automatically lead to outcomes that promote national welfare.
- The right incentives could improve the behaviour of biosecurity stakeholders, either by creating an efficient, compulsory biosecurity risk market (e.g. through risk-based insurance premiums charged to importers) or by creating incentive-compatible regulations.
- Technology can play an important role in incentive-compatible policies.
- Economic experiments can provide robust guidance on how incentive-compatible regulations can be implemented in the real world.

REFERENCES

Abdulkadiroğlu, A., and T. Sönmez. 2003. "School Choice: A Mechanism Design Approach." *American Economic Review* 93 (3):729–747. https://doi.org/10.1257/000282803322157061.

Alvarado-Quesada, I., L. Hein, and H. P. Weikard. 2014. "Market-Based Mechanisms for Biodiversity Conservation: A Review of Existing Schemes and an Outline for a Global Mechanism." *Biodiversity and Conservation* 23 (1):1–21. https://doi.org/10.1007/s10531-013-0598-x.

Arrow, K. J. 1969. *The Organization of Economic Activity: Issues Pertinent to the Choice of Market Versus Non-Market Allocations*. Washington DC: Joint Economic Committee of Congress.

Chui, M., M. Evers, J. Manyika, A. Zheng, and T. Nisbet. 2020. *The Bio Revolution: Innovations Transforming Economies, Societies and Our Lives*. McKinsey Global Institute.

Cramton, P. C. 1995. "Money Out of Thin Air: The Nationwide Narrowband PCS Auction." *Journal of Economics & Management Strategy* 4 (2):267–343. https://doi.org/10.1111/j.1430-9134.1995.00267.x.

Cramton, P., and G. Soros. 2007. "How Best to Auction Oil Rights." In *Escaping the Resource Curse*, edited by Macartan Humphreys, Jeffrey D. Sachs and Joseph E. Stiglitz, 114–152. New York: Columbia University Press.

Dodge, H. F. 1943. "A Sampling Inspection Plan for Continuous Production." *The Annals of Mathematical Statistics* 14 (3): 264–279.

Dodge, H. F., and M. N. Torrey. 1951. "Additional Continuous Sampling Inspection Plans." *Industrial Quality Control* 7 (5): 7–12.

Friedman, D. 2004. "Economics Lab: An Intensive Course in Experimental Economics". In *Routledge Advances in Experimental and Computable Economics*, edited by Alessandra Cassar and Reinhard Selten. London; New York: Routledge.

Guala, F. 2005. *The Methodology of Experimental Economics*. Cambridge: Cambridge University Press.

Hurwicz, L., and S. Reiter. 2006. *Designing Economic Mechanisms*. Cambridge: Cambridge University Press.

Jensen, M. C., and W. H. Meckling. 1976. "Theory of the Firm: Managerial Behavior, Agency Costs and Ownership Structure." *Journal of Financial Economics* 3 (4):305–360. https://doi.org/10.1016/0304-405X(76)90026-X.

McDaniel, T., and K. Neuhoff. 2004. Auctions to Gas Transmission Access: The British Experience. In *Cambridge Working Papers in Economics*: Cambridge University Faculty of Economics.

McMillan, J. 2002. *Reinventing the Bazaar: A Natural History of Markets*. New York: W.W. Norton. Book.

Nobel Prize Committee. 2002. *Foundations of Behavioral and Experimental Economics: Daniel Kahneman and Vernon Smith*. Nobel Prize Committee.

Plott, C. R., and V. L. Smith. 1978. "An Experimental Examination of Two Exchange Institutions." *The Review of Economic Studies* 45 (1):133–153. https://doi.org/10.2307/2297090.

Robinson, A., J. Bell, B. Woolcott, and E. Perotti. 2012. AQIS quarantine operations risk return: Imported plant-product pathways. Final Report for ACERA 1001 Study J. Australian Centre of Excellence for Risk Analysis, University of Melbourne.

Rossiter, A., and S. M. Hester. 2017. "Designing Biosecurity Inspection Regimes to Account for Stakeholder Incentives: An Inspection Game Approach." *Economic Record* 93 (301):277–301. https://doi.org/10.1111/1475-4932.12315.

Rossiter, A., S. M. Hester, C. Aston, J. Sibley, G. Stoneham, and F. Woodhams. 2016. Incentives for importer choices. Final Report for CEBRA 1304C, Centre of Excellence for Biosecurity Risk Analysis, University of Melbourne.

Rossiter, A., A. Leibbrandt, B. Wang, F. Woodhams, and S. M. Hester. 2018. Testing compliance-based inspection protocols. Final Report for CEBRA 1404C, Centre of Excellence for Biosecurity Risk Analysis, University of Melbourne.

Roth, A. E. 2002. "The Economist as Engineer: Game Theory, Experimentation, and Computation as Tools for Design Economics." *Econometrica* 70 (4):1341–1378. https://doi.org/10.1111/1468-0262.00335.

Roth, A. E. 2018. "Marketplaces, Markets, and Market Design." *American Economic Review* 108 (7):1609–1658. https://doi.org/10.1257/aer.108.7.1609.

Roth, A. E., Tayfun Sönmez, and M. Utku Ünver. 2004. "Kidney Exchange." *The Quarterly Journal of Economics* 119 (2):457–488. https://doi.org/10.1162/0033553041382157.

Roth, A. E., and Marilda Sotomayor. 1989. "The College Admissions Problem Revisited." *Econometrica* 57 (3):559–570. https://doi.org/10.2307/1911052.

Stoneham, G., V. Chaudhri, A. Ha, and L. Strappazzon. 2003. "Auctions for Conservation Contracts: An Empirical Examination of Victoria's BushTender Trial." *Australian Journal of Agricultural and Resource Economics* 47 (4):477–500. https://doi.org/10.1111/j.1467-8489.2003.t01-1-00224.x.

Stoneham, G., S. M. Hester, J. Siu, -H. Li, R. Zhou, and A. Chaudhry. 2021. "The Boundary of the Market for Biosecurity Risk." *Risk Analysis* 41 (8):1447–1462. https://doi.org/10.1111/risa.13620.

Teytelboym, A. 2019. "Natural Capital Market Design." *Oxford Review of Economic Policy* 35 (1):138–161. https://doi.org/10.1093/oxrep/gry030.

Wilson, R. 2002. "Architecture of Power Markets." *Econometrica* 70 (4):1299–1340. https://doi.org/10.1111/1468-0262.00334.

9 Resource Allocation
Using Economic Principles to Prioritise Projects and Allocate Biosecurity Budgets

Lucie M. Bland, Christine Li, Tom Kompas, Long Chu, Hoa-Thi-Minh Nguyen, and Susan M. Hester

ABSTRACT

Biosecurity budgets are unlikely to ever be adequate to reduce all threats to acceptable levels. Because of this, biosecurity agencies must decide on the relative importance or urgency of a large number of biosecurity risks, and which of these will be mitigated using the finite resources available. In this chapter, we outline methods for resource allocation based on the economic principle of value for money. Cost-efficient prioritisation involves understanding the costs and benefits of interventions across the range of pests and diseases that could be funded from a fixed budget. We outline the pros and cons of different prioritisation methods (including net present value, benefit-cost ratio, and optimisation) and assess when each method can lead to efficient budget allocation, depending on the decision context. Finally, we outline barriers and opportunities to increase the uptake of economic principles by biosecurity agencies. When economic principles are followed, biosecurity agencies can undertake the risk-reducing activities that provide the best "value for money" for taxpayers and the community.

GLOSSARY

Prioritisation The process of deciding which projects or activities are most important to fund and undertake. Prioritisation should be based on the economic principle of cost efficiency (i.e. value for money).

Cost efficiency vs cost effectiveness An activity is economically efficient if no other use of resources would yield a higher value to the community. Conversely, an activity is economically inefficient if its costs exceed its benefits or if it can be shown that resources could be used to produce something of higher value (Productivity Commission 2013). Cost efficiency should not be confused with cost effectiveness, which is the lowest cost approach to a specific output. In practice, cost efficiency is a higher standard to meet than cost effectiveness. A cost-effective activity is not necessarily cost-efficient, but a cost-efficient activity is also cost effective.

Costs and benefits The costs of a biosecurity activity refer to the human, monetary, or other resources need to implement the activity. The benefits of an activity refer to the value of the economic, environmental, and/or societal damages (impacts) that will be avoided if the activity is successful.

Discounting Allows costs and benefits that arise through time to be transformed into a comparable unit (namely their present value). Discounting accounts for the fact that a dollar received or spent today is worth more than a dollar in the future, as today's dollars could be invested and earn interest. Higher discount rates erode future benefits and costs more rapidly than lower discount rates.

Benefit-cost analysis (BCA) Encompasses a range of methods concerned with comparing the costs and benefits of activities. BCA outputs include the net present value (NPV) of benefits and benefit-cost ratios (BCR). Within a BCA, output measures are used to decide whether to invest in a project and/or prioritise projects.

Optimisation Describes the maximisation or minimisation of an objective, subject to one or more constraints. For example, a regulator may want to determine the combination of projects that minimises the total costs of impacts and activity costs (the objective) subject to a particular budget (the constraint). Optimisation problems are typically solved using mathematical programming techniques.

Returns to scale (RTS) Describe changes in extra benefits as expenditure increases. The three basic types of returns to scale are increasing, decreasing, and constant. For many biosecurity activities, returns to scale are expected to be decreasing (i.e. diminishing returns from extra expenditure).

INTRODUCTION

Biosecurity agencies have the difficult task of deciding how their budgets should be allocated to reduce, mitigate, or eliminate threats from pests and diseases. As biosecurity budgets are unlikely to ever be adequate to reduce all threats, choices must be made between many investment possibilities (Craik, Palmer, and Sheldrake 2017). The challenge for decision makers is deciding how to efficiently allocate limited monetary, human, or other resources to activities that will generate the best outcomes for the economy, society, and/or the environment (Epanchin-Niell 2017). Resource allocation therefore involves deciding on the relative importance or urgency of a large number of biosecurity risks.

The prioritisation task required to reduce biological harms is complex and occurs once biosecurity managers have identified, analysed, and evaluated risks (see Chapter 3. Anticipate). Managers are asked to prioritise budgets for a range of situations, such as prioritising consignments for screening at the border (see Chapter 5. Screen), selecting geographical areas for early detection surveillance (see Chapter 6. Detect) or allocating response budgets following an outbreak (see Chapter 7. Prepare, Respond, and Recover).

In practice, biosecurity agencies rely on historical legacies, simple rules-of-thumb, or the opinions of stakeholders (e.g. biosecurity managers, experts, or industry leaders) to select management activities or threats to address (Fox and Gordon 2004; Heikkilä 2011; Kompas et al. 2019). Biosecurity funds can be allocated in the same way they have been in previous years, as is typically the case in Australia's national biosecurity agency (Tongue 2020). Or, program funds can be allocated equally across government groups, as used to be the case in the Florida Department of Environmental Protection (Kim et al. 2007). Agencies also allocate funding as a result of political pressure, the visibility of the invasive species and its perceived impacts, co-funding availability, management experience, and pressure from lobby groups (Virtue 2007). Decisions made in these ways tend to lead to significant biases towards preventing some threats compared to others (Cook et al. 2011).

Some biosecurity agencies use risk assessment frameworks to allocate biosecurity budgets (see Chapter 3. Anticipate). Most frameworks incorporate a risk matrix in which likelihood and consequence (derived from published evidence, expert knowledge, or models) are combined to generate risk assessment results (e.g. negligible, low, medium, high, or extreme risk). Within these frameworks, the species deemed to pose the highest risk are often considered the highest priority for funding (see Hester and Mayo 2021 for a review). However, relying on risk rankings resulting from these assessments to allocate biosecurity budgets may not lead to desired outcomes, because risk assessments focus on threats and not on the activities used to manage those threats. Risk frameworks or rankings provide little to no information on the "value for money" of proposed projects— that is, the extent of risk reduction achieved from investing in a project, or whether one project should be funded over another. Risk assessments also do not give information on the total amount of expenditure needed to reduce risk to a particular level (Kompas et al. 2019). Because subjective opinions and risk rankings ignore value for money, resource allocation based on either of these methods results in economically inefficient decisions that do not achieve the highest benefits for the money spent.

Given increasing pressures from global trade and travel, human population growth, and climate change, resources invested in biosecurity activities will need to increase at an even faster rate to ensure absolute risk remains identical (Hulme 2009; IGB 2019). Predictions from the former Australian Government Department of Agriculture, Water and the Environment showed that increasing investment into biosecurity threefold by 2025 would not keep residual risk at 2014–2015 levels (Craik et al. 2017). This budgetary pressure does not result from a lack of knowledge about managing risks, but rather, from an inability for budgets to be increased enough to manage risks using the "business as usual" techniques described above. Government agencies also administer public funds under a culture of high scrutiny and accountability (Bossuyt, Shaxson, and Datta 2014; Chouinard 2013). Arguably, biosecurity agencies need to demonstrate that taxpayer dollars are spent in an economically efficient way to achieve the intended objectives of a biosecurity system—namely, to minimise the negative impacts of pests and diseases on the economy, environment, and community.

In this chapter, we summarise how economic efficiency can be achieved when allocating biosecurity budgets. We focus on the so-called *ex-ante* problems, that is, those comparing prospective projects rather than assessing the performance of past projects (*ex-post* problems). We summarise the types of data needed to prioritise biosecurity activities, including data on the costs of management activities, the benefits (avoided damages) they provide, their return to scale, the types of projects being compared, and economic discounting. We then review different approaches to prioritisation, including benefit-cost analysis (BCA) and optimisation. Based on data availability and the problem at hand, we outline how a decision maker can select a prioritisation method that achieves economic efficiency. Finally, we review current barriers and potential solutions to improving the uptake of economic principles in biosecurity.

INFORMATION UNDERPINNING RESOURCE ALLOCATION

Achieving cost-efficient allocation of biosecurity budgets involves understanding the costs of management activities and how much risk can be reduced by spending money on an activity.

COSTS

Generally, the costs of biosecurity activities can be classified into operating costs (e.g. personnel, consumables) and capital costs (e.g. infrastructure). This requires a breakdown of each activity into the time it is estimated to take, the geographical area over which it will be applied, the number of people required to carry it out, and the materials they will require. These estimates are then multiplied by their per-unit cost, and the costs of all activities are summed. For example, the administration of a weed eradication programme can involve funding a project coordinator, an administration officer, and on-site expenses such as office accommodation, computer equipment, telephone, internet use, and other minor consumables. Search-and-control processes involve labour to cover the potentially invaded area, flagging tape, herbicide, and the cost of travelling to and from sites (Hester et al. 2010).

Costs should be estimated for each year of the project lifespan, and in some cases beyond, to capture the entire timeframe over which damages may occur. When an agency needs to determine resource allocation across a set of candidate projects, only expenditures incurred from the same pool of funds should be accounted for. Local intervention or control costs (e.g. pesticide use) incurred by other parties such as producers and trade losses are not included in cost accounting, but they are included in the accounting of benefits in terms of avoided impacts.

BENEFITS

The benefits of an activity are the impacts (damages) that are likely to be avoided by implementing the activity. Benefits are compared to a base case or counterfactual—the situation that would most likely occur if no action was taken. The calculation of benefits considers the timing and extent of anticipated impacts (Summerson, Hester, and Graham 2018). This requires knowledge of the (often uncertain) relationship between the management activity and reduction in impacts. In recent decades, impact calculations have increased in sophistication by incorporating spread modelling. Pest or disease spread is simulated over time with and without the biosecurity activity and incorporated into benefits—here, the avoided impacts—and costs (see Box 9.1 and Chapter 14. Map).

Impacts may be classified as economic (e.g. production, revenue, and trade losses), environmental (e.g. native species declines), or societal (e.g. human health, infrastructure, or culture; see Chapter 3. Anticipate). Market impacts are comparably easy to calculate because prices reflecting value are readily available. The environment provides important ecosystem goods and services as well as cultural and recreational values often perceived to be priceless. Monetising environmental damages can be difficult, because environmental goods and services are not traded in markets and no market price is available. However, if the value of protecting the environment from the impacts of pests and diseases is not expressed in a common, monetary unit of measure, then this value can be implicitly discounted in resource allocation. Various non-market valuation methods are used to infer the value of non-market impacts (Emerton and Howard 2008). The monetised value of total

BOX 9.1. THE VALUE OF AUSTRALIA'S BIOSECURITY SYSTEM

A project by the Centre of Excellence for Biosecurity Risk Analysis (CEBRA) focused on estimating the value of Australia's biosecurity system (Dodd et al. 2020). The value of the Australian biosecurity system was defined as the difference between benefits under the status quo (biosecurity system "on") and a counterfactual (biosecurity system "off").

Dodd et al. (2020) estimated the value of benefits from 16 asset classes protected by the biosecurity system (namely primary production, ecosystem services, cultural assets, infrastructure, and other assets). The authors estimated the decline in asset values that would occur if exemplar pests and diseases across 40 functional groups enter, establish, and spread. The spatially explicit (2.5 km × 2.5 km resolution), bio-economic simulation model was run over 50,000 simulations for both the biosecurity system "off" and system "on" scenarios to model the arrival, spread, and impact of threats over 50 years, discounted to net present value (NPV).

The total flow of benefits arising from assets at risk was calculated as A\$5.696 trillion over 50 years. Without a biosecurity system, approximately \$672 billion in damages could be expected, but with a biosecurity system that costs \$10 billion, these damages are expected to be reduced to \$347 billion, resulting in a benefit of \$325 billion (= \$672 − \$347 billion; see Figure 9.1). The NPV of Australia's biosecurity system is therefore estimated to be around \$315 billion at an average benefit-cost ratio (BCR; return on investment [ROI]) of 30:1. Accounting for uncertainty, the 95% intervals of the NPV and BCR are \$156–467 billion and 15–45:1, respectively.

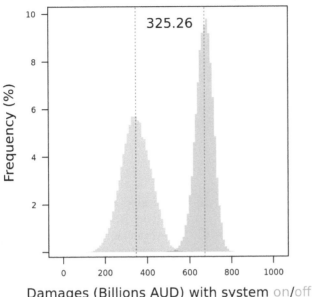

FIGURE 9.1   Damages expected over 50 years in Australia with the biosecurity system "on" (blue) or "off" (red). The dotted lines indicate median damages from 50,000 simulations for each scenario. Median avoided damages (benefits) were estimated to be A\$325 billion. (Reproduced with permission from Dodd et al. 2020).

benefits from market and non-market impacts is considered the amount that affected parties are willing to pay to avoid impacts or willingness to receive as compensation for impacts from the pest or disease, although cost-based approaches can also be used (e.g. based on restoration cost or avoidance cost). In some cases, it may only be necessary to assess the monetary value of a subset of impacts (e.g. market impacts) to demonstrate a project's cost-effectiveness relative to the costs of intervention (Cacho et al. 2008; Regan et al. 2006).

DISCOUNT RATE

Economic analysis often accounts for benefits and costs that occur over a time horizon spanning multiple years. When costs and benefits are compared at different points in time, costs and benefits must be normalised to 'present value' terms through *discounting* (Hester et al. 2013). The present value is the equivalent value today of a future benefit or cost. By discounting, we acknowledge that a dollar received today is worth more than a dollar received tomorrow (via inflation); today's dollar could be invested and earn interest. The formula for discounting when time is measured in discrete periods (e.g. years) is:

$$PV = FV \frac{1}{(1+r)^t} \qquad \text{(Eq. 9.1)}$$

where *PV* is the present value of a future value *FV* received in time period *t* at a discount rate *r*, which is assumed to be constant over time. Time is generally measured in years, with *t* = 0 representing the current year. Discrete discounting assumes all cash flows happen at the end of the year, while continuous discounting assumes cash flows occur continuously throughout the year (see Harrison 2010 for additional information about discounting).

Discounting reduces the value of costs or benefits of an activity, with the level of reduction depending on the discount rate and the number of years before society accrues the costs or benefits Harrison (2010). Higher discount rates erode future benefits and costs more rapidly than lower rates. Natural and social capital that are often counted in biosecurity analyses are not created or depleted in the same way as built capital, and their values may even increase over time, implying that very low discount rates may be appropriate (Costanza et al. 2021). Following, in the context of biosecurity economics, we recommend a discount rate of $r = 0.03$ for environmental assets and $r = 0.05$ for financial assets.

RETURNS TO SCALE

The costs and benefits of activities should be measured along the appropriate scale of effort for the project (e.g. time, geographic scale, and/or activity intensity). Many biosecurity activities can be scaled to achieve different levels of effort, for example to increase the area covered for early detection surveillance or to increase the number of traps per unit area. When deriving the costs of a biosecurity activity, the unit cost of control may vary with the scale of the project. These variations include where: projects have large up-front costs (e.g. infrastructure); decreases in unit costs may occur when activities are applied at large scales (e.g. bulk equipment purchases); or some activities may increase in unit cost when applied at very large scales (Epanchin-Niell and Hastings 2010).

Similarly, changes in benefits may vary with the scale of a project, such as when large benefits are only accrued after a certain level of investment or when extra benefits taper off as expenditure increases. For example, when inspecting air containers for the presence of an insect pest, a biosecurity agency will typically inspect the containers whose histories indicate they are more likely to be contaminated (Robinson et al. 2015). The cost of inspecting a container is fixed, but the benefit from each additional inspection declines as each extra container is less and less likely to be contaminated.

Increases or decreases in extra (marginal) costs and/or benefits with the scale of an activity lead to non-linearities in the relationship between costs and benefits. Returns to scale (RTS) describe changes in extra benefits as expenditure increases (i.e. the slope of the benefit-cost curve). The three basic types of RTS are constant, decreasing, or increasing, corresponding to whether increasing expenditure by a certain amount (e.g. 10%) leads to a marginal (extra) benefit of 10%, less than 10%, or more than 10%, respectively (Figure 9.2).

For many biosecurity activities, RTS is expected to be decreasing, leading to diminishing returns on extra expenditure, because early actions tend to focus on low-hanging fruit while further investment deals with more difficult tasks (Hauser and McCarthy 2009; Kompas, Chu, and Nguyen 2016). Furthermore, diminishing RTS occurs where pest damages irreversibly degrade assets and limit the benefits of additional control and restoration. An ecosystem may recover as the local pest population declines, but at a decreasing rate and never to its full pre-infestation function.

In practice, estimating the slope of a benefit-cost curve requires data on the "value-add" of benefits from additional spending at each level of expenditure. As these marginal benefits are usually sensitive to the specific threat, ecosystem, region, and management activity, quantitative estimation involves spread and population dynamics modelling (see Chapter 14. Map) or structured expert elicitation (see Chapter 12. Elicit). When data are not available to estimate the benefit-cost curve of an activity on a continuous scale, the costs, and benefits of individual project "blocks" (subsets) can be valued instead (see Table 9.4).

PROJECT MIX

When a regulator intends to optimise resource allocation across multiple projects, they should understand the relationship of projects and activities to one another. Projects may be mutually exclusive if

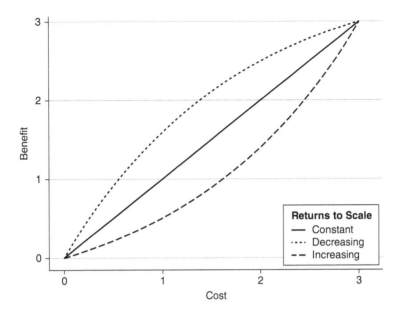

FIGURE 9.2 Benefit-cost curves for three basic types of returns to scale (RTS): Constant, decreasing, and increasing. Decreasing returns to scale is commonly observed for biosecurity activities.

alternative activities aim to achieve the same objective. For example, to contain the spread of foot-and-mouth disease (FMD), project A may involve implementing close-contact (ring) vaccination of cattle, sheep, and pigs, whereas project B may implement ring vaccination of cattle only, hence the two projects cannot be funded or conducted simultaneously. Projects may be independent (unrelated) when they achieve different objectives, and any combination of projects may be included in a prioritised list of projects. For example, project B (ring vaccination of cattle) may be implemented alongside project C, which focuses on improving diagnostic capacity. Due to the complexity of biosecurity systems, projects and activities can show complex relationships of mutual exclusion or independence, which must be accounted for when selecting the most appropriate method for allocating biosecurity budgets.

RESOURCE ALLOCATION METHODS

BENEFIT-COST ANALYSIS

BCA involves comparing the costs and benefits (avoided impacts) of a management activity. BCA is the standard method for evaluating the cost-effectiveness of response options when managing pest and disease incursions (Hester et al. 2013). For example, BCA has been used to compare control options for the marine carpet sea squirt (*Didemnum vexillum*; Coutts and Forrest 2007), to assess program funding to control the red imported fire ant (RIFA; *Solenopsis invicta*) in Australia (Kompas and Che 2001), and to compare different intervention strategies for RIFA (Hafi et al. 2014). Summerson et al. (2018) provide general guidelines to apply BCA to determine whether it is worth investing in a response to an invasive species (Table 9.1); these are demonstrated in Box 9.2.

Net Present Value

The sum of discounted costs and benefits is expressed as NPV as follows:

$$NPV = \sum_{t}^{T} \frac{B_t - C_t}{(1+r)^t}$$

(Eq. 9.2)

TABLE 9.1

Steps Required in a BCA of an Intervention to Reduce the Impacts of a Pest or Disease Incursion.

Step	Actions
1	Specify the intervention activity/activities and the base case (counterfactual) where the threat spreads unmanaged
2	Determine the costs of the intervention activity/activities
3	Identify the likely impacts from an unmanaged incursion
4	Predict impacts over time
5	Assign dollar values to impacts with market and non-market valuation
6	Discount and compare costs and benefits (avoided impacts) of activities
7	Calculate costs and benefits using net present value (NPV)
8	Run a sensitivity analysis for uncertain variables
9	Decide on an activity to implement

Source: Modified from Summerson et al. (2018) and Boardman et al. (2018).

where B_t and C_t represent the benefits and costs, respectively, that accrue in year t, r is the discount rate, and T is the time horizon of the evaluation. When NPV is positive, a proposal is considered to be economically efficient, and the community would be better off from its implementation. When two or more mutually exclusive projects are being considered, the project with the highest NPV is chosen.

BOX 9.2. HYPOTHETICAL EXAMPLE OF BCA USING NPV

The Asian green mussel (*Perna viridis*) is listed on Australia's Priority Marine Pest List (MPSC 2018) because of the significant economic (market) and environmental (non-market) impacts it is associated with, including direct impacts on vessel performance, fouling (e.g. industrial plants, power stations, desalination plants, water inlets, and sewage outlets), poisoning, and outcompeting native species (Murphy and Paini 2010).

In the hypothetical invasion scenario described in Summerson et al. (2018), Asian green mussels were discovered on a vessel arriving in Cairns in Queensland, Australia. The vessel returned to Cairns after travelling to other coastal locations in Queensland and in Darwin (Northern Territory, Australia) and Singapore. Subsequent investigations also found the mussel at the Gladstone port and on an artificial structure in Brisbane, Queensland.

Based on initial delimitation surveys, it was deemed technically feasible to eradicate the incursion. Only one intervention option (eradication) was selected in **step 1**, requiring five activities: in-water surveys of port infrastructure in (1) Cairns and (2) Gladstone to determine whether the mussel had established a founder population (and thus whether eradication at this port would be required), (3) delimitation surveys in Brisbane, (4) an eradication attempt in Brisbane, and (5) a post-eradication survey in Brisbane. The costs of the response action (**step 2**) were calculated as $137,500 (Table 9.2).

The costs of the response action (**step 2**) were compared to impacts from the unmanaged spread of the mussel (**step 3**), the base case. Predicted negative impacts of the mus-

TABLE 9.2

Costs of Response for the Asian Green Mussel in Queensland in a Hypothetical Scenario

Response Activities	Cost
(i) Cairns survey	$52,000
(ii) Gladstone survey	$25,000
(iii) Brisbane delimitation	$17,000
(iv) Brisbane eradication attempt	$20,000
(v) Brisbane post-eradication survey	$23,500
Total	$137,500

sel were identified on business activity (fouling on vessels, industrial plants, and power stations), the environment (outcompeting native species and changes to water quality), the public (fouling of recreational vessels), human health (poisoning), and urban infrastructure (fouling of water inlets, water and sewage outlets, and navigation buoys). The geographical extent of impacts (**step 4**) was estimated with a spread model and shipping data (see Chapter 14. Map), which predicted that the mussel could spread 60 km beyond Cairns within 5 years.

Market and non-market valuation estimated impacts of $24.6 million in present value terms (Table 9.3). These are the estimated value of damages (**step 5**) that will be avoided if the response is successful. The focus was on valuing the substantial environmental (non-market) impacts of this pest, in addition to a subset of direct impacts, so other impacts were ignored. It was unnecessary to discount impact and response costs (**step 6**) as they were already in present values.

The NPV (**step 7**) is $24.45 million (= $24,594,114 − $137,500). If non-market environmental impacts were excluded and only market impacts were estimated, then the NPV of the biosecurity response would be much lower ($656,614) and perhaps not be deemed large enough to invest in eradication. To protect substantial non-market values, eradication is the preferred option in this hypothetical scenario (**step 9**).

TABLE 9.3

Valuation of Hypothetical Impacts From Asian Green Mussels in Queensland

Impact	Value ($)
Fouling on vessels (additional fuel consumption for 23 cruise ships)	$174,875
In-water cleaning of fouling in Gladstone	$270,000
Investment in filtering and anti-fouling technology	$100,000
Environmental impact	$23,800,000
Treatment of ballast water	$249,239
Total	$24,594,114

Benefit-Cost Ratio

Rather than simply assessing a project based on comparing its benefits and costs using subtraction (Eq. 9.1), a project may be assessed in terms of its benefit-cost ratio (BCR), that is, its average ROI:

$$BCR = \frac{\sum_{t=1}^{T} B(1+r)^{-t}}{\sum_{t=1}^{T} C(1+r)^{-t}} \qquad \text{(Eq. 9.3)}$$

where the numerator and denominator represent the present value of benefits and costs, respectively, that accrue in year t, with r being the discount rate and T the time horizon. Projects are selected in order of decreasing BCR until the budget is exhausted (Epanchin-Niell 2017).

For example, the biosecurity Project Prioritization Protocol (Dodd et al. 2017) used the principles of BCR and Noah's Ark framework to allocate funding to eradicate 50 hypothetical plant incursions. Project benefit was defined as a combination of a species' weed risk score, the potential effectiveness of the intervention, and the probability of successfully intervention. Cost was the monetary cost of the project as NPV. Projects were then ranked by BCR, with resources allocated to projects according to their rank, until the budget was exhausted. In this hypothetical scenario, return on public expenditure using BCR ranking was improved by 25% compared to investing based on weed risk assessment scores alone (Dodd et al. 2017). However, investment decisions based on BCR ranking require strong assumptions that may not always be met in biosecurity (see section on *Allocating resources based on overall benefit-cost ratio*). The BCR approach is appropriate when the question is to invest in a project or not invest in a project. When faced with the question of where to allocate additional investments among projects, then we need to use the change in BCR, as discussed below.

Optimisation

Although the hypothetical Asian green mussel example above (Box 9.2) describes a problem of whether to fund a single project, in reality decision-makers are often faced with the question of selecting one (or multiple) project(s) to fund from a broad suite of proposed projects. The remainder of this chapter deals with principles for optimal allocation of resources across multiple projects and more complex examples. Complex decisions to identify the combination, scale, or timing of investments require *optimisation* approaches. Optimisation approaches aim to design the investments or management strategies that will provide the best value for money, which means achieving management objectives in the most economical way or providing the greatest overall net benefits (Epanchin-Niell 2017).

In biosecurity, optimisation approaches have been applied to determine optimal surveillance effort (Kompas et al. 2016; Yemshanov et al. 2015, 2019), optimal stopping time for eradication (Regan et al. 2006), optimal combinations of control measures (Odom et al. 2003), and optimal distribution and release of biological control agents (Nordblom et al. 2002). For example, Kompas et al. (2019) developed a generic optimisation algorithm in the MATLAB software to minimise expected total costs (damages and project costs) when managing invasive species. This tool requires a number of biological and economic parameters to be known, such as the average pest entry interval, size of initial pest entry, spread rate, surveillance and prevention effectiveness, damages, eradication costs, passive detection thresholds, and pest latency periods. To demonstrate the method, Kompas et al. (2019) applied their modelling framework to four invasive pests of concern in Australia: RIFA (*Solenopsis. invicta*), orange hawkweed (*Hieracium aurantiacum*), FMD, and papaya fruit fly (*Bactrocera papaya*). Results from the optimisation showed the size of the budget available influences both the priority given to invasive species and the specific biosecurity measures employed.

When prioritising biosecurity activities, optimisation approaches focus on the additional risk reduction (ΔB) achieved from additional expenditure (ΔC). If the shape of the benefit-cost curve for each candidate activity is known, the unit change in benefit (ΔB) per unit change in cost (ΔC) can be derived. For a single project or mutually exclusive projects, the optimal time to stop funding a project is when the extra benefits are no longer worth the extra costs ($\Delta B \leq \Delta C$). For a portfolio of candidate activities, the activity with the highest extra benefits (ΔB) for each increase in unit cost (ΔC) is iteratively identified until the budget is exhausted, or, for all projects, the extra benefits are not worth the extra costs ($\Delta B \leq \Delta C$). Using principles from economics, an optimal budget allocation is reached when the marginal return from investment ($\Delta B/\Delta C$) across all activities has been equalised.

In Figure 9.3, the exact optimum for resource allocation is identified by iteratively allocating every unit of cost (ΔC) to the action with the greatest additional benefit (ΔB). Project A exhibits constant RTS, meaning that a one-unit increase in cost results in a one-unit increase in benefits at any point along the benefit-cost curve. Projects B and C show diminishing returns, meaning that their marginal ROI ($\Delta B/\Delta C$, the slope of the pay-off curve at ΔC) differ at each level of investment. An optimal resource allocation outcome requires switching between projects with the steepest slope at each point along the curves.

Assuming the total funding available to a regulator is 50 units and considering discrete increases in project funding of 10 units, project C should be prioritised for the first 10 funding units (up to point 1 in Figure 9.3), as project C shows the highest marginal returns at that point. As the slope between 10 and 20 units of funding flattens for Project C, marginal returns from investing further diminish, and the regulator should switch to funding the first 10 units for project B (up to point 2 in Figure 9.3), as project B's slope between 0 and 10 units is steeper than project C's slope between 10 and 20 units. For the next 10 units, and until the budget of 50 units is exhausted, project A should be funded (points 3 to 5 in Figure 9.3). Overall, project A would be allocated 30 units of funding, and projects B and C would be allocated 10 units each.

True optimisation requires going beyond discrete changes in investment (e.g. 10 units of funding) to consider continuous changes along a curve (i.e. each additional dollar or cent). Even if data may not be available to estimate the slopes of all projects at all levels of investment (i.e. the entire shape of all benefit-cost curves), linear approximation can be used to estimate slopes for incremental changes in investment if increments are small enough. The optimisation method described in Figure 9.3 is easy to apply for projects that can be partially funded or scaled up and down as necessary. Projects that need to be funded fully to achieve any outcome will require additional optimisation constraints.

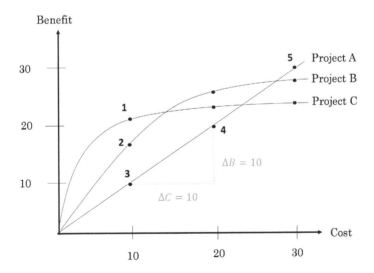

FIGURE 9.3 Conceptual representation of three biosecurity projects with varying benefit-cost curves. Project A shows constant (linear) returns to scale, while projects B and C both show diminishing returns to scale, albeit at different rates.

SELECTING A RESOURCE ALLOCATION METHOD TO ACHIEVE ECONOMIC EFFICIENCY

This chapter outlined different resource allocation methods for biosecurity budgets, including BCA (using the measures of NPV or BCR) and optimisation. In this section, we present the pros and cons of each method and outline when it may (or may not) be appropriate to cost-efficiently allocate budgets.

ALLOCATING RESOURCES BASED ON NET PRESENT VALUE

When NPV is used to allocate resources, funding is awarded when NPV is positive (i.e. benefits exceed costs) and when two or more mutually exclusive projects are being considered, the project with the highest NPV is typically chosen.

In the hypothetical example of an invasion by the Asian green mussel outlined in Box 9.2, eradication is the preferred option (compared to doing nothing) because the NPV of the eradication project is $24.45 million (= Benefits − Costs = $24,594,114 − $137,500). Here, it is clear that NPV does not account for the budget constraint (namely whether a budget of $137,500 is available) and whether resources would generate higher value for money if allocated to different projects.

The only circumstance under which it is economically efficient to allocate resources with NPV is when projects are mutually exclusive and the budget is limited (Pannell 2019). This is because NPV allows us to compare the opportunity costs of foregone project versions.

While it is possible to include opportunity costs in the calculation of project BCRs, the ranking of projects is sensitive to the budget level and opportunity costs, and BCRs would need to be recalculated for different budget levels (Pannell 2019). For projects with diminishing returns, it is best practice to first estimate benefit-cost curves to identify the level of investment at which more investment ceases to yield additional benefits ($\Delta B = \Delta C$). Once project budgets have been adjusted to reflect these diminishing returns, NPVs can be calculated and compared.

ALLOCATING RESOURCES BASED ON OVERALL BENEFIT-COST RATIO

When BCR is used to allocate resources, funding may be justified when BCR exceeds one (1:1) or another ratio selected by the decision maker. Given a set of projects ranked by BCR, decision makers typically proceed top-down, using a threshold where projects with a higher BCR get the entire budget they ask for and projects with lower BCR receive nothing. This all-or-nothing process is often known as the Noah's Ark solution (Dodd et al. 2017). Allocating resources based on the highest BCR makes the implicit assumption that projects' RTS are constant, which is usually untrue or at least should be tested rather than assumed a priori. Noah's Ark solution requires strong assumptions about the shapes and relationships of projects' benefit-cost curves, which are likely to be false in many biosecurity projects.

In Figure 9.4 a, b, two alternative projects, each with declining RTS curves originating from O_1 and O_2, respectively, are proposed against a total budget of 10 cost units. For the sake of illustration, in these examples Project 1 has an overall BCR of 12.6 and Project 2 has an overall BCR of 15.5. Under an overall BCR ranking, Project 2 would secure the entire budget because it has a higher overall BCR (a larger area below the curve). However, allocating the entire budget to the project with the highest overall BCR is not optimal. The optimal allocation that maximises the total benefit from both projects is the intersection of the two RTS curves, where the proportional allocation of the budget is 2:8 in favour of Project 2. This is because the extra benefit that would arise from the 9th And 10th cost units of Project 2 is less than the extra benefit that would arise from the 1st and 2nd cost units of Project 1.

In Figure 9.4 a, b, the BCR ranking is not optimal but is still indicatively right, in that the project with the higher BCR is also the one that is proportionally favoured under optimisation

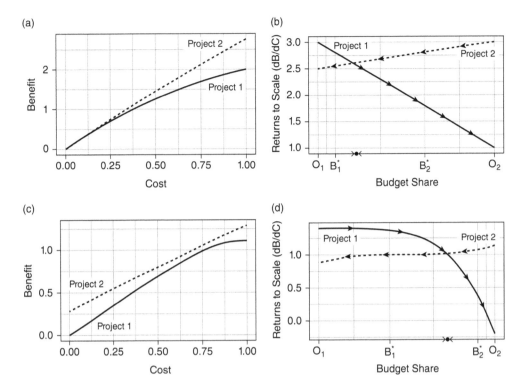

FIGURE 9.4 (a) and (c) Two examples of benefit-cost curves for two competing projects (Projects 1 and 2). (b) and (d) Two examples of returns to scale curves for Projects 1 and 2, with origins at O_1 and O_2, respectively (n.b. the curve for Project 2 starts on the right-hand of the x-axis). The position of the circle on the x-axis indicates when optimal investment switches between Projects 1 and 2. B_1^* and B_2^* indicate optimal budget shares. (Adapted from Kompas et al. (2019) with permission.)

(Project 2). However, in some cases BCR rankings can give an incorrect signal regarding budget shares.

Figure 9.4 c, d shows a situation where the extra benefit arising from the extra expenditure on Project 1 starts relatively flat before decreasing. Project 1 has an overall BCR of 7.1 and Project 2 has an overall BCR of 8.8. In this case, Project 1 has a lower BCR, but it should have 60% of the budget, directly contrasting its BCR ranking. In this case, allocating by overall BCR is neither optimal nor indicatively right.

The only time it is correct to use the highest BCR to allocate resources is when projects are unrelated (independent) and all of their benefit-cost curves are linear (i.e. constant RTS). The only reason overall BCR rankings are economically efficient in this instance is because the slopes of the curves of each project are identical at each level of effort (cost) and equal to the overall BCR (i.e. BCR = $\Delta B/\Delta C$).

ALLOCATING RESOURCES BASED ON BLOCKS OF BENEFIT-COST RATIO

Data may not be readily available to estimate benefit-cost curves on a continuous scale to allow for a full optimisation. To approximate benefit-cost curves for unrelated (independent) projects, splitting projects into distinct "blocks" of funding and ranking each block can result in more economically efficient prioritisation compared to using BCR ranks for entire projects. The smaller the funding blocks, the more accurately RTS can be estimated for each additional unit of funding.

Table 9.4 provides an example of BCR block ranking and illustrates its potential superiority compared to overall BCR ranking. Table 9.4a shows the rankings of three projects (C, D, and E)

TABLE 9.4

Comparison of Ranking by (a) Overall Benefit-Cost Ratio (BCR) and (b) Splitting Projects into Blocks and Ranking Individual Blocks by BCR.

(a) Ranking by Overall BCR

	Project C	Project D	Project E
Cost	1	2	7
Benefit	5	7	8
Overall BCR	5	3.5	1.1
Overall BCR ranking	**1**	**2**	**3**

(b) Ranking by BCR Blocks

	Project C	Project D		Project E						
Block	Block 1	Block 1	Block 2	Block 1	Block 2	Block 3	Block 4	Block 5	Block 6	Block 7
Cost	1	1	1	1	1	1	1	1	1	1
Benefit	5	6	1	7	0.5	0.1	0.1	0.1	0.1	0.1
BCR per block	5	6	1	7	0.5	0.1	0.1	0.1	0.1	0.1
Block BCR ranking	**3**	**2**	**4**	**1**	**5**	**6**	**6**	**6**	**6**	**6**

Source: Derived from Kompas et al. (unpub.) with permission.

according to their overall BCR. In Table 9.4b, the three projects are split into individual blocks (1 block for Project C, 2 blocks for Project D, and 7 blocks for Project E). Rankings based on the individual costs, benefits, and BCRs of these blocks show that the first unit of budget should be allocated to block 1 of Project E (and the second unit allocated to block 1 of Project D), despite Project E having the lowest overall BCR (and Project D the second lowest overall BCR). Ranking by BCR blocks accounts, to some extent, for non-constant RTS in projects. The BCR for block 1 of Project E is high (BCR = 7), whereas BCRs for blocks 2 to 7 are less than 1, indicating that funding these blocks would be economically inefficient. These low BCR blocks contribute to the low BCR of project E as a whole.

Under a hypothetical budget of 3 cost units and using overall BCR rankings, a regulator would choose to fund Projects C and D. By splitting projects into blocks and assessing these separately using BCR blocks, a regulator would choose to invest in block 1 of Project E, block 1 of Project D, and block 1 of Project C.

Given a hypothetical budget of 4 cost units, a regulator using overall BCR would still fund projects C and D and would have 1 unit of cost left over, whereas a regulator using a block BCR approach would be able to fully fund projects C and D and block 1 of Project E, generating additional benefit. This example thus illustrates how blocking may also make better use of all funding available.

ALLOCATING RESOURCES BASED ON OPTIMISATION

When RTS is not constant, it is always best to formally optimise resource allocation by minimising the total costs of impacts and management actions. Even when RTS is constant, optimisation methods can still be used.

The costs of misallocation using other methods such as ranking by overall (average) BCR or NPV can be high when dealing with pest impacts that increase non-linearly with pest population density and/or with time and where management actions are also likely to exhibit nonlinear RTS. Some pests can cause high impacts at low density (e.g. those that trigger export prohibitions at low thresholds), such as weeds affecting wheat crops or FMD (Cook et al. 2012; Yokomizo et al. 2009). Because impacts are incurred in a highly nonlinear way, they are also not likely to be reduced at a constant rate proportional to additional spending. Incorrect allocation by applying simple ranking rules can result in underinvestment towards control and eradication early in the outbreak or underinvestment in measures preventing arrival and establishment. Such resource allocation may lead to impacts that cost more to rehabilitate than the money originally saved (Yokomizo et al. 2009). Conversely, some pests have almost no impacts at low density and with impacts growing only when population densities become large. Incorrectly assuming constant RTS can lead to allocation decisions that waste resources by overinvesting in unnecessary actions early in the outbreak or in early eradication (Yokomizo et al. 2009). Optimisation approaches should thus be used when benefits (avoided damages) are sensitive in variable and non-linear ways to pest density, maximum economic impacts, ecological parameters (such as environmental fluctuations in mortality, density-dependent population recovery, and difficulty of eradication), and management parameters (such as the time-horizon of actions and discount rates; Yokomizo et al. 2009).

Optimisation methods may also be required when combinations of independent projects or combinations of independent and mutually exclusive projects are being considered—the technical detail may be beyond the interest of some readers. Two of this chapter's authors (TK and CL) applied optimisation to assist an Australian State Government in project prioritisation. They developed a spreadsheet-based BCA model for New South Wales Biosecurity & Food Safety that provides an integer programming optimisation approach for prioritising a combination of independent and mutually exclusive projects. If a set of projects are all mutually exclusive, optimisation methods can still be used but may not be necessary (see section on *Allocating resources based on net present value*).

SELECTING A RESOURCE ALLOCATION METHOD BASED ON THE DECISION CONTEXT

Using optimisation methods always ensures resource allocation is cost-efficient, regardless of the specifics of the problem at hand. However, under specific decision contexts, it may be appropriate to use simpler approaches, as these will provide the same results as (or an approximation of) optimisation.

To select an appropriate method, Figure 9.5 shows how a biosecurity resource allocation problem can be broken down into core questions relating to the project mix, assumptions about RTS, and data availability:

- *Projects are mutually exclusive*: Distinct projects could achieve a set objective. Ranking by NPV can be used, regardless of the RTS of projects, as this will provide the same results as optimisation.
- *Projects are unrelated (independent)*: If the RTS of all projects is constant, ranking by overall BCR is correct and will provide the same results as optimisation. If the RTS of any of the projects is not constant (which is likely to be the case in biosecurity), then ranking with overall BCR will lead to inefficient resource allocation (and ranking may be indicatively wrong; see Figure 9.4b). If RTS is not constant, only optimisation will lead to economically efficient decisions. If data are unavailable to calibrate benefit-cost curves on a continuous scale, ranking by BCR blocks can provide a useful approximation of optimisation results and will be superior to ranking by overall BCR.
- *Projects are a combination of mutually exclusive and independent*: When the project mix includes a combination of mutually exclusive and independent projects, only optimisation leads to economically efficient decisions. If the RTS of all projects is constant (or assumed to be), basic integer programming can be used to solve the optimisation problem.

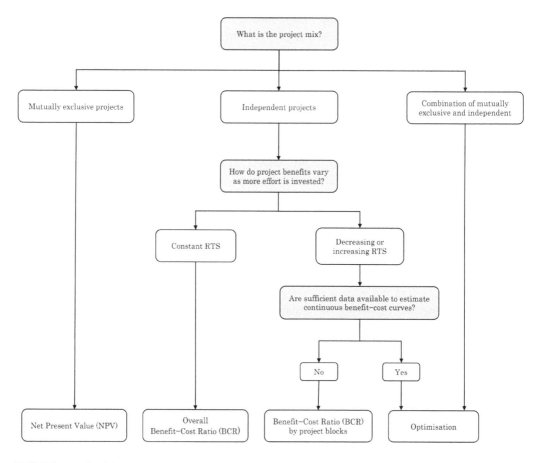

FIGURE 9.5 Decision tree to inform the selection of a cost-efficient resource allocation method in biosecurity. *Abbreviation*: RTS: Returns to scale.

BARRIERS AND OPPORTUNITIES FOR COST-EFFICIENT RESOURCE ALLOCATION IN BIOSECURITY

Under the World Trade Organisation (WTO) Agreement on the Application of Sanitary and Phytosanitary Measures (SPS Agreement), signatory countries are entitled to set an appropriate level of protection (ALOP) to protect their economy, society, and environment from pests and diseases (see Chapter 2. Biosecurity Systems and International Regulations). The WTO Agreement aims to achieve open and non-discriminatory trade policies, including policies aimed at mitigating biosecurity risks. Although principles of economic efficiency would dictate an assessment of the benefits and costs of risk mitigation measures, this is not a requirement under the SPS Agreement. Rather, the language of the SPS Agreement used to describe the choice and application of ALOP (e.g. Article 5.3; WTO 1995) leads biosecurity agencies to choose measures that reduce risk to negligible levels and to focus on the direct risk-related costs of imports (rather than on foregone trade benefits to domestic consumers).

At present, many methods used by biosecurity agencies to prioritise threats and activities and allocate budgets do not make use of economic principles. Funding allocations based on subjective opinions, historical legacies, the visibility of a threat, or industry pressure do not account for the value for money of projects and are thus economically inefficient. As a perceived improvement, many agencies allocate funding based on risk assessment results or lists of priority species. Although these methods give insight into the relative risks posed by species, they give no insight into the value for money of the activities that could be implemented to manage those risks. Failure by biosecurity agencies to incorporate economic principles wastes taxpayers' money, as implementing unnecessary or ineffective activities does not provide the highest value for the money spent. Budget misallocations can also contribute to increasing risk throughout the biosecurity system as a whole, as underperforming or ineffective programs detract resources from other risk-reducing activities (see Chapter 10. Monitoring, Evaluation, and Reporting).

The basic premise of value for money holds regardless of whether allocation decisions are made to respond to a single threat, across a portfolio of threats, or at different stages of the biosecurity system (namely pre-border, border, or post-border). In an ideal world, a decision maker would consider the costs of prevention and management activities and their effects on reducing damages for all pests and diseases simultaneously. Optimal effort would be identified for each activity, and funds would be allocated to the activities and pests where the value of risk reduction is greatest, subject to a budget constraint (Heikkilä 2011; Moffitt and Osteen 2006).

In practice, it is very difficult for biosecurity agencies to obtain all the data required to compute and select the most economically efficient set of activities across all threats. One of the suggested reasons for the limited uptake of economic principles in biosecurity is the complex nature of biosecurity funding decisions, rather than the simple decision of whether to fund or not (Epanchin-Niell 2017). Policymakers are often faced with large numbers of threats, and each can be associated with a complex range of biosecurity activities pre-border, at the border, and post-border.

For these reasons, resource allocation methods in biosecurity should be practically implementable, even with limited input data. Biosecurity activities are often designed at the early stages of an incursion or risk mitigation planning process, where information about a threat is limited or uncertain (see Chapter 7. Prepare, Respond, and Recover). This chapter has shown that project costs can be broken down into measurable components (e.g. operational, infrastructure, or per-unit costs) and benefits can be estimated using well-established market and non-market evaluation methods and, where necessary, the assistance of an economist. Benefit and cost information compiled for previous projects (e.g. for the purposes of monitoring and evaluation; see Chapter 10. Monitoring, Evaluation, and Reporting) should be collated and used to inform budget allocation. In some cases, using benefit transfer techniques (e.g. from other projects or geographic areas) can be useful; in other cases, structured expert elicitation of potential benefits will be necessary. In some cases, estimating a subset of impacts or benefits (e.g. monetary benefits) might be sufficient (as in the hypothetical example of the Asian green mussel in Box 9.2).

Even when the ecological dynamics of threats and the effects of management activities are uncertain, advances in spread modelling (and associated sensitivity analyses; see Chapter 14. Map) have enabled more reliable valuation and prioritisation in biosecurity (Kompas et al. 2019). Similarly, using structured expert elicitation to gather information from experts (see Chapter 12. Elicit) can fill data gaps and be subject to high standards of scientific enquiry and sensitivity analyses.

When a project mix contains mutually exclusive and independent projects, or when independent projects are expected to show non-constant RTS (Figure 9.5), estimating benefit-cost curves is essential to achieving cost-efficiency. Because diminishing RTS is expected for most biosecurity applications, assuming a generic diminishing returns benefit-cost curve based on sparse data is likely to lead to more efficient prioritisation than assuming constant RTS and using overall BCR to rank independent projects (Figure 9.5). When data are sparse, RTS can also be accounted for by using simple project blocks (Table 9.4). If data are too sparse to make appropriate assumptions or calibrate sensitivity analyses regarding project costs and benefits, the utility of conducting projects with very uncertain costs or benefits should be questioned.

Beyond data availability, another possible explanation for the lack of uptake of economic principles by biosecurity agencies is the perceived lack of practical and readily applicable prioritisation methods. While it may not be possible to develop and apply sophisticated optimisation models to assist in all budget allocations, this chapter has shown that simple prioritisation problems can be solved using pen and paper by following basic economic principles (Figure 9.3 and Table 9.4). In practice, biosecurity agencies could implement BCAs to allocate funding across candidate projects by using simple spreadsheet tools, thereby greatly improving the benefits reaped from taxpayer money. When prioritisation problems are complex and spreadsheet tools are no longer suitable, consultation with appropriately qualified economists (e.g. natural resource or agricultural economists) is advisable. This chapter is intended as a palatable guide for biosecurity practitioners to understand and commission the development of prioritisation tools. The decision tree outlining the suitability of prioritisation methods (Figure 9.5) focuses on the different decision-making contexts faced by a regulator and aims to support communication and consultation with economists.

The lack of application of well-established economics methods in biosecurity constitutes a breakdown in the research uptake process—one with potentially high costs to society. Strong leadership support in favour of cost-efficient resource allocation, basic economics training for biosecurity staff, appointing knowledge brokers to bridge organisational silos, and creating social opportunities for knowledge sharing are strategies that could be employed by biosecurity agencies to shift organisational culture and attitudes (see Chapter 11. Research Uptake). Creating a strong monitoring and evaluation culture (see Chapter 10. Monitoring, Evaluation, and Reporting) may not only contribute to an increased information base on the costs and benefits of past projects, but such an approach may also encourage critical assessment of projects in terms of their value for money.

Many government policies are politically driven rather than based on the best evidence or methods available (Bray, Gray, and Stanton 2020). Government decision making is constrained by past choices, established programs and their funding, and ongoing risk avoidance (Matheson 1998). In Australia, two high-profile reviews have questioned the current allocation of biosecurity budgets (Australian National Audit Office 2021; Inspector-General of Biosecurity 2021). Policymakers may choose to ignore methods that put in question past decisions, or they may disregard resource allocation results that contradict their aspirations or that are unpalatable to vested interest groups. However, there is a strong argument that value for money should be the overarching criterion driving budget allocations when taxpayer dollars are being spent.

Public sector agencies are subject to extreme top-down budgetary pressures. As long as economic principles remain on the periphery of biosecurity decision making, biosecurity agencies will forgo the opportunity to understand (and credibly ask for) the total budget that is required to reduce biosecurity risks to a socially acceptable level (as has been investigated in species conservation (Wintle et al. 2019). Biological threats from pests and diseases are just one of many hazards confronting societies (e.g. climate change, cybersecurity threats, terrorism; World Economic Forum 2022). The high-level decision about how many resources to devote to reducing biological threats compared to other hazards is largely a political decision, which implicitly reflects the relative priority given to this type of hazard (Heikkilä 2011). The costs of managing COVID-19 (in the order of billions of dollars for many economies) and the massive impacts of the disease are a cogent reminder of the potential size of biosecurity impacts if threats are not managed effectively.

IN A NUTSHELL

- Budgets will never be adequate to eliminate or reduce all risks. Because of this, biosecurity agencies must choose which projects to implement to manage threats.
- An activity is cost-efficient if no other use of resources would yield a higher value to the community. When taxpayer dollars are being spent, cost-efficiency (i.e. "value for money") should be the overarching criterion driving budget allocations.
- Funding allocations based on subjective opinions, historical legacies, industry pressure, or risk assessments do not account for the "value for money" of competing projects and are economically inefficient.

- To prioritise biosecurity activities, regulators should obtain data on the costs of management activities and their potential benefits (avoided damages) and use an appropriate economic method for the problem at hand.
- As biosecurity agencies become more familiar with economics methods, the process of identifying and closing key data gaps can also become more targeted.

REFERENCES

Australian National Audit Office. 2021. Responding to Non-Compliance with Biosecurity Requirements: Department of Agriculture, Water and the Environment.

Boardman, A. E., D. H. Greenberg, A. R. Vining, and D. L. Weimer. 2018. *Cost-Benefit Analysis: Concepts and Practice.* 5th ed. Cambridge: Cambridge University Press.

Bossuyt, J., L. Shaxson, and A. Datta. 2014. Study on the uptake of learning from EuropeAid's strategic evaluations into development policy and practice. Final Report. European Commission.

Bray, J. R., M. C. Gray, and D. I. Stanton. 2020. "Performance Management and Evaluation Meets Culture and Politics: Australia's Experience." *Asia Pacific Journal of Public Administration* 42 (4):290–302. https://doi.org/10.1080/23276665.2020.1808491.

Cacho, O. J., R. M. Wise, S. Hester, and J. Sinden. 2008. "Bioeconomic Modeling for Optimal Control of Weeds in Natural Environments." *Ecological Economics* 65: 559–568.

Chouinard, J. A. 2013. "The Case for Participatory Evaluation in an Era of Accountability." *American Journal of Evaluation* 34 (2):237–253. https://doi.org/10.1177/1098214013478142.

Cook, D., R. Fraser, J. Waage, and M. Thomas. 2011. "Prioritising Biosecurity Investment between Agricultural and Environmental Systems." *Journal für Verbraucherschutz und Lebensmittelsicherheit* 6 (1):3–13. https://doi.org/10.1007/s00003-011-0689-0.

Cook, D. C., G. Long, H. P. Possingham, L. Failing, M. Burgman, R. Gregory, R. Estevez, and T. Walshe. 2012. Potential methods and tools for estimating biosecurity consequences for primary production, amenity, and the environment. Australian Centre of Excellence for Risk Analysis.

Costanza, R., I. Kubiszewski, N. Stoeckl, and T. Kompas. 2021. "Pluralistic Discounting Recognizing Different Capital Contributions: An Example Estimating the Net Present Value of Global Ecosystem Services." *Ecological Economics* 183:106961. https://doi.org/10.1016/j.ecolecon.2021.106961.

Coutts, A, and B. M. Forrest. 2007. "Development and Application of Tools for Incursion Response: Lessons Learned from the Management of the Fouling Pest *Didemnum vexillum.*" *Journal of Experimental Marine Biology and Ecology* 342 (1):154–162. https://doi.org/10.1016/j.jembe.2006.10.042.

Craik, W., D. Palmer, and R. Sheldrake. 2017. *Priorities for Australia's Biosecurity System. An Independent Review of the Capacity of the National Biosecurity System and Its Underpinning Intergovernmental Agreement.* Canberra: Department of Agriculture and Water Resources.

Dodd, A. J., N. Ainsworth, C. E. Hauser, M. A. Burgman, and M. A. McCarthy. 2017. "Prioritizing Plant Eradication Targets by Re-Framing the Project Prioritization Protocol (PPP) for Use in Biosecurity Applications." *Biological Invasions*:1–15. https://doi.org/10.1007/s10530-016-1335-7.

Dodd, A., N. Stoeckl, J. B. Baumgartner, and T. Kompas. 2020. *Key Result Summary: Valuing Australia's Biosecurity System, CEBRA Project 170713.* Centre of Excellence for Biosecurity Risk Analysis.

Emerton, L., and G. Howard. 2008. *A Toolkit for the Economic Analysis of Invasive Species.* Nairobi: Global Invasive Species Programme, CABI/IUCN.

Epanchin-Niell, R. S. 2017. "Economics of Invasive Species Policy and Management." *Biological Invasions*:1–22. https://doi.org/10.1007/s10530-017-1406-4.

Epanchin-Niell, R. S., and A. Hastings. 2010. "*Controlling Established Invaders: Integrating Economics and Spread Dynamics to Determine Optimal Management.*" 13 (4):528–541. https://doi.org/10.1111/j.1461-0248.2010.01440.x.

Fox, A. M., and D. R. Gordon. 2004. "Criteria for Listing Invasive Plants." *Weed Technology* 18: 1309–1313.

Harrison, M. 2010. Valuing the Future: the Social Discount Rate in Cost-Benefit Analysis. In *Visiting Researcher Paper.* Canberra: Productivity Commission.

Hauser, C. E., and M. A. McCarthy. 2009. "Streamlining 'search and destroy': Cost-Effective Surveillance for Invasive Species Management." *Ecology Letters* 12 (7):683–692. https://doi.org/10.1111/j.1461-0248.2009.01323.x.

Heikkilä, J. 2011. "Economics of Biosecurity Across Levels of Decision-Making: a Review." *Agronomy for Sustainable Development* 31 (1):119. https://doi.org/10.1051/agro/2010003.

Hester, S. M., S. J. Brooks, O. J. Cacho, and F. D. Panetta. 2010. "Applying a Simulation Model to the Management of an Infestation of *Miconia calvescens* in the Wet Tropics of Australia." *Weed Research* 50 (3):269–279. https://doi.org/10.1111/j.1365-3180.2010.00778.x.

Hester, S. M., O. J. Cacho, F. Dane Panetta, and C. E. Hauser. 2013. "Economic Aspects of Post-Border Weed Risk Management." *Diversity and Distributions* 19 (5-6):580–589. https://doi.org/10.1111/ddi.12053.

Hester, S., and J. Mayo. 2021. Improving the Methodology for Rapid Consequence Assessment of Amenity and Environmental Pests. Centre of Excellence for Biosecurity Risk Analysis.

Hulme, P. E. 2009. "Trade, Transport and Trouble: Managing Invasive Species Pathways in an Era of Globalization." *Journal of Applied Ecology* 46 (1):10–18. https://doi.org/10.1111/j.1365-2664.2008.01600.x.

IGB. 2019. *Pest and Disease Interceptions and Incursions in Australia*. Canberra: Inspector-General of Biosecurity. Department of Agriculture and Water Resources.

Inspector-General of Biosecurity. 2021. *Adequacy of Department's Operational Model to Effectively Mitigate Biosecurity Risks in Evolving Risk and Business Environments*. Canberra: Department of Agriculture and Water Resources.

Kim, C. S., D. Lee, G. Schaible, and U. Vasavada. 2007. "Multiregional Invasive Species Management: Theory and an Application to Florida's Exotic Plants." *Journal of Agricultural and Applied Economics* 39 (s1):111–124. https://doi.org/10.1017/S1074070800028984.

Kompas, T., and T. N. Che. 2001. An Economic Assessment of the Potential Costs of Red Imported Fire Ants in Australia. ABARES.

Kompas, T., L. Chu, A. Dodd, and H. T. M. Nguyen. unpub. Misusing Cost-Benefit Analysis. *Centre for Environmental and Economic Research*, University of Melbourne.

Kompas, T., L. Chu, and H. T. M. Nguyen. 2016. "A Practical Optimal Surveillance Policy for Invasive Weeds: An Application to Hawkweed in Australia." *Ecological Economics* 130:156–165. http://doi.org/10.1016/j.ecolecon.2016.07.003.

Kompas, T., L. Chu, P. Van Ha, and D. Spring. 2019. "Budgeting and Portfolio Allocation for Biosecurity Measures." *Australian Journal of Agricultural and Resource Economics*. https://doi.org/10.1111/1467-8489.12305.

Matheson, C. 1998. "Rationality and Decision-Making in Australian Federal Government." *Australian Journal of Political Science* 33 (1):57–72. https://doi.org/10.1080/10361149850723.

Moffitt, L. J., and C. D. Osteen. 2006. "Prioritizing Invasive Species Threats Under Uncertainty." *Agricultural and Resource Economics Review* 35 (1): 41–51.

MPSC. 2018. *Australian Priority Marine Pest List: Process and Outcomes*. Canberra: ABARES. [Report].

Murphy, B., and D. Paini. 2010. CCIMPE Review: UPDATE 2007–2009.

Nordblom, T. L., M. J. Smyth, A. Swirepik, A. W. Sheppard, and D. T. Briese. 2002. "Spatial Economics of Biological Control: Investing in New Releases of Insects for Earlier Limitation of Paterson's Curse in Australia." *Agricultural Economics* 27 (3):403–424. https://doi.org/10.1016/S0169-5150(02)00069-5.

Odom, D. I. S., O. J. Cacho, J. A. Sinden, and G. R. Griffith. 2003. "Policies for the Management of Weeds in Natural Ecosystems: the Case of Scotch Broom (*Cytisus scoparius*, L.) in an Australian National Park." *Ecological Economics* 44: 119–135.

Pannell, D. J., H. T. M. Nguyen, H. L. Chu, T. Kompas, and A. Rogers. 2024. "Benefit-Cost Analysis Decision Criteria: Reconciling Conflicting Advice." *Applied Economics Teaching Resources* 6 (1):13–28. https://doi.org/10.22004/ag.econ.341811.

Productivity Commission. 2013. On Efficiency and Effectiveness: Some Definitions. In *Productivity Commission Staff Research Note*. Canberra: Productivity Commission. [Report].

Regan, T., M. A. McCarthy, P. W. J. Baxter, F. D. Panetta, and H. P. Possingham. 2006. "Optimal Eradication: When to Stop Looking for an Invasive Plant." *Ecology Letters* 9: 759–766.

Robinson, A., M. Chisholm, R. Mudford, and R. Maillardet. 2015. "Ad Hoc Solutions to Estimating Pathway non-Compliance Rates Using Imperfect and Incomplete Information." In *Biosecurity Surveillance: Quantitative Approaches*, edited by F. Jarrad, S. Low-Choy and K. Mengersen, 167–180. CABI International.

Summerson, R., S. Hester, and S. Graham. 2018. *Methodology to Guide Responses to Marine Pest Incursions Under the National Environmental Biosecurity Response Agreement*. Centre of Excellence for Biosecurity Risk Analysis.

Tongue, A. 2020. *Future Department Review: Revitalising the Organisation to Deliver for the Government and Australians*. Canberra: Department of Agriculture, Water and the Environment.

Virtue, J. G. 2007. "Weed risk assessment in South Australia." 2nd International weed risk assessment workshop., Perth, Australia.

Wintle, B. A., N. C. R. Cadenhead, R. A. Morgain, S. M. Legge, S. A. Bekessy, M. Cantele, H. P. Possingham, J. E. M. Watson, M. Maron, D. A. Keith, S. T. Garnett, J. C. Z. Woinarski, and D. B. Lindenmayer. 2019. "Spending to Save: What Will It Cost to Halt Australia's Extinction Crisis?" *Conservation Letters* 12 (6):e12682. https://doi.org/10.1111/conl.12682.

World Economic Forum. 2022. Global Risks Report 2022. Geneva: World Economic Forum.

WTO. 1995. "The WTO agreement on the application of Sanitary and Phytosanitary Measures (SPS Agreement)." World Trade Organization, accessed 13 January 2023. https://www.wto.org/english/tratop_e/sps_e/spsagr_e.htm.

Yemshanov, D., R. G. Haight, F. H. Koch, B. Lu, R. Venette, D. B. Lyons, T. Scarr, and K. Ryall. 2015. "Optimal Allocation of Invasive Species Surveillance With the Maximum Expected Coverage Concept." *Diversity and Distributions* 21 (11):1349–1359. https://doi.org/10.1111/ddi.12358.

Yemshanov, D., R. G. Haight, F. H. Koch, R. C. Venette, T. Swystun, R. E. Fournier, M. Marcotte, Y. Chen, and J. J. Turgeon. 2019. "Optimizing Surveillance Strategies for Early Detection of Invasive Alien Species." *Ecological Economics* 162:87–99. https://doi.org/10.1016/j.ecolecon.2019.04.030.

Yokomizo, H., H. P. Possingham, M. B. Thomas, and Y. M. Buckley. 2009. "Managing the Impact of Invasive Species: the Value of Knowing the density–Impact Curve." 19 (2):376–386. https://doi.org/10.1890/08-0442.1.

10 Monitoring, Evaluation, and Reporting
Assessing the Performance of Biosecurity Programs

Edith Arndt

ABSTRACT

Monitoring, evaluation, and reporting are activities that allow managers to know whether program objectives are being met. Monitoring, evaluation, and reporting also help fulfil public accountability requirements and help inform future decision making and resource allocation. In this chapter, we outline the role of monitoring and evaluation in the biosecurity planning cycle and summarise the elements that underpin successful performance appraisal in biosecurity systems. Effective monitoring and evaluation rely on appropriate resourcing, a robust evaluation framework and program theory, relevant performance indicators, and a positive organisational culture that favours learning and transparency. Challenges remain for monitoring, evaluation, and reporting in the public sector in Australia and elsewhere, but shifts in organisational culture can help to overcome those challenges.

GLOSSARY

Monitoring The routine or continuous measurement of an object or a process over time. Monitoring is about generating information that can be used to assess whether standards are being met (Alexander 2013).

Evaluation The systematic and periodic process of judging whether undertaking a program is worthwhile, meaning that the program achieves its intended outcomes and fits the policy environment and stakeholder needs. Evaluation uses, analyses, and interprets the information gathered through monitoring to inform higher level strategic decisions about the direction of a program (Funnell and Rogers 2011; Markiewicz and Patrick 2016).

Reporting The formal communication of monitoring and/or evaluation outcomes to different audiences, such as members of the government, program managers, and the public (Markiewicz and Patrick 2016).

Program theory A description of how a program is expected to work, focusing on a program's outputs and intended outcomes (Funnell and Rogers 2011).

Logic model Illustrates a program theory as a diagram. Logic models usually depict program inputs, activities, outputs, outcomes, and impacts (see Figure 10.2; Funnell and Rogers 2011).

DOI: 10.1201/9781003253204-14

Performance indicators Measurable characteristics of a program that reflect the level of success (or performance) of a management intervention.
Biosecurity risk material (BRM) Living organisms exotic to a country, or biological material carrying an exotic pest or disease. BRM includes live animals and plants, animal
and plant material (e.g. fur, feathers, pollen, and flowers), food scraps, and soil.

INTRODUCTION

Government agencies use taxpayer funds to coordinate a multitude of programs across public sectors,
while being constantly challenged by a scarcity of resources. Government agencies administer public
funds and coordinate service delivery within a culture of accountability that demands evidence of
efficiency and effectiveness (Bossuyt, Shaxson, and Datta 2014; Chouinard 2013). Monitoring and
evaluation can be used to satisfy accountability requirements by providing evidence on the performance of government policies and activities. Monitoring and evaluation are therefore fundamental
elements of good public sector governance and budget management (Podger et al. 2018).

In Australia, interest in public sector performance and effectiveness (or how managers spend public
money) has increased since the 1980s, when the focus started to shift from assessing compliance to
assessing performance (Hoque and Pearson 2018). Compliance is about adhering to rules and regulations and proper reporting of finances and activities, whereas performance evaluation determines the
merit or worth of an intervention, with the aim to use that information to improve policies and programs (Hoque and Pearson 2018; Peersman et al. 2016). With the *Public Governance, Performance
and Accountability Act 2013*, performance monitoring and evaluation have become anchored in public
sector management in Australia. However, significant concerns remain in Australia about the extent to
which public sector performance is assessed (Gray and Bray 2019). In the United States, the *Foundations
for Evidence-Based Policymaking Act of 2018* prescribes government agencies to make agency data
publicly accessible and to develop statistical evidence to support policy making (Ryan 2019).

Some public sectors have actively embedded performance evaluation in their management processes.
The Australian health sector, for example, strives for performance reporting using a set of national indicators to derive strategies for system-level improvement to enable patients and carers to make more
informed decisions (AIHW 2009; Dixit and Sambasivan 2018). The Australian healthcare system follows international standards for quality and performance measurement (Dixit and Sambasivan 2018),
and Australia has been recognised for its significant progress in developing performance frameworks
and using indicators in healthcare (Braithwaite et al. 2017). Similarly, New Zealand introduced a new
approach to measuring and monitoring healthcare performance in 2016. The Systems Level Measure
Framework applies a whole-of-system approach, focusing on system-level outcomes (e.g. patient experience of care and acute hospital bed days per capita) and requiring all parts of the health system to work
together (Chalmers, Ashton, and Tenbensel 2017; Ministry of Health 2018).

The fire management sector is another area under high public scrutiny because of the substantial
damages that result from destructive bushfires. The Victorian State Government in Australia has
developed and implemented a monitoring, evaluation, and reporting framework that focuses on the
effectiveness of planned burning in minimising the impact of bushfires and improving ecosystem
resilience (DELWP 2015). As another example, Californian government researchers have advo-
cated for the use of performance measurement systems to support decision making in fire management (Thompson et al. 2018).

Monitoring, evaluation, and reporting allow decision makers to know whether program objectives are met. These methods can be applied to projects, programs, and systems of any size and
in any sector, whether it be healthcare, natural resource management, education, or international
development. Whatever the purpose of management activities may be or what outcomes they intend

to deliver, decisions should always be guided by how well these activities achieve their objectives. The present chapter focuses on biosecurity programs and their main objectives, namely, to minimise the negative impacts of pests and diseases on the economy, environment, and community.

Recognising the value of performance evaluation in biosecurity is a natural step in the process of biosecurity system maturity. Essentially, performance evaluation adds a second lens through which to view a biosecurity system. The first lens focuses on the exotic pests, diseases, and entry pathways that need to be identified and managed so that adverse effects on national biosecurity can be mitigated—the main business of biosecurity agencies. The second lens focuses on the performance or effectiveness of the regulating organisation itself, and, importantly, identifies management areas that do not achieve their objectives. For example, an early surveillance program that has deployed insufficient traps or employed insufficient staff to check them regularly may not detect an incursion early enough to stop pests from establishing or spreading (see Chapter 6. Detect).

If program managers do not monitor and evaluate the performance of their program, they do not know how well the program achieves its intended outcomes. They are "flying blind", with no knowledge about whether the program is creating value or steering off course. They also do not know where risks occur, and they therefore cannot act to prevent or mitigate those risks. Without insight into a program's past performance and weak spots, managers can struggle to make decisions about future investments and resource allocation.

In this chapter, we outline the role of monitoring and evaluation in the biosecurity planning cycle and summarise the elements that underpin successful performance appraisal in biosecurity systems. Such performance appraisal can in turn support transparent and efficient decision making when biosecurity funding is drawn from the public purse.

MONITORING, EVALUATION, AND REPORTING IN THE PLANNING CYCLE

Monitoring and evaluation are not conducted in isolation of other program management activities—rather, monitoring and evaluation are integrated into what is known as the planning cycle (Figure 10.1). Program management consists of three fundamental phases:

1. *Learning and Planning*: Assess and define the problem the program addresses, ideally based on learnings, knowledge, and input from stakeholders. Planning includes defining

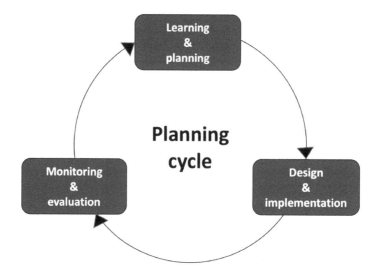

FIGURE 10.1 Program management activities forming a planning cycle. (Adapted from Williams and Brown 2014.)

the goals and objectives of the program, identifying alternative actions, predicting their consequences, and evaluating trade-offs. The outcome of the planning phase provides a clear direction for program management. Structured decision-making puts emphasis on this initial phase, where both facts and values are considered during collaborative deliberation (see Chapter 12. Elicit).

2. *Design and implementation*: The planning work is put into practice by designing program activities and developing monitoring and evaluation protocols. Program managers develop operational work plans and coordinate and allocate resources (namely human, financial, and physical) to undertake planned activities.

3. *Monitoring and evaluation*: Collect data on the performance of program activities and periodically analyse whether the overall program is achieving its objectives. Outcomes of monitoring and evaluation provide lessons on previous steps (namely planning, design, and implementation).

Definitions and labels for these phases vary among sources, and different planning approaches emphasise different steps within the planning cycle. Monitoring and evaluation are sometimes depicted as separate activities (e.g. adaptive management cycle; Williams and Brown 2014) or as activities grouped with other activities, such as implementation and reporting (e.g. MER framework for bushfire management, DELWP 2015; structured decision making, Failing, Gregory, and Harstone 2007; Gregory et al. 2012).

The phases of the planning cycle are performed iteratively to adapt activities based on learning. Monitoring and evaluation provide the information necessary for learning, thereby closing the loop of the planning cycle. When a new iteration of the planning cycle starts, model predictions produced during the previous planning phase are compared to observed outcomes to generate learnings that can inform the current planning phase and any adjustments to project objectives and predictive models (Williams and Brown 2014). The outcomes of program evaluation may also inform higher-level prioritisation of investment choices, for example, when decision makers need to decide on allocating resources among different pests or programs and need to weigh up alternative options (see Chapter 9. Resource Allocation and Chapter 7. Prepare, Respond and Recover).

In practice, linking the outcomes of monitoring and evaluation to decision making is often difficult. This can be due to a lack of guidance on how to make management decisions (Gregory et al. 2012), risk aversion (Walters 1997), or a poor evaluation culture. A structured decision making approach can increase the chances of results being taken up, because it clearly accounts for how subsequent decisions should be made (Gregory et al. 2012).

ELEMENTS OF SUCCESSFUL MONITORING, EVALUATION, AND REPORTING

Monitoring and evaluation are integral to the long-term success of management programs. Ideally, a monitoring and evaluation framework is developed during the initial planning of a program and implemented jointly with other program activities. Below, we outline how a manager can develop a successful performance evaluation approach for a biosecurity program by using:

- Appropriate resourcing.
- Monitoring and evaluation frameworks.
- Performance indicators.
- Consistent methodologies.
- Data accessibility.
- Continuous learning.
- Culture of monitoring and evaluation.

Appropriate Resourcing

Evaluation outcomes are only as good as the effort invested in performance appraisal, and the level of investment needed to monitor and evaluate program activities should not be underestimated. Monitoring and evaluation require extensive planning in the design phase and are also resource intensive during data collection, analysis, and management. It can be difficult to sustain an institutional commitment to monitoring and evaluation beyond the initial set-up of a program, for example, when decision makers believe that limited resources are better invested in on-the-ground activities with visible impact (Williams and Brown 2014). The cost of monitoring and evaluation activities depends on the complexity of the program and existing capacity within an organisation, but an allocation of 5%–10% of the overall program budget can usually cover the costs of monitoring and evaluation (OECD 2011).

Monitoring and Evaluation Frameworks

Program evaluation should be based on a comprehensive monitoring and evaluation framework. Developing such a framework starts with defining the purpose, scope, and management of the evaluation (i.e. a document detailing the purpose and scope of the evaluation and how the evaluation will be conducted and managed; Roberts, Khattri, and Wessal 2011). Evaluators also need to develop a conceptual framework, evaluative criteria, key evaluation questions, performance indicators, a strategy for data collection and analysis, and an outline of reporting approaches (Peersman et al. 2016).

Many government organisations apply a theory-driven approach to evaluation (Schneider and Arndt 2020), which means that evaluators create a model of the evaluation subject (e.g. a biosecurity program) to show how interventions are expected to work and how they contribute to a chain of intended outcomes (Bickman 1987; Funnell and Rogers 2011; Vogel 2012). These theoretical models are often described as a program theory or theory of change, and they can be represented in the form of a diagram, which is then referred to as a logic model or program logic.

A logic model displays relationships among program resources, planned activities, and the changes the program aims to achieve (i.e. outputs, outcomes, and impacts; Figure 10.2). Resources are the inputs needed to operate the program, for example, staff time, funding, equipment, and office and laboratory facilities. Activities are planned actions that are implemented using inputs. Outputs are the direct and measurable products, results, or services delivered by the program, often expressed in terms of volumes or units delivered. Outcomes are specific changes to people's behaviours or to the status and functioning of program targets (e.g. natural resources or organisational processes). Outcomes are often the result of multiple outputs over time, and they can be categorised into short-term, intermediate, and long-term outcomes.

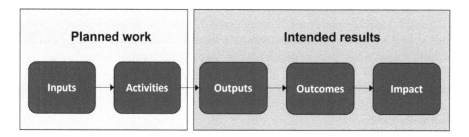

FIGURE 10.2 Basic elements of a logic model. (Adapted from W. K. Kellogg Foundation 2004. Reprinted from the W. K. Kellogg Foundation Logic Model Guide with permission from the W. K. Kellogg Foundation.)

Impacts are fundamental intended or unintended changes at a broader level (e.g. organisations, communities, or systems) over longer time frames, for example, 7 to 10 years (Mertens and Wilson 2012; W. K. Kellogg Foundation 2004).

For example, in biosecurity surveillance (see Chapter 6. Detect), inputs include on-the-ground practitioners (e.g. from government, industry, research, and community groups) who have appropriate skills for undertaking surveys and diagnostic testing; government and industry funding and in-kind contributions; and field and laboratory equipment, taxonomic collections, information systems, and laboratory facilities. Surveillance activities involve undertaking on-the-ground surveys of particular areas in a targeted effort or in an unplanned manner, using visual detection approaches or traps. Surveillance can also include the collection of biological specimens for further testing in diagnostic laboratories. The outputs of surveillance activities include the numbers of hectares surveyed and diagnostic tests completed. The desired outcome of surveillance is the rapid detection of incursions or outbreaks of priority pests and diseases. In the long term, the impacts of surveillance activities include maintaining healthy ecosystems and communities and a strong economy because surveillance activities can prevent or mitigate the consequences of incursions or outbreaks of pests and diseases (Schneider and Arndt 2020).

In Australia, different organisations have developed monitoring and evaluation frameworks to keep track of their programs. In the research sector, the Australian Commonwealth Scientific and Industrial Research Organisation (CSIRO) has established an impact evaluation framework for all of its flagship projects (CSIRO 2020; Morgan 2014), and Australian universities have implemented an engagement and impact framework (Gunn and Mintrom 2018). In the biosecurity sector, the Victorian State Government uses the Biosecurity Evidence Framework to monitor, evaluate, and report on the performance of biosecurity programs and services. This framework is built around an overarching logic model of how the Victorian biosecurity system aims to achieve intended outcomes, and the framework contains performance indicators for different business units (DEPI 2014).

PERFORMANCE INDICATORS

Indicators are integral to any monitoring and evaluation framework because they allow consistent tracking of program performance over time, thereby supporting policy makers' decision making. Ideally, the development of indicators is guided by a program theory to ensure that selected indicators are relevant and sensitive to the actual performance of the program. The United Nations Development Programme has outlined the following SMART criteria for the development of indicators (UNDP 2015):

- *Specific*: The indicator should capture expected changes.
- *Measurable*: Evaluators should be able to retrieve or collect data for the indicator at reasonable cost and effort.
- *Attainable*: The indicator should produce results that are realistic and achievable.
- *Relevant*: The indicator should point to the intended outputs and outcomes.
- *Time-bound and trackable*: The indicator should capture performance or progress over time at a reasonable cost and effort.

Although not based on a program theory or being part of an overall evaluation framework, the development of performance indicators for the Australian border allows inspection program managers to consistently track the performance of inspection activities over time and to compare performance among import pathways and points of first entry (Box 10.1).

A program theory can support the development of meaningful performance indicators, but government agencies may, at times, use performance indicators without an underlying concept of how

BOX 10.1. PERFORMANCE INDICATORS FOR BORDER COMPLIANCE

Quarantine inspection refers to any activities that aim to detect biosecurity risk material (BRM) in the belongings of international travellers, mail, air and sea cargo, containers, ships, and aircraft. Following a recommendation by the Australian National Audit Office (ANAO 2001), the Australian Quarantine Inspection Service began using inspection effectiveness (i.e. the rate of detection of existing contamination) as a performance indicator. Inspection effectiveness is a useful performance indicator, but it does not provide insight as to how to manage the relative risks of different pathways. To better align performance measurement with a risk-based approach to biosecurity management, the Centre of Excellence for Biosecurity Risk Analysis (CEBRA) developed a suite of performance indicators to account for pathway inspection effort, cleanliness (i.e. the proportion of non-compliant units), risk, and workload (i.e. number of units; Robinson, Cannon, and Mudford 2011; Robinson et al. 2013).

For example, for international passenger and mail pathways arriving to Australia, manual inspections are undertaken behind screening tools (e.g. manual inspections, detector dogs, and X-ray units) to determine the amount of BRM passing undetected through an intervention point (also called the leakage rate) when particular screening tools are used. Figure 10.3 shows two leakage sources: leakage from inspected units (V_I) and leakage from previously released units (V_{NI}). To obtain the leakage rate for a screening tool on a pathway, the counts of non-compliant units from both sources (NC_1 and NC_2) are added, considering the incoming volume of units and the number of endpoint inspections (also called leakage or slippage inspections).

An estimate of leakage is required to calculate the following indicators of border performance:

- Before-intervention compliance (to measure the inherent risk of the pathway).
- Post-intervention compliance (to measure overall pathway risk management performance).

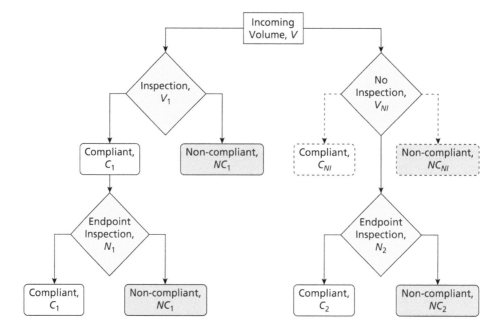

FIGURE 10.3   Endpoint inspections along an import pathway. (Adapted from Lane, Baumgartner, and Robinson 2020.)

- Non-compliance effectiveness (to measure the quality of inspection methods in detecting non-compliance).
- Hit rate (to measure the efficiency or accuracy of the profiling process).

These four performance indicators have been implemented for the international traveller and mail pathways by the Australian Government Department of Agriculture, Fisheries and Forestry (DAFF). Developing quantitative performance indicators for other activities across the biosecurity continuum may be more challenging, as there are no equivalent datasets to the interception data used to evaluate border activities (Figure 10.3). When developing high-level indicators, the inherent complexity of biosecurity systems can also confound clear attribution of observed outcomes to specific activities (see Chapter 2. Biosecurity Systems and International Regulations).

activities link to outputs and outcomes, making it difficult to understand cause-and-effect relationships (Box 10.2). Indicators that are not based on a theoretical framework may lead to evaluation outcomes that are biased or not reflective of the performance of the program. For example, Olsson et al. (2009) described how sector interests can bias the selection of indicators in a sustainable development context. To assess a policy aiming to reduce nitrogen leaching, the water management sector may suggest reducing nitrogen surplus in the soil (e.g. composting manure and slurry). However, composting can release ammonia in the air, which can be washed into the soil, leading to

BOX 10.2. UNSTRUCTURED AND OUTPUT-FOCUSED PROGRAM EVALUATION

In Australia, federal government departments report on a list of performance indicators set out in Portfolio Budget Statements and corporate plans, and performance results are published in annual reports. Most departments provide a list of performance indicators and link government programs and activities to high-level outcomes, but program outputs are usually not clearly identified and aligned with activities and outcomes (see Figure 10.2). Indicators can be arbitrary because it is not clear what they assess, and they may focus on outputs instead of outcomes. For example, the following performance indicators track outputs, not outcomes:

- Engagement with stakeholders (e.g. the number of meetings held across sectors).
- Number of partnerships developed and maintained.
- Attendance at international events.
- Number of regions where businesses or communities have received assistance.
- Number of large projects contracted.

These performance indicators do not assess the quality and effectiveness of management activities (outcomes), instead they give confirmation that planned activities are being undertaken and of the level of effort involved (outputs). If evaluation is focused on outputs without linking them to outcomes, then evaluation only measures the scope and scale of activities and whether implementation targets are being met (Figure 4.2; Schneider and Arndt 2020). Output-focused evaluation is short-sighted because it ignores the intended results that truly matter (i.e. outcomes and impacts). Only performance indicators that link activities with outcomes and impacts provide answers about whether management activities achieve their objectives.

soil acidification and subsequent release of greenhouse gas (N_2O). In this example, it is important to understand trade-offs between sectors to select indicators that cover both soil acidification and greenhouse gases (Olsson et al. 2009).

CONSISTENT METHODOLOGIES

Standardisation is key to comparing monitoring and evaluation results at different spatial scales and over time (Beard, Scott, and Adamson 1999). Standardisation prescribes that all participants, whether they be local offices, states, or countries, apply the same strictly defined data collection methods. For example, in the United Kingdom, the National Health System developed six outcome indicators to assess the effectiveness and efficiency of new care model vanguards (i.e. pioneer organisations) across the country. The evaluation strategy for this program planned to equip all local vanguards with guidance documents to ensure consistent data collection, which would then produce high quality local data that could be scaled up to the national level (NHS 2016).

In another example, an interdisciplinary research team developed a national framework for evaluating the economic impacts of climate change and demonstrated its utility in Austria (Steininger et al. 2015). The methodology expanded on sector-specific studies and combined them into a consistent modelling approach to estimate economic impacts at the national level. Steps were taken to avoid double counting and to ensure consistency, including clearly assigning and dividing impact chains among sectors, translating physical impacts into economic indicators, and assessing the costs of climate change for each sector using the same tools and methods. To ensure consistent application of the framework across sectors, researchers developed a common language and use of terms (Steininger et al. 2015).

The development of performance indicators for inspection activities at the Australian border (Box 10.1) is an example of an evaluation approach that uses a consistent methodology to compare inspection performance among inspection methods and points of entry. State and territory governments in Australia have developed their own frameworks for evaluating the overall performance of biosecurity activities, but evaluation results cannot be scaled up to the national level because the frameworks use different methods (Craik, Palmer, and Sheldrake 2017). To remedy this, DAFF and biosecurity agencies in all states and territories are jointly developing performance indicators to assess the effectiveness and efficiency of the national biosecurity system, under a national performance evaluation framework (DAWE 2020).

DATA ACCESSIBILITY

Monitoring and evaluation activities produce a lot of information. Knowledge management systems provide solutions for end users to search, retrieve, and apply relevant knowledge in their daily work (see Chapter 11. Research Uptake). The careful stewardship of monitoring and evaluation data not only helps improve agency operations; if information is available to a broad range of stakeholders, it can also strengthen relationships, increase collaboration among organisations, and increase public trust in government (Wirtz, Weyerer, and Rösch 2019).

Government agencies now report on the status of natural resources and the scale and impacts of management activities by presenting information in interactive, online formats (FFM 2021; SoE 2016), sometimes with the option to download raw data (DAWE 2021; SoE 2016). Agriculture Victoria has resolved the issue of managing performance data and other information related to biosecurity by developing a software platform called MAX. The MAX software supports performance evaluation under the Biosecurity Evidence Framework and provides data collection solutions for field surveillance (DEDJTR 2016; DEPI 2014). These are welcome developments to support transparency, reproducibility, and accountability.

Data rehearsals (whereby scenarios are developed using hypothetical data before real data are collected) can help in deciding what monitoring and evaluation data to collect, how to collect them, and how to analyse and report them (Rogers and Macfarlan 2018). The availability, quality, and accessibility of existing data can also be tested through targeted data rehearsals, such as pilot studies (Schneider and Arndt 2020).

CONTINUOUS LEARNING

Most programs are dynamic in nature, which means that they can be influenced by external and internal factors, such as changes in funding, policies, legislations, and environmental conditions (see Chapter 2. Biosecurity Systems and International Regulations). Continuous learning and review ensure that program goals and approaches remain relevant under changing conditions.

Performance assessment requires a comparison of what was achieved with what was expected (i.e. benchmarking). In the public sector, benchmarking has been used to improve the performance of higher education, employment services, and library facilities. Benchmarking in the public sector can be constrained by the lack of direct competitors against which to assess results or negative perceptions of benchmarking by managers (Magd and Curry 2003; Tillema 2010). However, performance expectations can be set by establishing a baseline over time or by consulting with stakeholders (Mayne 2004).

Evaluation frameworks (including performance indicators and program theories) themselves should be reviewed regularly and revised where necessary, as their implementation is iterative and adaptive (Peersman et al. 2016; Schneider and Arndt 2020). Managers will not get everything right the first time around, so reviewing the evaluation framework is an important step to close the loop of the planning cycle (Figure 10.1).

CULTURE OF MONITORING AND EVALUATION

Senior management supporting and promoting program evaluation is an essential part of evidence-based decision making. A culture shift has been observed in Australian government agencies towards more effective and transparent monitoring, evaluation, and reporting (Box 10.3). For example, the most recent Australian and Victorian State of the Environment reports provide information on the underlying conceptual framework and methods used as well as an indication of data quality for each dataset (CES 2018; DAWE 2021).

BOX 10.3. THE EVOLUTION OF BIOSECURITY EVALUATION IN AUSTRALIA

In the Australian biosecurity sector, the significance of performance evaluation has increased over the past 15–20 years to a point where accountability is viewed from a more holistic perspective. In 1996, a review of quarantine policies and procedures recommended that the Australian government establish performance objectives and indicators for border programs (Nairn et al. 1996). A further review in 2008 expanded on these recommendations and suggested that under a new national biosecurity authority, all program planning should include measurable performance indicators (Beale et al. 2008). The Intergovernmental Agreement on Biosecurity (an agreement between the Commonwealth and states and territories; COAG 2019) also recommended that a performance framework be developed to capture the performance of the national biosecurity system (Craik et al. 2017).

Stakeholder engagement is an integral part of developing this national evaluation framework, as its implementation requires program managers to engage and work collaboratively with their counterparts across jurisdictions. At the national level, all jurisdictions are collaboratively developing a national evaluation framework for the biosecurity system (including performance indicators) to establish a performance baseline. A recent project by CEBRA further proposed a performance evaluation framework for the national biosecurity system, including a comprehensive description of the biosecurity system, evaluative criteria, key evaluation questions, and a set of quantitative and qualitative indicators (Schneider and Arndt 2020).

However, the public sector often grapples with the "gaming" of performance results (Podger 2018) and a poor evaluation culture (Wanna 2018). The gaming of performance data compromises the quality and integrity of evaluation and can take place when there are financial rewards for better performance. For example, universities can attract students through higher rankings, and hospitals with shorter waiting lists are more appealing to patients (Podger 2018). The gaming of evaluation data can also be driven by fear of negative consequences, leading staff to report false data during data collection (Donaldson, Gooler, and Scriven 2002; Chapter 8. Incentives).

In an organisational culture that disregards the benefits of evaluation, managers and practitioners can see monitoring and evaluation activities as a burden that needs to be carried out on top of everything else. They may also consider monitoring and evaluation to be disruptive to program operations and a diversion of limited resources (Williams and Brown 2014). Many of these fears can be alleviated through appropriate resourcing. In addition, many strategies that can be used to foster a positive knowledge culture in an organisation (see Chapter 11. Research Uptake) can also be applied to improve evaluation culture.

If stakeholders are not engaged from the start, then a cascade of unwanted consequences can occur, such as a poor understanding of the evaluation process, diminished trust in findings, and low evaluator credibility (Taut and Alkin 2003). A poor evaluation culture can also result in evaluation findings not being communicated to or taken up by policy makers. This can lead to ineffective use of resources and possibly the failure of the overall program. Without continuous learning, managers re-invent the wheel and repeat the same mistakes. If performance results are unknown or not shared, it is also difficult to maintain community support for a program because public accountability responsibilities are not fulfilled (Craik et al. 2017).

The next frontier for monitoring and evaluation is to overcome these impediments and incorporate performance appraisal activities and outcomes more firmly into the entire lifecycle of biosecurity programs (Figure 10.1). Monitoring and evaluation should be considered during program design, and avenues should be defined for how evaluation results are to inform different areas of organisational decision making.

IN A NUTSHELL

- Performance monitoring, evaluation, and reporting are used to assess whether biosecurity interventions achieve their intended outcomes.
- Evaluation activities are integral to a program's planning cycle and should be considered at the outset of a biosecurity program.
- Performance appraisal relies on appropriate resourcing for monitoring and evaluation, a robust evaluation framework and program theory, and relevant performance indicators.
- A positive organisational culture that favours learning and transparency is key to evaluation results being taken up in decision making.

REFERENCES

AIHW. 2009. Towards National Indicators of Safety and Quality in Health Care. Canberra, Australia: Australian Institute of Health and Welfare.

Alexander, M. 2013. "Survey, Surveillance, Monitoring & Recording." In *Management Planning for Nature Conservation. A Theoretical Basis & Practical Guide*, 53–68. Dordrecht, The Netherlands: Springer.

ANAO. 2001. Managing for Quarantine Effectiveness: Department of Agriculture, Fisheries and Forestry, Biosecurity Australia. Audit Report No. 47, 2000-01 Performance Audit. Canberra, Australia: Australian National Audit Office.

Beale, R., J. Fairbrother, A. Inglis, and D. Trebeck. 2008. One Biosecurity. A Working Partnership. The Independent Review of Australia's Quarantine and Biosecurity Arrangements. Report to the Australian Government: Quarantine and Biosecurity Review Panel, Department of Agriculture, Fisheries and Forestry (Australia).

Beard, G. R., W. A. Scott, and J. K. Adamson. 1999. "The Value of Consistent Methodology in Long-Term Environmental Monitoring." *Environmental Monitoring and Assessment* 54: 239–258. https://doi.org/10.1023/A:1005917929050.

Bickman, L. 1987. "The Functions of Program Theory." *New Directions for Program Evaluation* 1987 (33): 5–18. https://doi.org/10.1002/ev.1443.

Bossuyt, J., L. Shaxson, and A. Datta. 2014. Study on the Uptake of Learning from EuropeAid's Strategic Evaluations into Development Policy and Practice. Final Report. European Commission.

Braithwaite, J., P. Hibbert, B. Blakely, J. Plumb, N. Hannaford, J. C. Long, and D. Marks. 2017. "Health System Frameworks and Performance Indicators in Eight Countries: A Comparative International Analysis." *SAGE Open Medicine* 5: 1–10. https://doi.org/10.1177/2050312116686516.

CES. 2018. Interim Victorian State of the Environment Report 2018. Victoria, Australia: Commissioner for Environmental Sustainability Victoria, Victoria State Government.

Chalmers, L. M., T. Ashton, and T. Tenbensel. 2017. "Measuring and Managing Health System Performance: An Update from New Zealand." *Health Policy* 121: 831–835. https://doi.org/10.1016/j.healthpol.2017.05.012.

Chouinard, J. A. 2013. "The Case for Participatory Evaluation in an Era of Accountability." *American Journal of Evaluation* 34 (2): 237–253. https://doi.org/10.1177/1098214013478142.

COAG. 2019. Intergovernmental Agreement on Biosecurity. Council of Australian Governments, Canberra.

Craik, W., D. Palmer, and R. Sheldrake. 2017. Priorities for Australia's Biosecurity System. An Independent Review of the Capacity of the National Biosecurity System and Its Underpinning Intergovernmental Agreement. Canberra: Department of Agriculture and Water Resources.

CSIRO. 2020. Impact Evaluation Guide. Canberra, Australia: Commonwealth Scientific and Industrial Research Organisation.

DAWE. 2018. Australia's State of the Forests Report 2018. Canberra, Australia: Department of Agriculture, Water and the Environment.

DAWE. 2020. Corporate Plan 2020-21. Canberra, Australia: Department of Agriculture, Water and the Environment.

DAWE. 2021. Australia State of the Environment 2021. Canberra, Australia: Department of Agriculture, Water and the Environment.

DEDJTR. 2016. Submission to the Intergovernmental Agreement on Biosecurity (IGAB) Review. Victorian Government, Department of Economic Development, Jobs, Transport and Resources.

DELWP. 2015. Monitoring, Evaluation and Reporting Framework for Bushfire Management on Public Land. Victoria, Australia: Victoria Department of Environment, Land, Water and Planning.

DEPI. 2014. Biosecurity Evidence Framework (internal Document). Victoria, Australia: Department of Environment and Primary Industries.

Dixit, S. K., and M. Sambasivan. 2018. "A Review of the Australian Healthcare System: A Policy Perspective." *SAGE Open Medicine* 6: 1–14. https://doi.org/10.1177/2050312118769211.

Donaldson, S. I., L. E. Gooler, and M. Scriven. 2002. "Strategies for Managing Evaluation Anxiety: Toward a Psychology of Program Evaluation." *American Journal of Evaluation* 23 (3): 261–273. https://doi.org/10.1177/109821400202300303.

Failing, L., R. Gregory, and M. Harstone. 2007. "Integrating Science and Local Knowledge in Environmental Risk Management: A Decision-Focused Approach." *Ecological Economics* 64 (1): 47–60. https://doi.org/10.1016/j.ecolecon.2007.03.010.

FFM. 2021. "Managing Victoria's Bushfire Risk: Fuel Management Report 2019–20, the Eight Annual Report of Victoria's Fuel Management Program on Public Land." Forest Fire Management Victoria. Victoria Department of Environment, Land, Water and Planning. https://www.ffm.vic.gov.au/fuel-management-report-2019-20.

Funnell, S. C., and P. J. Rogers. 2011. *Purposeful Program Theory: Effective Use of Theories of Change and Logic Models*. San Francisco: Jossey-Bass, A Wiley Imprint.

Gray, M., and J. R. Bray. 2019. Evaluation in the Australian Public Service: Current State of Play, Some Issues and Future Directions. Australia & New Zealand School of Government.

Gregory, R., L. Failing, M. Harstone, G. Long, T. McDaniels, and D. Ohlson. 2012. "Structuring Environmental Management Choices." In *Structured Decision Making: A Practical Guide to Environmental Management Choices*, 1–20. West Sussex, UK: John Wiley & Sons, Incorporated.

Gunn, A., and M. Mintrom. 2018. "Measuring Research Impact in Australia." *Australian Universities' Review* 60 (1): 9–15.

Hoque, Z., and D. Pearson. 2018. "Accountability Reform, Parliamentary Oversight and the Role of Performance Audit in Australia." In *Value for Money. Budget and Financial Management Reform in the People's Republic of China, Taiwan and Australia*, edited by Andrew Podger, Tsai-Tsu Su, John Wanna, Hon S. Chan, and Meili Niu, 175–374. Canberra, Australia: ANU Press.

Lane, S. E., J. B. Baumgartner, and A. P. Robinson. 2020. Evaluating the Health of Australia's Biosecurity System: Summary of Technical Report: Key Performance Indicators of "Anticipate" and "Prevent" Activities. Melbourne, Australia: Centre of Excellence for Biosecurity Risk Analysis.

Magd, H., and A. Curry. 2003. "Benchmarking: Achieving Best Value in Public-Sector Organisations." *Benchmarking: An International Journal* 10 (3): 261–286. https://doi.org/10.1108/14635770310477780.

Markiewicz, A., and I. Patrick. 2016. "Introduction to Developing and Monitoring and Evaluation Frameworks." In *Developing Monitoring and Evaluation Frameworks*, 1–27. Thousand Oaks, CA: Sage Publications.

Mayne, J. 2004. "Reporting on Outcomes: Setting Performance Expectations and Telling Performance Stories." *The Canadian Journal of Program Evaluation* 19 (1): 31–60.

Mertens, D. M., and A. T. Wilson. 2012. "Data Collection Strategies and Indicators." In *Program Evaluation Theory and Practice. A Comprehensive Guide*, edited by Donna M. Mertens and Amy T. Wilson, 353–408. New York: The Guilford Press.

Ministry of Health. 2018. Guide to Using the System Level Measures Framework for Quality Improvement. Wellington, New Zealand: New Zealand Ministry of Health.

Morgan, B. 2014. "Income for Outcome." *Nature* 511 (7510): S72–S75. https://doi.org/10.1038/511S72a.

Nairn, M. E., P. G. Allen, A. R. Inglis, and C. Tanner. 1996. Australian Quarantine: A Shared Responsibility. Canberra: Department of Primary Industries and Energy.

NHS. 2016. Evaluation Strategy for New Care Model Vanguards. England: Operational Research and Evaluation Unit, NHS England.

OECD. 2011. "Section 10: Monitoring and Evaluation." In *The OECD DAC Handbook on Security System Reform: Supporting Security and Justice*. Paris, France: OECD Publishing.

Olsson, J. A., C. Bockstaller, L. M. Stapleton, F. Ewert, R. Knapen, O. Therond, G. Geniaux, S. Bellon, T. P. Correira, N. Turpin, and I. Bezlepkina. 2009. "A Goal Oriented Indicator Framework to Support Integrated Assessment of New Policies for Agri-Environmental Systems." *Environmental Science & Policy* 12: 562–572. https://doi.org/10.1016/j.envsci.2009.01.012.

Peersman, G., P. Rogers, I. Guijt, S. Hearn, T. Pasanen, and A. L. Buffardi. 2016. When and How to Develop an Impact-Oriented Monitoring and Evaluation System. London, UK: A Methods Lab Publication. Overseas Development Institute.

Podger, A. 2018. "Making 'accountability for results' Really Work?" In *Value for Money. Budget and Financial Management Reform in the People's Republic of China, Taiwan and Australia*, edited by Andrew Podger, Tsai-Tsu Su, John Wanna, Hon S. Chan, and Meili Niu, 95–125. Canberra, Australia: ANU Press.

Podger, A., T.-t. Su, J. Wanna, M. Niu, and H. S Chan. 2018. "How Political Institutions, History and Experience Affect Government Budgeting Processes and Ways of Achieving 'Value for Money." In *Value for Money. Budget and Financial Management Reform in the People's Republic of China, Taiwan and Australia*, edited by Andrew Podger, Tsai-Tsu Su, John Wanna, Hon S. Chan, and Meili Niu, 1–16. Canberra, Australia: ANU Press.

Roberts, D., N. Khattri, and A. Wessal. 2011. Writing Terms of Reference for an Evaluation: A How-to Guide. Washington, DC: World Bank, Independent Evaluation Group.

Robinson, A., R. Cannon, and R. Mudford. 2011. DAFF Biosecurity Quarantine Operations Risk Return – Study I: Performance Indicators. Australian Centre of Excellence for Risk Analysis.

Robinson, A., R. Mudford, K. Quan, P. Sorbello, and M. Chisholm. 2013. Adoption of Meaningful Performance Indicators for Quarantine Inspection Performance. Australian Centre of Excellence for Risk Analysis.

Rogers, P. J., and A. Macfarlan. 2018. Innovations in Evaluation: How to Choose, Develop and Support Them. Briefing Paper from Joint UNICEF-BetterEvaluation – EVALSDGs Webinar Held May 2018.

Ryan, P., D. 2019. H.R.4174 - 115th Congress (2017-2018): Foundations for Evidence-Based Policymaking Act of 2018.

Schneider, K., and E. Arndt. 2020. Evaluating the Health of Australia's Biosecurity System. Final Report for CEBRA Project 170714. Melbourne, Australia: Centre of Excellence for Biosecurity Risk Analysis.

Steininger, K. W., M. König, B. Bednar-Friedl, and H. Formayer. 2015. "Climate Impact Evaluation at the National Level: The Interdisciplinary Consistent Framework." In *Economic Evaluation of Climate Change Impacts: Development of a Cross-Sectoral Framework and Results for Austria*, edited by Karl W. Steininger, Martin König, Birgit Bednar-Friedl, Lukas Kranzl, Wolfgang Loibl, and Franz Prettenthaler, 45–54. Cham, Switzerland: Springer International Publishing.

Taut, S. M., and M. C. Alkin. 2003. "Program Staff Perceptions of Barriers to Evaluation Implementation." *American Journal of Evaluation* 24 (2): 213–226. https://doi.org/10.1016/S1098-2140(03)00028-6.

Thompson, M. P., D. G. MacGregor, C. J. Dunn, D. E. Calkin, and J. Phipps. 2018. "Rethinking the Wildland Fire Management System." *Journal of Forestry* 116 (4): 382–390. https://doi.org/10.1093/jofore/fvy020.

Tillema, S. 2010. "Public Sector Benchmarking and Performance Improvement: What Is the Link and can It Be Improved?" *Public Money & Management* 30 (1): 69–75. https://doi.org/10.1080/09540960903492414.

UNDP. 2015. Why, What and How to Measure? A User's Guide to Measuring Rule of Law, Justice and Security Programmes. United Nations Development Programme.

Vogel, I. 2012. Review of the Use of 'Theory of Change' in International Development. Draft Review Report and Practical Resource. UK: Department of International Development.

W. K. Kellogg Foundation. 2004. *Logic Model Development Guide*. Michigan: W. K. Kellogg Foundation.

Walters, C. 1997. "Challenges in Adaptive Management of Riparian and Coastal Ecosystems." *Conservation Ecology* 1 (2): 1.

Wanna, J. 2018. "Government Budgeting and the Quest for Value-for-Money Outcomes in Australia." In *Value for Money. Budget and Financial Management Reform in the People's Republic on China, Taiwan and Australia*, edited by Andrew Podger, Tsai-Tsu Su, John Wanna, Hon S. Chan, and Meili Niu, 17–42. Canberra, Australia: ANU Press.

Williams, B. K., and E. D. Brown. 2014. "Adaptive Management: From More Talk to Real Action." *Environmental Management* 53: 465–479. https://doi.org/10.1007/s00267-013-0205-7.

Wirtz, B. W., J. C. Weyerer, and M. Rösch. 2019. "Open Government and Citizen Participation: An Empirical Analysis of Citizen Expectancy Towards Open Government Data." *International Review of Administrative Sciences* 85 (3): 566–586. https://doi.org/10.1177/0020852317719996.

11 Research Uptake
Improving Knowledge Management in Biosecurity Agencies

Susan M. Hester, Edith Arndt, Les Kneebone,
Lucie M. Bland, and Jocelyn Cranefield

ABSTRACT

For agencies managing biosecurity risks, failure to find, obtain, and exploit relevant research findings can result in negative impacts, such as exotic pests and diseases entering a country and significantly damaging the environment, economy, and social amenities. Repeated failures to take up relevant research outcomes represent missed opportunities for innovation, as benefits from improved efficiency in decision making do not accrue over time. Impediments to the smooth flow of knowledge include a culture that doesn't value knowledge and learning (e.g. where organisational silos develop or where decision making is driven by political agendas). Tools are available to assess an organisation's 'knowledge culture' and many solutions are available to improve the flow of knowledge within systems. In this chapter, we discuss managing and communicating knowledge in the context of agencies that are charged with managing biosecurity risk, so that opportunities provided by a healthy knowledge culture can be fully exploited. We provide a knowledge audit framework and an example of its use in the biosecurity context to show how a biosecurity agency might assess and improve its culture around the use and uptake of knowledge.

GLOSSARY

Knowledge Information that has been internally validated by individuals or a group by relating it to beliefs, ideas, procedures, observations, expectations, or methods (Alavi and Leidner 2001; Newell 2015; Wiig 1993). Knowledge is transformed into information when it is expressed as text, graphics, or other symbolic forms (Alavi and Leidner 2001).

Knowledge management The set of processes an organisation uses to create, store, disseminate, and apply information assets. The knowledge management discipline aims to improve organisational performance by identifying, capturing, and leveraging knowledge so people can use knowledge effectively, when and where it is needed (Alavi and Leidner 2001).

Knowledge culture Describes how an organisation relates to knowledge. A positive knowledge culture understands and values knowledge management, whereas a negative culture devalues it. Knowledge culture is a component of organisational culture, besides cultural values, power structures, control systems, and trust (Baskerville and Dulipovici 2006).

DOI: 10.1201/9781003253204-15

Silos Describe the physical, structural, or mental divisions that occur in different parts of an organisation (de Waal et al. 2019).

Knowledge audit A methodical examination and review of knowledge assets and systems within an organisation.

Biosecurity risk material Living organisms exotic to a country, or biological material carrying an exotic pest or disease. Biosecurity risk material includes live animals and plants, animal and plant material (e.g. fur, feathers, pollen, and flowers), food scraps, and soil.

INTRODUCTION

Biosecurity agencies often commission research to inform decision making and improve the effectiveness of biosecurity policies and practices. In many cases, the uptake of research outcomes leads to measurable improvements in the efficiency of biosecurity activities. For example, the Centre of Excellence for Biosecurity Risk Analysis (CEBRA) in Australia devised a new border inspection algorithm, based on methods originally developed for quality control (Box 11.1). The implementation of the algorithm under a rewards-based inspection approach, called the Compliance-Based Intervention Scheme (CBIS), has allowed the national regulator to reduce border inspections on specific plant product pathways without significantly increasing biosecurity risk (Rossiter et al. 2019; Rossiter and Hester 2017).

BOX 11.1 UPTAKE OF BIOSECURITY RESEARCH IN PRACTICE: THE COMPLIANCE-BASED INTERVENTION SCHEME

In 2008, an influential review of Australia's quarantine and biosecurity arrangements encouraged the adoption of a risk-based approach to biosecurity regulation (Beale et al. 2008). In practice, a risk-based approach implies focussing border activities on pathways in which consignments have a higher likelihood of containing biosecurity risk material rather than treating all consignments as posing the same risk (see Chapter 5. Screen). To fill this research gap, CEBRA investigated the use of a continuous sampling plan, an algorithm originally developed for quality control. In a biosecurity context, a continuous sampling plan can help determine whether to inspect a consignment based on the recent inspection history of the pathway and parameters set by the pathway manager, thereby rewarding consistently compliant importers with reduced inspection rates (Dodge 1943; Dodge and Torrey 1951). Key parameters are the clearance number (CN; the number of successive consignments that must pass inspection for the importer to be eligible for a reduced inspection frequency) and the monitoring fraction (MF; the reduced inspection frequency and probability that a given consignment is inspected in 'monitoring mode'; see Chapter 4. Prevent).

Following a recommendation by CEBRA, the Australian Government Department of Agriculture, Fisheries and Forestry (DAFF) implemented the resulting CBIS in 2013. Implementation was initially via three-month pilot trial on the green coffee beans pathway (Andrew Robinson, personal communication) with three plant product pathways subsequently added to the trial (Figure 11.1).

Various pieces of research were subsequently commissioned by DAFF to improve the rollout of CBIS. Research into the incentive properties of CBIS (2014) and a subsequent field trial (2017) recommended supplying regular feedback reports to importers on their compliance history. DAFF also found that setting parameter values was problematic, slowing the rollout.

FIGURE 11.1 Timeline of the development and rollout of the CBIS by the Department of Agriculture, Fisheries and Forestry (DAFF). *Abbreviations*: CEBRA, Centre of Excellence for Biosecurity Risk Analysis; CSP, continuous sampling plan.

In 2020, CEBRA research and a subsequent tool (CBIS Sensitivity Module; Chapter 4. Prevent) demonstrated how information on a commodity's supply chain (e.g., production method, treatment activities, and producing country) could help set CBIS parameters.

While these research projects were commissioned by DAFF and co-produced with agency staff, there has not yet been the anticipated widespread uptake of recommendations. Feedback reports are still being trialled by DAFF and have not been incorporated into business as usual and a lack of knowledge on how to set CBIS parameters remains (Hester et al. 2020). Departmental IT systems have required updating to enable the rollout of CBIS.

Nevertheless, the implementation of CBIS continues, with 68 low-risk plant-based products (out of more than 360 import cases) and one animal product now being subject to CBIS (DAFF 2022) with a per-consignment compliance cost saving of around $400 per inspection (Rossiter et al. 2019).

While agencies acknowledge the importance of making decisions based on evidence, this does not always occur in practice. The lack of uptake of policy-relevant information is a problem across many public sectors (Daviter 2015; Hyder et al. 2010; Walt 1994), including biosecurity (Sutcliffe et al. 2018). A recent report for New Zealand's national biosecurity agency, the Ministry for Primary Industries (MPI) found that despite more than five years of collaboration with CEBRA, knowledge of CEBRA research outputs was much lower than expected and could be improved (Hester and Reed 2021). Several reviews of the Australian biosecurity system also revealed issues with implementing research recommendations and potential flow-on effects on management (ANAO 2021). When DAFF ceased using a methodology developed by CEBRA to assess how mail and traveller detection activities were targeted, it left Australia with no knowledge of the effectiveness of its targeting and thus vulnerable to changes in riskiness of these pathways (ANAO 2021).

Failure to take up relevant research findings not only constitutes a waste of taxpayer dollars that fund the research but also a possible misallocation of resources into inefficient biosecurity activities (van der Graaf et al. 2021). Disregarding research findings can have real and tangible impacts, such as exotic pests and diseases entering and establishing in a country and significantly damaging the environment, economy, and social amenity. Failures to take up relevant research outcomes represent missed opportunities for innovation and continuous learning, as benefits from improved decision making and management do not accrue over time (Newell 2015). If valid research outcomes are not adopted or become invisible due to poor knowledge retainment strategies, then there is a danger that future projects may unnecessarily reinvent the wheel or repeat mistakes (Newell 2015).

Unfortunately, not all knowledge available to government agencies is automatically taken up, even when it would be of benefit to national economies to do so. Evidence shows that having the best science and policy advice does not guarantee that the processes of biosecurity agencies or stakeholders will change in any way (Hester and Reed 2021). To understand why and to provide potential solutions for improved research uptake in biosecurity, we review the knowledge management literature and summarise key concepts as they apply to biosecurity. We develop a practical audit tool that can assist biosecurity agencies and research providers in reflecting on and improving their research culture. Finally, we propose solutions to common knowledge management problems diagnosed via the audit.

KNOWLEDGE MANAGEMENT IN ORGANISATIONS

Knowledge management describes the ideas and practices involved in managing and communicating knowledge so that knowledge can be exploited to fulfil organisational objectives and take on new opportunities (Nawab et al. 2015; Scarbrough and Swan 2001; Schultze and Leidner 2002). Innovation and efficiency gains in an organisation rely on having access to, and utilising, multiple forms of knowledge. Explicit knowledge can be written down or articulated in language or other symbolic form (e.g. newsletters, conferences, refresher courses, websites, databases, manuals, and patents; Farnese et al. 2019; Newell 2015). Tacit knowledge refers to the people in an organisation and what they know. This type of knowledge is acquired through personal experience and can be difficult to articulate (Hansen, Nohria, and Tierney 1999; Nonaka 1994).

Keeping in mind the different forms of knowledge an organisation may hold, knowledge management is typically conceptualised as a set of four iterative and related processes (Table 11.1):

- *Knowledge creation and/or acquisition*: The process of developing, acquiring, and accumulating new insights, skills, products, services, and systems. Generating new knowledge or acquiring knowledge from outside sources are critical enablers of innovation. While knowledge is created by biosecurity practitioners in their day-to-day work, suppliers of

---

TABLE 11.1

The Four Knowledge Management Processes and Their Application in Biosecurity.

	Knowledge Creation and/or Acquisition	Knowledge Capture and/or Storage	Knowledge Dissemination	Knowledge Application
Definition	Discovering or acquiring new knowledge	Storing knowledge to allow its retrieval	Distributing, transferring, and sharing knowledge	Applying knowledge to generate value
Examples	• Method • Concept • Algorithm	• Document • Database • Website • Archive • Meeting minutes • Research paper	• Meeting • Conference • Conversation • Secondment • Newsletter • Journal	• Regulation • Standard • Procedure • Instruction • Task team
Biosecurity application	Developing an algorithm that rewards compliant behaviour by importers (Box 11.1)	Creating a database of research outputs	Holding workshops with biosecurity agency staff and researchers	Implementing a scheme that prescribes interventions based on the previous compliance of importers (Box 11.1)

Source: Developed based on Alavi and Leidner (2001), Baskerville and Dulipovici (2006), Nawab et al. (2015), Newell (2015), and Shin, Holden, and Schmidt (2001).

biosecurity research outputs are often universities, research institutes, and government departments. Some suppliers of knowledge may only focus on the creation of new knowledge and give little consideration to its access, use, or application.

- *Knowledge capture and/or storage*: The process of storing knowledge in explicit form to allow its retrieval and use for decision making. Databases and repositories are developed to support organisational memory and reduce access time for users. Examples of knowledge storage in biosecurity include horizon scanning tools, knowledge banks on plant pests, and compendia on invasive species (CABI 2021).

- *Knowledge dissemination*: Involves sharing, distributing, and transferring knowledge with a focus on the tools and mechanisms needed to share knowledge within and between organisational settings. For example, a biosecurity agency may need to consider how new employees can discover knowledge generated by past research projects—the availability of stored knowledge (e.g. a database) does not necessarily guarantee this knowledge will be accessed or used by interested parties. Likewise, the dissemination of tacit knowledge will not occur without social knowledge sharing.

- *Knowledge application*: The use and integration of knowledge into organisational operations to create capability or value. Integration can take place through regulations, standards, and procedures, which may require integrating knowledge from different areas. To apply knowledge, organisations need to combine 'business as usual' procedures with new knowledge. Organisations may need to 'unlearn' or stop applying knowledge when their current ways of thinking or operating become obsolete.

BARRIERS TO RESEARCH UPTAKE IN BIOSECURITY

Many factors relating to an organisation's knowledge management can impede the flow and use of knowledge. Based on the knowledge management literature, we identify four main impediments to research uptake in biosecurity. Solutions are provided in a subsequent section.

LEARNING BARRIERS

A pre-requisite for effective knowledge management is that an organisation encourages learning and provides space for individuals to reflect and internalise information (Baskerville and Dulipovici 2006). Policy makers without a technical background, who are not supported in developing their understanding of science, may be unfamiliar with scientific methodology and therefore lack the necessary skills to interpret and apply scientific results. These policy makers may struggle to assess the merit of scientific claims because of their inexperience with concepts such as measurement error, confirmation bias, sample size, uncertainty, extrapolation, correlation, and replication (Sutherland, Spiegelhalter, and Burgman 2013; also see Chapter 5. Screen and Chapter 12. Elicit). These problems are likely to be compounded when scientific results are presented in an opaque or jargon-laden way. Learning barriers might present themselves in the form of competency traps, where internal views of an agency's core strengths stifle innovation and the integration of new knowledge (Leonard-Barton 1992).

ORGANISATIONAL SILOS

Organisational silos can develop when organisations are structured into divisions or departments for the purpose of achieving efficiency (Zyngier 2008). A 2002 survey of American companies revealed that 83% of respondents reported the existence of silos (Stone 2004). Silos can also appear because of legacy structures, differing professional values, or organisational cultures that are competitive rather than collaborative (de Waal et al. 2019). While some internal structures are beneficial for organisational performance, silos can become an obstacle to knowledge sharing and collaboration (de Waal et al. 2019; Stone 2004). Getting people to work together across discipline-focused silos is a challenge.

This is made more difficult if an organisation does not value knowledge. Organisational silos (as well as other obstructions to informal and formal networks) deprive people of opportunities to learn and innovate and can lead to turf wars, power struggles, and conflicts (de Waal et al. 2019; Stone 2004). Interviewees in New Zealand's national biosecurity agency reported the presence of organisational silos, which led to difficulties in communicating research activities and outcomes, and thus reduced the ability of staff to use new knowledge generated by research projects (Hester and Reed 2021).

BIASES

Many different types of bias can inhibit the use of knowledge (e.g. motivational or confirmation biases; see Chapter 12. Elicit). Of particular importance is the "not-invented-here" bias, which occurs when knowledge and ideas from external sources are viewed negatively by default (Antons and Piller 2015). The not-invented-here bias leads to external knowledge being devalued or rejected, even when it is potentially useful to the organisation. For example, a study of NASA's Jet Propulsion Laboratory showed that organisational hubris, and the organisation's belief it was world class in all aspects of its operations, led to reinvention of a knowledge base that could have been obtained by collaborating with external groups (Majchrzak, Neece, and Cooper 2001).

POLITICAL DECISION MAKING

Many government policies are politically driven rather than based on the best evidence available (Bray, Gray, and Stanton 2020). Government decision making is constrained by past choices, established programs and their funding, and ongoing risk avoidance (Matheson 1998). Policy makers may choose to ignore knowledge that contradicts their aspirations or that draws criticism. If decision makers exclude knowledge (or evaluation results) for political reasons, management decisions may lead to sub-optimal outcomes (see Chapter 10. Monitoring, Evaluation, and Reporting).

OPPORTUNITIES TO IMPROVE RESEARCH UPTAKE IN BIOSECURITY

The barriers to research uptake identified above can be remediated via a range of activities that emphasise research leadership and buy-in, many of which can also contribute to a strong monitoring and evaluation culture (see Chapter 10. Monitoring, Evaluation, and Reporting).

ESTABLISH LEADERSHIP SUPPORT AND CLEAR GOVERNANCE

Top-level support and leadership are critical to improving knowledge management (Burstein et al. 2010; Cvitanovic, McDonald, and Hobday 2016). In this context, leadership is the willingness of senior decision makers to support organisational learning and implement a knowledge management strategy. For example, to improve knowledge culture in MPI, Hester and Reed (2021) suggested appointing an executive sponsor of biosecurity research and identifying knowledge brokers to improve research culture (Box 11.2).

APPOINT KNOWLEDGE BROKERS TO BRIDGE SILOS

Knowledge brokers focus on sharing knowledge and creating connections between researchers and the intended end users of the research (Meyer 2010). Given the size of most biosecurity agencies and the likely presence of silos, identifying existing knowledge brokers who can cut across organisational boundaries (such as team or department silos) may be a useful first step in improving research uptake. Knowledge brokers can be identified through social network analysis and tasked with improving and rewarding knowledge sharing and application (Newell 2015).

BOX 11.2 FOSTERING KNOWLEDGE CULTURE IN A BIOSECURITY AGENCY

One of the few published examples of internal knowledge management practices in biosecurity is an investigation into knowledge transfer between the MPI (the national biosecurity agency) and one of its key suppliers of research, CEBRA (Hester and Reed 2021). The investigation was prompted due to the perception that the ministry had not yet fully capitalised on existing research outputs from CEBRA, despite a formal collaboration having been in place for seven years.

To understand knowledge management practices in MPI and assess awareness of CEBRA research outputs, staff interviews and a qualitative analysis were conducted. Staff responses showed organisational silos had developed in the ministry, leading to issues with communicating research activities and outcomes. Accessing the ministry's internal database of research outputs was also difficult. Staff interviews were used to inform the design and evaluation of a new CEBRA research database, with the hope that improved accessibility would result in increased research uptake by ministry staff. However, key findings from the project showed that the CEBRA research database would not in and of itself lead to increased knowledge reuse; improved visibility of CEBRA staff members within the ministry would also be key to improving research uptake.

The review concluded that the ministry's knowledge management capability would determine the benefits obtained from investing in CEBRA's research. Interviewee responses indicated emerging themes around research culture that were likely to hinder staff when undertaking or interpreting research, including competing time pressures, a perceived lack of support from 'the top' for research activities (e.g., conferences), and issues with staff turnover. Specific recommendations for improving the ministry's research and knowledge culture included:

- Appointing new or existing knowledge brokers to focus on improving values and behaviours around research, with a view to promoting the ministry's capacity for creating, sharing, and using knowledge and maintaining links with researchers.
- Promoting a positive knowledge culture, including by appointing an executive sponsor with an overall responsibility for improving knowledge culture in the ministry, developing 'internet yellow pages', and setting up social initiatives to foster knowledge sharing within the ministry and between the ministry and CEBRA.

Foster Relationships and Social Networks

Many factors influence research uptake in organisations, including workplace culture, staff motivation, timeframes, personal relationships, and communication (Arndt et al. 2020; Gibbons et al. 2008). Evidence from the literature suggests that people, including scientists, do not evaluate knowledge solely based on objective criteria (such as quality and content) but also with subjective criteria, including attitudes and social relationships (Burcharth and Fosfuri 2014). If knowledge producers are not part of a biosecurity agency's social network, it is unlikely their work will have the necessary visibility for uptake to occur, even if that work is of high quality (Hester and Reed 2021). The following actions can improve the visibility of knowledge producers and the uptake of their research:

- Regular in-person or online meetings.
- Regular news items about current, past, and upcoming projects.
- Exit seminars to present project outcomes.
- Opportunities for agency staff to share and showcase the re-use of knowledge (e.g. seminars, knowledge fairs, mentoring, and secondments).

Adopt a Co-production Model

Knowledge co-production refers to a collaborative approach where the development and delivery of research are shared between policy makers and scientists (Arndt et al. 2020; Burgman 2015) with the aim of creating tangible benefits for each partner (Cooke et al. 2021). In a survey of Canadian scientists and policy makers, respondents considered collaborative approaches as the ideal way of conducting evidence-informed policy making in the future, marking a shift from knowledge dissemination towards knowledge generation (Choi et al. 2016). Australia's national biosecurity agency (DAFF) and the University of Melbourne have successfully used a shared governance model for solving biosecurity problems with the creation of CEBRA (Box 11.3). The shared governance model is based on close interactions between scientists and policy makers, a configuration that enhances the likelihood of research uptake because decision makers develop a sense of ownership of the research (Burgman 2015; Cvitanovic et al. 2016). The co-production of research has a range of benefits for policy makers and scientists when requirements for the implementation of the working model are met (Burgman 2015). For CEBRA and DAFF, mutual respect and trust have been key to the success of this unique working arrangement in the biosecurity research sector (Arndt et al. 2020).

Stimulate Demand for Knowledge

A commonly suggested approach to counteract distrust of science is to improve communication—the so-called "knowledge deficit" model, where better explanations lead to better understanding and greater acceptance of scientific arguments. However, this view is limited and tends to fail in affecting peoples' judgements (Toomey, Knight, and Barlow 2017). Many databases created for the purpose of knowledge management in fact have supply-demand issues, where knowledge supply

BOX 11.3 CO-PRODUCING BIOSECURITY RESEARCH

To solve biosecurity management problems, Australia's national biosecurity agency (DAFF) and the University of Melbourne have successfully used a shared governance model since the inception of CEBRA in 2006.

Under this co-production model, biosecurity policy makers and scientists jointly identify research topics and develop proposals, which requires both partners to think about research uptake from the outset, given the aim is to fund projects that demonstrate tangible management outcomes. The prioritisation of proposals is guided by the available budget and a Steering Committee consisting of senior members of the Australian and New Zealand governments (Arndt et al. 2020). A Scientific Review Panel takes part in a peer-review process to ensure that proposals and outputs meet academic standards, thus increasing the likelihood that government end-users will adopt research outcomes into operational processes and policy decisions (Arndt et al. 2020).

While these governance arrangements have supported research uptake, interpersonal relationships within the co-production model have had the most profound impact. The close relationship between scientists and policy makers increases knowledge transfer, in that participants form their own social networks where knowledge is disseminated and shared to achieve project objectives. Because knowledge is shared more easily and funding arrangements exist for the longer term, the CEBRA/DAFF co-production model has led to short projects extending into multiple-year projects. This more in-depth pursuit of biosecurity solutions has allowed government and university teams not only to achieve policy impact but also to develop a deeper understanding of each other's working environments (Arndt et al. 2020).

exceeds demand (Newell 2015; Scarbrough and Swan 2001). The presence of a database or repository cannot in and of itself lead to an increase in knowledge use—even a well-designed repository is unlikely to stimulate knowledge demand and sharing, unless it is accompanied by valuable social interactions and appropriate incentives for its use (Bergquist, Ljungberg, and Lundh-Snis 2001; Newell 2015).

Sharing knowledge through social networks and collaborative peer-review can help legitimise knowledge and increase its use (Bergquist et al. 2001; Liu, Ray, and Whinston 2010). Demand for knowledge can also be stimulated by improving an organisation's capability, in that a workforce is more likely to use available knowledge if they understand it and know how to apply it (Burcharth, Knudsen, and Søndergaard 2014; Ji et al. 2012). Technical capability can be increased through technical briefings, courses, secondments, or by appointing knowledge brokers (Hyder et al. 2010). Knowledge brokers can further stimulate demand for knowledge as they facilitate interactions between researchers and policy makers (Markus 2001; Rajić, Young, and McEwen 2013).

CREATE INCENTIVES AND OPPORTUNITIES FOR STAFF

Changes in staff behaviours can be facilitated by incentives such as promotions, bonuses, pay rises, and official recognitions, with the intent to increase the uptake of work practices such as knowledge sharing (Bartol and Srivastava 2002; Burcharth and Fosfuri 2014; Ji et al. 2012; Lyu and Zhang 2017; Newell 2015). In the public sector, the intention to share knowledge can be high, suggesting that knowledge sharing may be less constrained by individual motivation than by available opportunities or existing skill levels, perhaps calling for management practices that enhance opportunities instead of personal incentives (Fischer 2022). Incentivising engagement between researchers and policy makers may help improve research uptake into policy (Hyder et al. 2010).

COMMUNICATE RESEARCH RESULTS

Research is unlikely to be translated into action when the communication needs of end-users (e.g. policy makers) are not met (Hennink and Stephenson 2005). In effect, poor research communication fails to link knowledge supply to demand. Research findings that are not worded appropriately to fit policy makers' needs (e.g. when reports are too technical or too detailed) may be stored in a knowledge repository but are unlikely to be shared among users and reach relevant decision makers. Trying to reach a broader audience involves making content more accessible and easier to understand for audiences with different professional backgrounds and expertise. This includes choosing the right communication channel and repackaging research as succinct blog content, opinion pieces, summaries, press releases, or policy briefs (i.e. a document that addresses a single policy issue of interest to policy makers, including potential solutions and implementation considerations; Arndt et al. 2020; Rajić et al. 2013). Effective science communication also requires high levels of trust and mutual understanding, as fostered in social networks (Holmes and Lock 2010; Schröder-Hinrichs, Hebbar, and Alamoush 2020).

DESIGN EFFECTIVE REPOSITORIES THAT SUPPORT USER TASKS

Databases and repositories may be considered as material agents within the social networks responsible for communicating research (Newell 2015). Any identified design shortfalls in research repositories should be addressed, as these systems support a range of common information seeking tasks—*user tasks*. The International Federation of Library Associations and Institutions (IFLA) Library Reference Model (LRM) defines common user tasks in the context of information systems. In the IFLA LRM (Riva, Le Boeuf, and Žumer 2018), the five common user tasks are to: find, identify, select, obtain, and explore (Box 11.4). Support for uptake of research via repository services should include

BOX 11.4 USER TASKS

Research repositories support a range of common information seeking tasks, for example, find, identify, select, obtain, and explore. These should be considered in repository design. Example initiatives include following:

- *Find*: Tagging research with synonym-augmented vocabularies, such as the Biosecurity Thesaurus (CEBRA 2022), and enhancing repository search indexes with synonym relationships. Users should be able to find research about pests, threats, controls, treatments, methodologies, impacted commodities, and environmental assets without knowing the exact terminology used in research literature.
- *Identify*: Including resource type (report, figure, presentation, etc.) and attribution (personal author and agency) metadata, so that the user may better identify a retrieved work.
- *Select*: Indicative, non-evaluative summaries, or abstracts that support selection of works from disparate contexts, such as in search result lists.
- *Obtain*: Licensing and access rights information that support use and reuse of works in new contexts.
- *Explore*: Key biosecurity concepts that form the basis for consistent, intuitive topic navigation throughout repositories, and form links between similar research works.

The IFLA LRM (Riva et al. 2018) can guide repository design, as well as design of projects and workflows that harvest and collect biosecurity research into accessible databases. Simply accounting for and obtaining all relevant research outputs from research, industry, and government institutions itself remains a challenge for improving research uptake.

consideration of these tasks in repository design (Box 11.4). Hester and Reed (2021) found that simply establishing a repository of CEBRA research was in itself insufficient for ensuring research uptake. Internal research at DAFF showed that established research databases failed to support basic tasks, where users needed to know in advance "the exact title" of a work in order to retrieve it. A mismatch between search terms and research language, or lack of repository capability to translate or predict key concepts within a query, will reduce the probability of relevant research being found. Further, repackaged or summarised extracts may accompany research, but if these are not rendered through database discovery layers, the systems fail to support selection of the most appropriate research in search result contexts.

Monitor and Evaluate

When an organisation implements a knowledge management strategy (e.g. adopts any of the solutions proposed here), the outcomes of such strategy should be reflected on and reviewed through monitoring and evaluation (see Chapter 10. Monitoring, Evaluation, and Reporting). Program evaluation provides decision makers with evidence of the status of knowledge management in their organisation, helping to establish a baseline prior to changes and develop performance benchmarks and targets. To measure progress with knowledge management activities over time, Maag et al. (2018) recommend tracking specific metrics (e.g. the number of internal meetings held to improve a process) and regularly checking in on particular aspects of knowledge management using surveys and interviews.

ASSESSING AND IMPROVING KNOWLEDGE MANAGEMENT WITH AUDITS

To shift away from inadequate knowledge management strategies, an organisation must first investigate its status quo. A knowledge audit is a formal examination and review of knowledge assets, systems, and processes in an organisation (Leung et al. 2010). The purpose of a knowledge audit is to evaluate an organisation's knowledge resources; identify strengths, weaknesses, and gaps; and use these insights to plan better-functioning knowledge management systems (Ayinde et al. 2021; Rodríguez-Elias, Rose, and Martíenz-Garcia 2010; Tiwana 2000). Audits are recommended before embarking on any new knowledge management initiatives, as without a well-designed audit, new initiatives are unlikely to succeed (Hylton 2002). A knowledge audit comprises a range of activities similar to those used in program evaluation (see Chapter 10. Monitoring, Evaluation, and Reporting). Below and in Table 11.2, we provide a summary of audit activities to investigate research uptake in a biosecurity agency.

1. PREPARING FOR A KNOWLEDGE AUDIT

An important first part of the audit is assessing an organisation's readiness to examine its knowledge processes and implement cultural changes (Choy, Lee, and Cheung 2004). An organisation that disregards learning, knowledge sharing, and performance evaluation (i.e. has a poor knowledge or evaluation culture) is unlikely to successfully conduct a knowledge audit, let alone make cultural changes on the basis of audit results. As mentioned earlier, top-level support and buy-in are critical to improving knowledge management, including through knowledge audits (Burstein et al. 2010; Cvitanovic et al. 2016). Leung et al. (2010) stressed the need for involving all people affected by an audit early in the process and informing them of audit objectives and methods, as this engagement increases the likelihood the audit will produce useful outcomes.

TABLE 11.2
Framework for Conducting a Knowledge Audit in a Biosecurity Agency

Audit Step

1. **Prepare for the knowledge audit**
    * Get top-level support and engage with staff affected by the audit.
2. **Define the purpose of the knowledge audit**
    * Identify any barriers (and potential solutions) to research uptake within a biosecurity agency.
3. **Identify knowledge assets, processes, and people**
    * Access biosecurity agency's internal repository of research projects;
    * Collate list of additional research outputs; and
    * Obtain organisational charts.
4. **Determine the 'target state' of knowledge management**
    * Define what 'good' knowledge management looks like (e.g. leadership is supportive of research uptake initiatives, awareness and use of research outputs is high, internal research is stored in an easily accessible repository, and a positive knowledge culture is in place).
5. **Select audit method(s)**
    * Semi-structured interviews of 20–30 biosecurity research staff.
    * Analyse interview transcripts with thematic analysis and, where possible, derive quantitative metrics.
    * Quantify database access logs.
6. **Implement audit**
    * Create a detailed knowledge inventory, knowledge map, summary table (see Table 11.3), and audit report.
7. **Improve knowledge management and knowledge culture**
    * Identify and take action steps to improve knowledge management and knowledge culture in the organisation (e.g., appoint executive sponsor of research, appoint knowledge broker, create research 'yellow pages', and improve database accessibility).

---

2.   Defining the Purpose of a Knowledge Audit

The purpose of a knowledge audit will relate to an organisation's knowledge management ambitions and may range from creating an inventory of knowledge assets to checking on progress with past knowledge management initiatives. Since it is not possible to manage all knowledge with equal emphasis, it is important to identify the specific knowledge processes or areas that need be managed effectively, based on the organisation's overall strategy and goals. Focal processes may relate to the uptake of research in policies and practices or to effective communication among actors in a system (see Chapter 2. Biosecurity Systems and International Regulations). A comprehensive knowledge audit will investigate existing knowledge types and assets; knowledge needs, strengths, and gaps; and the extent to which organisational culture is conducive to knowledge sharing.

3.   Identify Knowledge Assets, Processes, and People

Additional preparatory activities may entail collecting information on key knowledge assets (e.g. databases and repositories), processes (e.g. procedures and usage logs), and people (e.g. organisational charts) that are relevant to the audit's purpose. Such information can be assembled in a preliminary knowledge inventory or knowledge map, which is a graphical depiction of how knowledge is disseminated and used in an organisation.

4.   Determine a Target State (i.e. Benchmark) for Knowledge Management

Based on the key assets, processes, and people identified earlier, specific targets (or benchmarks) are defined to provide a measure of what "good" knowledge management would entail (see Chapter 10. Monitoring, Evaluation, and Reporting). For example, in a biosecurity agency, target states may involve staff being aware of and using an internal research repository, staff reporting on a positive knowledge culture, or top-level management being supportive of research uptake initiatives (Table 11.2).

5.   Select Audit Method(s)

Knowledge audits can be undertaken using a range of methods, which should be adapted to the audit's purpose, the specific items under investigation, and the organisational context. Setting up an audit methodology involves developing or collating materials (e.g., surveys, interview questions, and databases), identifying the target personnel of the audit (e.g., research staff and top-level management), and selecting a method of analysis. An audit based on interviews or surveys is likely to involve knowledge management theory, qualitative analysis, and sometimes quantitative analysis (i.e. a mixed methods approach; Jefferson et al. 2014; Patton 2015). Thematic analysis involves transcribing data and assigning 'tags' to blocks of text (i.e. coding) according to a set of emerging themes (Jackson and Bazeley 2019).

For example, Hester and Reed's (2021) investigation of knowledge management in MPI used semi-structured interviews with agency staff (Box 11.2). Interview transcripts were analysed using thematic analysis and some quantitative metrics were derived (e.g. the number of research projects recalled by staff). When looking at a social service organisation, Leung et al. (2010) complemented a survey questionnaire with in-depth interviews, direct observations, and reviews of document inventories and databases. As such, a mix of methods can produce a more comprehensive audit (Liebowitz et al. 2000).

6.   Implement Audit

An example audit that could be undertaken to investigate research uptake in a biosecurity agency is given in Table 11.3 (adapted from Hester and Reed 2021). Here, the audit is framed around the four core knowledge management processes: knowledge creation, knowledge storage, knowledge

TABLE 11.3

An Example Knowledge Audit Developed for a Biosecurity Agency

Survey Questions	Survey Answers		Potential Solutions and Opportunities
	Strengths	Barriers	
Knowledge creation and/or acquisition			
• Do you have a process for bringing new staff up-to-speed with past and current research? • Are you aware of any current or past research projects or outputs?	*"We have a formal induction process which includes research projects and research networks."* *"Our 'go-to guide' is comprehensive."*	*"Our induction is very high-level. It isn't data-specific or research-specific."* *"There aren't enough opportunities to talk to researchers."* • *Quantitative metric*: number/proportion of projects recalled is low.	Modify the completion process for research projects to include: • An email to research networks listing the new research outputs. • Implement social initiatives that better link key knowledge suppliers, brokers, and users within the organisation.
Knowledge capture and/or storage			
• How do you store and retrieve research outputs?	*"Our internal database is very useful. Not only do we routinely load one-page impact summaries onto the repository, but we also review summaries via a keyword search."* *"Our librarian is excellent at pushing relevant research our way."*	*"Have tried a lot of things over the years. Internal research is in the internal database. If you know the exact title of the work then it's ok, otherwise the database is a black hole."* *"When people leave, knowledge goes with them."*	• Focus resources on improving existing internal database capacity for information storage and retrieval. • Continue including short, plain-language summaries for project outputs.
Knowledge dissemination			
• Once a project is completed, how do you distribute outputs throughout the organisation or to a wider audience?	*"The important thing is to share how to find the information, and I think we do this well."* *"We have an excellent Research and Technical Innovation Manager. S/He helps people get support outside of our research division."*	*"To be honest, research is at the bottom of the list because there are so many other things to get their heads around."* *"Dissemination is a once-off."*	• Identify and use knowledge brokers, including those who span organisational divisions. • Develop a knowledge 'yellow pages' containing profiles of employees and documenting their technical knowledge, past projects, and research networks.

(Continued)

TABLE 11.3 (Continued)

An Example Knowledge Audit Developed for a Biosecurity Agency

Survey Questions	Survey Answers		Potential Solutions and Opportunities
	Strengths	Barriers	
Knowledge application • When you develop new projects, do you make your staff aware of relevant internal and external research? • Have you used internal or external research to develop new projects?	*"My team works closely with an external agency. We work collaboratively—they are on top of the literature."* *"Yes, after reviewing external project outputs, I am aligning my modelling with the findings."*	*"I'm struggling to keep track of the work that is going on, just because the team is pretty busy."* *"Things are happening in silos."* *"We tried to build on a methodology but came up against barriers with the data."*	• To avoid re-inventing the wheel, modify the project development process to require searches in the internal database. • Develop a method to record the use of internal and external research and record changes to biosecurity practices deriving from the research. • Implement social initiatives to connect researchers and policy makers and ensure semi-regular contact between these groups.
Knowledge culture • How could knowledge management be improved in the organisation?	*"In my team, business as usual work includes adequate time to develop new research projects."* *"I regularly present at conferences and as a result I am well connected to international experts."*	*"Top-level management doesn't care about research. Conferences are not perceived as useful by managers/the system."*	• Appoint an 'executive sponsor' who has overall responsibility for improving knowledge culture. • Identify existing knowledge brokers and/or appoint new brokers. • Establish and maintain engagements between researchers and policy makers (e.g. through conferences, internal seminars, and networks).

Source: Adapted from Hester and Reed (2021).

dissemination, and knowledge application. A fifth category covers general knowledge culture. Through quantitative and thematic analysis, survey responses can be analysed to reveal strengths, weaknesses, gaps, and opportunities to improve knowledge management.

Further audit outputs can include comprehensive knowledge inventories (e.g. asset functions, formats, creators, and users) and knowledge maps illustrating the flow of information in an organisation (e.g. through people, tasks, or documents). Knowledge maps can be used to identify sources and sinks of knowledge and reveal constraints in information flow and transfer, for example, through social network analyses. The main output of an audit is a report summarising current knowledge processes, observed problems, and proposed solutions (Rodríguez-Elias et al. 2010).

7. IMPROVE KNOWLEDGE MANAGEMENT AND KNOWLEDGE CULTURE

The audit may suggest problems with a particular aspect of knowledge management (e.g. research uptake) as well as potential solutions that may be feasible within the organisational context. Low research uptake generally reflects a lack of recognition of the value of knowledge, poor handling of knowledge within an organisation, and inadequate provisions for managing knowledge flow (Baskerville and Dulipovici 2006).

Establishing an organisational culture that understands and values knowledge management will make it easier to implement the potential solutions outlined in this chapter, and vice versa, such solutions will contribute to improving knowledge culture. A knowledge-sharing culture values trust, free sharing of information, and working closely with others (Handzic and Agahari 2004; Ribière and Sitar 2003). Organisational culture is a complex system and the product of shared and accumulated learning as a group, including the beliefs, values, and behavioural norms of its members. Culture change can be easier in emerging organisations than in more mature ones, where past changes have created established structures and values (Schein and Schein 2017). Often, organisational culture as such does not need to be changed, only elements of the culture need to be aligned, which can be achieved through training, technology changes, and purposeful leadership (Corfield and Paton 2016; Schein and Schein 2017).

IN A NUTSHELL

- Because biosecurity systems are complex, knowledge acquired in one part of the system or developed by an outside party (e.g. university or research institute) may not reach or be taken up by its intended users.
- Low research uptake generally reflects issues with an organisation's knowledge management processes (e.g. learning barriers, organisational silos, and biases).
- Effective ways to improve knowledge management include top-level leadership support, appointing knowledge brokers, improved repository design, and fostering a positive knowledge culture.
- Knowledge audits can help identify barriers to research uptake (and potential solutions) in an organisation, and thereby help accrue benefits from the research and improved decision making.

REFERENCES

Alavi, M., and D. E. Leidner. 2001. "Review: Knowledge Management and Knowledge Management Systems: Conceptual Foundations and Research Issues." *MIS Quarterly* 25 (1): 107–136. https://doi.org/10.2307/3250961.

ANAO. 2021. *Responding to Non-Compliance with Biosecurity Requirements.* Canberra: Australian National Audit Office.

Antons, D., and F. T. Piller. 2015. "Opening the Black Box of "Not Invented Here": Attitudes, Decision Biases, and Behavioral Consequences." *Academy of Management Perspectives* 29 (2): 193–217. https://doi.org/10.5465/amp.2013.0091.

Arndt, E., M. Burgman, K. Schneider, and A. Robinson. 2020. "Working With Government — Innovative Approaches to Evidence-Based Policy-Making." In *Conservation Research, Policy and Practice*, edited by William J. Sutherland, Peter N. M. Brotherton, Zoe G. Davies, Nancy Ockendon, Nathalie Pettorelli, and Juliet A. Vickery, 216–229. Cambridge, UK: Cambridge University Press.

Ayinde, L., I. O. Orekoya, Q. A. Adepeju, and A. M. Shomoye. 2021. "Knowledge Audit as an Important Tool in Organizational Management: A Review of Literature." *Business Information Review* 38 (2): 89–102. https://doi.org/10.1177/0266382120986034.

Bartol, K. M., and A. Srivastava. 2002. "Encouraging Knowledge Sharing: The Role of Organizational Reward Systems." *Journal of Leadership & Organizational Studies* 9 (1): 64–76. https://doi.org/10.1177/107179190200900105.

Baskerville, R., and A. Dulipovici. 2006. "The Theoretical Foundations of Knowledge Management." *Knowledge Management Research & Practice* 4 (2): 83–105. https://doi.org/10.1057/palgrave.kmrp.8500090.

Beale, R., J. Fairbrother, A. Inglis, and D. Trebeck. 2008. One Biosecurity. A Working Partnership. The Independent Review of Australia's Quarantine and Biosecurity Arrangements. Report to the Australian Government: Quarantine and Biosecurity Review Panel, Department of Agriculture, Fisheries and Forestry (Australia).

Bergquist, M., J. Ljungberg, and U. Lundh-Snis. 2001. "Practising Peer Review in Organizations: A Qualifier for Knowledge Dissemination and Legitimization." *Journal of Information Technology* 16 (2): 99–112. https://doi.org/10.1080/02683960110054780.

Bray, J. R., M. C. Gray, and D. I. Stanton. 2020. "Performance Management and Evaluation Meets Culture and Politics: Australia's Experience." *Asia Pacific Journal of Public Administration* 42 (4): 290–302. https://doi.org/10.1080/23276665.2020.1808491.

Burcharth, A. L. d. A., and A. Fosfuri. 2014. "Not Invented Here: How Institutionalized Socialization Practices Affect the Formation of Negative Attitudes Toward External Knowledge." *Industrial and Corporate Change* 24 (2): 281–305. https://doi.org/10.1093/icc/dtu018.

Burcharth, A. L. d. A., M. P. Knudsen, and H. A. Søndergaard. 2014. "Neither Invented Nor Shared Here: The Impact and Management of Attitudes for the Adoption of Open Innovation Practices." *Technovation* 34 (3): 149–161. https://doi.org/10.1016/j.technovation.2013.11.007.

Burgman, M. 2015. "Governance for Effective Policy-Relevant Scientific Research: The Shared Governance Model." *Asia & the Pacific Policy Studies* 2 (3): 441–451. https://doi.org/10.1002/app5.104.

Burstein, F., S. Sohal, S. Zyngier, and A. S Sohal. 2010. "Understanding of Knowledge Management Roles and Responsibilities: A Study in the Australian Context." *Knowledge Management Research & Practice* 8 (1): 76–88. https://doi.org/10.1057/kmrp.2009.18.

CABI. 2021. "Open Access Services: Making Tools and Research Available to Solve Problems in Agriculture and the Environment." Accessed 7 May 2022. https://www.cabi.org/products-and-services/cabi-open-access/.

CEBRA. 2022. "Biosecurity Thesaurus." Australian Research Data Commons. Accessed 21 July 2023. https://vocabs.ardc.edu.au/viewById/646.

Choi, B. C. K., L. Li, Y. Lu, L. R. Zhang, Y. Zhu, A. W. P. Pak, Y. Chen, and J. Little. 2016. "Bridging the Gap Between Science and Policy: An International Survey of Scientists and Policy Makers in China and Canada." *Implementation Science* 11 (1): 16. https://doi.org/10.1186/s13012-016-0377-7.

Choy, S. Y., W. B. Lee, and C. F. Cheung. 2004. "A Systematic Approach for Knowledge Audit Analysis: Integration of Knowledge Inventory, Mapping and Knowledge Flow Analysis." *Journal of Universal Computer Science* 10 (6): 674–682.

Cooke, S. J., V. M. Nguyen, J. M. Chapman, A. J. Reid, S. J. Landsman, N. Young, S. G. Hinch, S. Schott, N. E. Mandrak, and C. A. D. Semeniuk. 2021. "Knowledge Co-Production: A Pathway to Effective Fisheries Management, Conservation, and Governance." *Fisheries* 46 (2): 89–97. https://doi.org/10.1002/fsh.10512.

Corfield, A., and R. Paton. 2016. "Investigating Knowledge Management: Can KM Really Change Organisational Culture?" *Journal of Knowledge Management* 20 (1): 88–103. https://doi.org/10.1108/JKM-12-2014-0502.

Cvitanovic, C., J. McDonald, and A. J. Hobday. 2016. "From Science to Action: Principles for Undertaking Environmental Research That Enables Knowledge Exchange and Evidence-Based Decision-Making." *Journal of Environmental Management* 183: 864–874. https://doi.org/10.1016/j.jenvman.2016.09.038.

DAFF. 2022. "Compliance-Based Intervention Scheme (CBIS)." Accessed 4 August 2022. https://www.agriculture.gov.au/biosecurity-trade/import/goods/plant-products/risk-return.

Daviter, F. 2015. "The Political Use of Knowledge in the Policy Process." *Policy Sciences* 48 (4): 491–505. https://doi.org/10.1007/s11077-015-9232-y.

de Waal, A., M. Weaver, T. Day, and B. van der Heijden. 2019. "Silo-Busting: Overcoming the Greatest Threat to Organizational Performance." *Sustainability* 11 (23): 6860. https://doi.org/10.3390/su11236860.

Dodge, H. F. 1943. "A Sampling Inspection Plan for Continuous Production." *The Annals of Mathematical Statistics* 14 (3): 264–279.

Dodge, H. F., and M. N. Torrey. 1951. "Additional Continuous Sampling Inspection Plans." *Industrial Quality Control* 7 (5): 7–12.

Farnese, M. L., B. Barbieri, A. Chirumbolo, and G. Patriotta. 2019. "Managing Knowledge in Organizations: A Nonaka's SECI Model Operationalization." *Frontiers in Psychology* 10. https://doi.org/10.3389/fpsyg.2019.02730.

Fischer, C. 2022. "Incentives Can't Buy Me Knowledge: The Missing Effects of Appreciation and Aligned Performance Appraisals on Knowledge Sharing of Public Employees." *Review of Public Personnel Administration* 42 (2): 368–389. https://doi.org/10.1177/0734371X20986839.

Gibbons, P., C. Zammit, K. Youngentob, H. P. Possingham, D. B. Lindenmayer, S. Bekessy, M. Burgman, M. Colyvan, M. Considine, A. Felton, R. J. Hobbs, K. Hurley, C. McAlpine, M. A. McCarthy, J. Moore, D. Robinson, D. Salt, and B. Wintle. 2008. "Some Practical Suggestions for Improving Engagement Between Researchers and Policy-Makers in Natural Resource Management." *Ecological Management & Restoration* 9 (3): 182–186. https://doi.org/10.1111/j.1442-8903.2008.00416.x.

Handzic, M., and D. Agahari. 2004. "Knowledge Sharing Culture: A Case Study." *Journal of Information & Knowledge Management* 03 (2): 135–142. https://doi.org/10.1142/S0219649204000754.

Hansen, M. T., N. Nohria, and T. Tierney. 1999. "What's Your Strategy for Managing Knowledge?" *Harvard Business Review* 77 (2): 106–187.

Hennink, M., and R. O. B. Stephenson. 2005. "Using Research to Inform Health Policy: Barriers and Strategies in Developing Countries." *Journal of Health Communication* 10 (2): 163–180. https://doi.org/10.1080/10810730590915128.

Hester, S., and C. Reed. 2021. *CEBRA Research: Harnessing Past and New Work to Improve Uptake and Impact of Best Practice Risk Analysis Approaches in MPI Analysis.* Centre of Excellence for Biosecurity Risk Analysis. Final report for CEBRA Project 180702.

Hester, S. M., A. Rossiter, A. Robinson, J. Sibley, C. Woolcott, C. Aston, and A. Hanea. 2020. *CBIS/CSP Sensitivity: Incorporating Pre-Border Information Analysis.* Centre of Excellence for Biosecurity Risk Analysis. Final Report for CEBRA Project 170608.

Holmes, J., and J. Lock. 2010. "Generating the Evidence for Marine Fisheries Policy and Management." *Marine Policy* 34 (1): 29–35. https://doi.org/10.1016/j.marpol.2009.04.004.

Hyder, A. A., A. Corluka, P. J. Winch, A. El-Shinnawy, H. Ghassany, H. Malekafzali, M.-K. Lim, J. Mfutso-Bengo, E. Segura, and A. Ghaffar. 2010. "National Policy-Makers Speak Out: Are Researchers Giving Them What They Need?" *Health Policy and Planning* 26 (1): 73–82. https://doi.org/10.1093/heapol/czq020.

Hylton, A. 2002. *A KM Initiative Is Unlikely to Succeed Without a Knowledge Audit.* Hylton Associates.

Jackson, K., and P. Bazeley. 2019. *Qualitative Data Analysis with NVivo.* 3rd ed. London: Sage.

Jefferson, T., S. Austen, R. Sharp, R. Ong, G. Lewin, and V. Adams. 2014. "Mixed-Methods Research: What's in It for Economists?" *The Economic and Labour Relations Review* 25 (2): 290–305. https://doi.org/10.1177/1035304614530819.

Ji, L., J. Huang, Z. Liu, H. Zhu, and Z. Cai. 2012. "The Effects of Employee Training on the Relationship Between Environmental Attitude and Firms' Performance in Sustainable Development." *The International Journal of Human Resource Management* 23 (14): 2995–3008. https://doi.org/10.1080/09585192.2011.637072.

Leonard-Barton, D. 1992. "Core Capabilities and Core Rigidities: A Paradox in Managing New Product Development." *Strategic Management Journal* 13 (S1): 111–125. https://doi.org/10.1002/smj.4250131009.

Leung, Z. C. S., C. F. Cheung, K. F. Chu, Y.-c. Chan, W. B. Lee, and R. Y. W. Wong. 2010. "Assessing Knowledge Assets: Knowledge Audit of a Social Service Organization in Hong Kong." *Administration in Social Work* 34 (4): 361–383. https://doi.org/10.1080/03643107.2010.512843.

Liebowitz, J., B. Rubenstein-Montano, D. McCaw, J. Buchwalter, C. Browning, B. Newman, and K. Rebeck. 2000. "The Knowledge Audit." *Knowledge and Process Management* 7 (1): 3–10. https://doi.org/10.1002/(SICI)1099-1441(200001/03)7:1<3::AID-KPM72>3.0.CO;2-0.

Liu, D., G. Ray, and A. B. Whinston. 2010. "The Interaction Between Knowledge Codification and Knowledge-Sharing Networks." *Information Systems Research* 21 (4): 892–906. https://doi.org/10.1287/isre.1080.0217.

Lyu, H., and Z. Zhang. 2017. "Incentives for Knowledge Sharing: Impact of Organisational Culture and Information Technology." *Enterprise Information Systems* 11 (9): 1416–1435. https://doi.org/10.1080/17517575.2016.1273393.

Maag, S., T. J. Alexander, R. Kase, and S. Hoffmann. 2018. "Indicators for Measuring the Contributions of Individual Knowledge Brokers." *Environmental Science & Policy* 89: 1–9. https://doi.org/10.1016/j.envsci.2018.06.002.

Majchrzak, A., O. E. Neece, and L. P. Cooper. 2001. "Knowledge Reuse for Innovation – the Missing Focus in Knowledge Management: Results of a Case Analysis at the Jet Propulsion Laboratory." *Academy of Management Proceedings* 2001 (1): A1–A6. https://doi.org/10.5465/apbpp.2001.6133668.

Markus, L. M. 2001. "Toward a Theory of Knowledge Reuse: Types of Knowledge Reuse Situations and Factors in Reuse Success." *Journal of Management Information Systems* 18 (1): 57–93. https://doi.org/10.1080/07421222.2001.11045671.

Matheson, C. 1998. "Rationality and Decision-Making in Australian Federal Government." *Australian Journal of Political Science* 33 (1): 57–72. https://doi.org/10.1080/10361149850723.

Meyer, M. 2010. "The Rise of the Knowledge Broker." *Science Communication* 32 (1): 118–127. https://doi.org/10.1177/1075547009359797.

Nawab, S., T. Nazir, M. M. Zahid, and S. M. Fawad. 2015. "Knowledge Management, Innovation and Organizational Performance." *International Journal of Knowledge Engineering* 1 (1): 43–48. https://doi.org/10.7763/IJKE.2015.V1.7.

Newell, S. 2015. "Managing Knowledge and Managing Knowledge Work: What We Know and What the Future Holds." *Journal of Information Technology* 30 (1): 1–17. https://doi.org/10.1057/jit.2014.12.

Nonaka, I. 1994. "A Dynamic Theory of Organizational Knowledge Creation." *Organization Science* 5 (1): 14–37. https://doi.org/10.1287/orsc.5.1.14.

Patton, M. Q. 2015. *Qualitative Research & Evaluation Methods.* 4th ed. Thousand Oaks, CA: Sage Publications.

Rajić, A., I. Young, and S. A. McEwen. 2013. "Improving the Utilization of Research Knowledge in Agri-Food Public Health: A Mixed-Method Review of Knowledge Translation and Transfer." *Foodborne Pathogens and Disease* 10 (5): 397–412. https://doi.org/10.1089/fpd.2012.1349.

Ribière, V. M., and A. S. Sitar. 2003. "Critical Role of Leadership in Nurturing a Knowledge-Supporting Culture." *Knowledge Management Research & Practice* 1 (1): 39–48. https://doi.org/10.1057/palgrave.kmrp.8500004.

Riva, P., P. Le Boeuf, and M. Žumer. 2018. "IFLA Library Reference Model: A Conceptual Model for Bibliographic Information." International Federation of Library Associations and Institutions (IFLA), accessed 21 July 2023. https://repository.ifla.org/handle/123456789/40.

Rodríguez-Elias, O., C. Rose, and A. Martíenz-Garcia. 2010. "Integrating Current Practices and Information Systems in KM Initiatives – A Knowledge Management Audit Approach." International Conference on Knowledge Management and Information Sharing (KMIS 2010), Valencia, Spain, 25–28 October, 2010.

Rossiter, A., and S. M. Hester. 2017. "Designing Biosecurity Inspection Regimes to Account for Stakeholder Incentives: An Inspection Game Approach." *Economic Record* 93 (301): 277–301. https://doi.org/10.1111/1475-4932.12315.

Rossiter, A., F. Mody, J. Whyte, B. Wang, C. Brent, J. Vandenbroek, E. Miech, S. Ryan, and H. Susie. 2019. *Testing Incentive-Based Drivers for Importer Compliance. Final Report for CEBRA 1608C.* Centre of Excellence for Biosecurity Risk Analysis. https://cebra.unimelb.edu.au/__data/assets/pdf_file/0006/3464754/1608C_Final-Report_for-endorsement-20200713.pdf

Scarbrough, H., and J. Swan. 2001. "Explaining the Diffusion of Knowledge Management: The Role of Fashion." *British Journal of Management* 12 (1): 3–12. https://doi.org/10.1111/1467-8551.00182.

Schein, E. H., and P. A. Schein. 2017. *Organizational Culture and Leadership.* 5th ed. Hoboken, New Jersey: Wiley.

Schröder-Hinrichs, J.-U., A. A. Hebbar, and A. S. Alamoush. 2020. "Maritime Risk Research and Its Uptake in Policymaking: A Case Study of the Baltic Sea Region." *Journal of Marine Science and Engineering* 8 (10): 742. https://doi.org/10.3390/jmse8100742.

Schultze, U., and D. E. Leidner. 2002. "Studying Knowledge Management in Information Systems Research: Discourses and Theoretical Assumptions." *MIS Quarterly* 26 (3):213–242. https://doi.org/10.2307/4132331.

Shin, M., T. Holden, and R. A. Schmidt. 2001. "From Knowledge Theory to Management Practice: Towards an Integrated Approach." *Information Processing & Management* 37 (2): 335–355. https://doi.org/10.1016/S0306-4573(00)00031-5.

Stone, F. 2004. "Deconstructing Silos and Supporting Collaboration." *Employment Relations Today* 31 (1): 11–18. https://doi.org/10.1002/ert.20001.

Sutcliffe, C., C. H. Quinn, C. Shannon, A. Glover, and A. M. Dunn. 2018. "Exploring the Attitudes to and Uptake of Biosecurity Practices for Invasive Non-Native Species: Views Amongst Stakeholder Organisations Working in UK Natural Environments." *Biological Invasions* 20 (2): 399–411. https://doi.org/10.1007/s10530-017-1541-y.

Sutherland, W. J., D. Spiegelhalter, and M. Burgman. 2013. "Policy: Twenty Tips for Interpreting Scientific Claims." *Nature* 503 (7476): 335–337. https://doi.org/10.1038/503335a.

Tiwana, A. 2000. *The Knowledge Management Toolkit: Practical Techniques for Building a Knowledge Management System.* Upper Saddle River, NJ: Prentice Hall PTR.

Toomey, A. H., A. T. Knight, and J. Barlow. 2017. "Navigating the Space Between Research and Implementation in Conservation." *Conservation Letters* 10 (5): 619–625. https://doi.org/10.1111/conl.12315.

van der Graaf, P., M. Cheetham, S. Redgate, C. Humble, and A. Adamson. 2021. "Co-Production in Local Government: Process, Codification and Capacity Building of New Knowledge in Collective Reflection Spaces. Workshops Findings from a UK Mixed Methods Study." *Health Research Policy and Systems* 19 (1): 12. https://doi.org/10.1186/s12961-021-00677-2.

Walt, G. 1994. "How Far Does Research Influence Policy?" *European Journal of Public Health* 4 (4): 233–235. https://doi.org/10.1093/eurpub/4.4.233.

Wiig, K. M. 1993. *Knowledge Management Foundations: Thinking About Thinking.* Arlington, Texas: Schema Press.

Zyngier, S. 2008. "Risk Management: Strengthening Knowledge Management." *International Journal of Knowledge Management (IJKM)* 4 (3): 19–32. http://doi.org/10.4018/jkm.2008070102.

Section V

Innovative Methods

12 Elicit
Using Structured Elicitation in Biosecurity

*Sana Bau, Anca Hanea, Andrew P. Robinson,
and Mark Burgman*

ABSTRACT

Expert elicitation plays a critical role in risk analysis when data are unavailable and when decisions are imminent, for example when predictions about facts such as the outcomes of future events, or value judgements—preferences for an objective or choice—are required. However, these subjective judgements are often faulty due to uncertainty and the influences of overconfidence, and other cognitive and motivational biases. Techniques have been developed that obtain predictions and value judgements that anticipate and mitigate against these influences. In this chapter, we review contemporary methods used for expert elicitation in biosecurity and describe their challenges and limitations. For the challenge of deriving predictions of outcomes to use in a scientific setting, we recommend the IDEA protocol, which comprises four steps, namely: Investigate, Discuss, Estimate, and Aggregate. For eliciting value-based preferences, we recommend swing weights, a technique that supports accounting for the range of consequences that could arise from a decision. Using repeatable, systematic, and transparent methods for elicitation helps to ensure that data derived from experts are subject to the same standards of rigour as empirical or modelled data. This in turn allows for relatively accurate estimates of risk and reliable and representative preferences to be incorporated in decision making across the biosecurity continuum.

GLOSSARY

Technical prediction (also called scientific prediction or estimate) Prediction of matters of a factual or scientific nature, forecasting.

Value judgement A subjective evaluation assigning importance or preference for an objective or option in decision making.

Expert An individual with specialist knowledge in a field relevant to the topic under investigation.

Structured elicitation Systematic methods for facilitating groups to address a complex problem using structured, multiple-round group communication processes.

Linguistic uncertainty Uncertainty that arises from people using and interpreting language differently (Regan, Colyvan, and Burgman 2002).

Consensus The endorsement or acceptance of a prediction or choice by all parties involved in the elicitation.

Subjective scale A category scale that uses verbal descriptors for each category level (e.g. rating likelihoods on a scale ranging from "rare" to "almost certain").

Four-point elicitation The elicitation of quantities using four features (i.e. best guess, plausible lower bound, plausible upper bound and confidence that the true value lies within those bounds; Speirs-Bridge et al. 2010).

DOI: 10.1201/9781003253204-17

Credible interval An uncertainty interval capturing the degree of one's belief (i.e. confidence) in the elicited estimate of a numerical value (Hemming et al. 2018).

Calibration The frequency of expert interval estimates enclosing the true value relative to their reported confidence (Burgman 2015). For instance, 90% confidence intervals should capture the true value nine times out of ten if they are well calibrated.

Structured decision making (SDM) A framework for making decisions incorporating both facts and values that combines decision theory and the practice of decision analysis. A systematic, deliberative process for working sequentially through steps including defining objectives and performance measures, identifying and evaluating alternatives, and making explicit trade-offs (Gregory et al. 2012).

Objectives The interests that could be affected by policy and management decisions.

Fundamental objectives The overarching, ends objectives of a decision that are important in and of themselves. Fundamental objectives describe a set of independent, fundamentally important consequences that could be affected by a decision (Keeney 2007).

Means objectives Intermediate objectives which serve to help achieve fundamental objectives—a means to an end. Meeting means objectives is important because they are required to meet other, more fundamental objectives.

Importance The perceived significance of an outcome; how an outcome is valued.

Value function A mathematical function used in decision analysis to scale the performance of alternatives against criteria by combining measurable attributes of objectives with preference weights.

Performance measure A specific metric corresponding to a decision-making criterion intended to consistently estimate and report the anticipated consequences for a given objective under a management alternative. Performance measures can be natural (direct) measures, indicators, or constructed scales for reporting hard-to-measure consequences (Gregory et al. 2012).

INTRODUCTION

Decision making in biosecurity deals with complex and uncertain problems that require rapid responses based on the best available knowledge. It is rare that all data are available to predict what will happen under different scenarios, such as increasing surveillance efforts or implementing different response strategies to contain an incursion. In the absence of adequate data to support decision making, it may be necessary to turn to the judgement of topic specialists, such as veterinarians, biologists, researchers, land managers, consultants, people with extensive local knowledge, and others, all of whom we commonly refer to as experts, to fill critical knowledge gaps.

Data deficiencies and the uncertain nature of knowledge inputs that inform decisions (e.g. data, models, and causal reasoning) are not the only impediments to decision making. Making good biosecurity decisions is also challenging due to multiple, competing objectives in biosecurity management and differences in what people hope to achieve. Formulating decision problems involves working with values (what stakeholders deem important) as much as facts. Conventional ideas about science serving as a purely objective method of understanding the world are no longer widely accepted, because scientific judgement cannot be divorced from social and cultural values (Longino 1990). Bias and social context are inescapable in data collection and analyses; thus, facts and values often appear to be entangled.

Good decision making also relies on a clear understanding of what decision makers and stakeholders stand to gain or lose, such as avoiding or limiting economic damages or minimising the impacts of invasive species on native species. Historically, there has been a tendency for decision analysts to take charge of decision-making processes, marginalising stakeholders and thereby ignoring plausible options or important objectives (O'Brien, 2000). With the aim of countering these shortcomings, an alternative, more nuanced strategy of democratising policy-relevant science includes participatory decision making. This approach engages stakeholders from the wider community in setting research priorities and policy agendas, as well as in deliberative decision processes. However, capturing stakeholder preferences is fraught with difficulties, as value judgements are often hidden, highly personal and informed by feelings about what is important. Many of the complex problems tackled in biosecurity affect a wide range of stakeholders with competing interests, which may result in conflicting preferences.

Howard (1988) describes decision making as having three main components: *alternatives* (the set of choices), *information* (relevant models and technical judgements that predict the outcomes of alternatives), and *preferences* (value judgements relevant to the problem). This chapter introduces commonly used techniques for eliciting information (factual or technical judgements or estimates of the consequences of actions, see Chapter 3. Anticipate) and preferences (value judgements). These aspects are important considerations in biosecurity risk management but are dealt with in very different ways. In this chapter, we describe effective tools for eliciting technical estimates (facts) with wide applicability to biosecurity prediction and forecasting, and separately, elicitation methods to capture value judgements using criteria weights—measures of individual preferences for different outcomes of a decision.

ELICITING FACTS IN BIOSECURITY

In biosecurity, decisions are often made after considering the risk of future events, such as the potential entry, establishment, and spread of exotic organisms, for which direct data can never be available. Similarly, the potential effectiveness of a strategy to eradicate an incursion is often informed by limited data. Under these circumstances, the elicitation of technical judgements is crucial.

Eliciting technical judgements (also referred to as technical elicitation) includes prediction, forecasting, and intelligence gathering and deals with estimating well-defined facts (which may be verifiable or not). In technical elicitations, experts are asked to estimate facts, predict future events, or advise on likely outcomes of different courses of action (Burgman 2015). Experts are assumed to have relevant knowledge and reasoning skills that can help meet the needs of reliable estimation and prediction. Conventionally, expertise is identified by qualifications, experience, and peer recognition; however, in this chapter, we take a wider and more inclusive view of expertise, challenging traditional norms of who qualifies as an "expert". Obtaining useful technical estimates from experts relies on asking clear and specific questions about facts (e.g. "what is the contamination rate of a particular import?"). In practice, vague questions asked in informal, discussion contexts frequently make elicitations ineffective, resulting in unreliable estimates (Hanea and Robinson 2020). The quality of elicited judgements may also be limited by short study timelines and other resource constraints.

Eliciting unbiased estimates from experts is not always straightforward, as experts are prey to a host of biasing influences, such as cognitive biases, motivational influences, linguistic uncertainty, and group dynamic issues (Burgman 2015). Cognitive biases and heuristics (i.e. mental shortcuts engaged in intuitive, as opposed to deliberative reasoning) include overconfidence (excessive confidence in one's knowledge), confirmation bias (a tendency to seek and interpret information to confirm one's beliefs), the availability heuristic (basing estimates on the most easily recalled information), and anchoring (an initial, potentially irrelevant or inaccurate piece of information provides an "anchor" for numerical estimates; see Böhm and Tanner (2018) and Burgman (2004)). Motivational bias is thinking founded on perceptions of what is to be gained or lost personally and can influence experts to interpret evidence in a way that affirms their desired outcomes, even inadvertently (Burgman 2004). For example, in the 2007 trade dispute over risks of introducing fire blight via import of New Zealand apples into Australia (Fraser, Cook, and Haddock-Fraser 2019), expert advisors engaged by both Australian and New Zealand growers used scientific arguments to advance their opposing positions, giving higher or lower estimates of risk depending on which side they represented (Higgins and Dibden 2011). In group elicitations, individual experts are susceptible to unconsciously anchoring on the judgements of highly regarded or vocal individuals who dominate discussions. In-group social pressures can lead to conformity in thinking and limit consideration of alternatives, narrowing the range of information elicited through a phenomenon known as "groupthink" (Janis 1982). Groups lacking in demographic diversity have a smaller pool of cognitive resources (i.e. perspectives on the questions at hand, task-relevant knowledge and skills), limiting group performance (Steel et al. 2021).

Knowing about the limitations of expert judgement and designing elicitation techniques to counter biases is important because decision makers and the public often defer to the authority of experts and uncritically accept their conclusions (Burgman 2015). Decisions based on flawed, compromised, or otherwise biased judgements may result in the misallocation of limited biosecurity resources and may generate undesirable biosecurity outcomes. As such, a number of techniques for eliciting technical judgements have been developed and adopted for use in biosecurity contexts (reviewed in Hanea and Robinson 2020). These range from simple, unstructured elicitations through to more systematic and prescriptive tools intended to counter the common influences biasing human judgement. New innovations in elicitation methods have emerged in line with the changing needs of biosecurity agencies, with improvements in expert elicitation likely motivated by increasing demand for public accountability in government service delivery (see Chapter 10. Monitoring, Evaluation, and Reporting).

Labelling an elicitation method as unstructured or structured is challenging, because an agreed definition of a structured protocol has yet to be established. Cooke (2008) proposed that to qualify as "structured", elicited data and their uncertainty representations must meet the rigour and transparency standards of the scientific method. This implies that the elicitation questions should have clear operational meanings, the elicitation should be thoroughly documented, and provide opportunities for empirical evaluation (which can be achieved using calibration questions). Hemming et al. (2018) expand on this definition by stipulating that the processes should also pre-empt and mitigate potential judgement biases. Here, we define structured approaches as adhering to a systematic method for facilitating groups of experts to collectively address a complex problem through a formal process. We recommend observing the following key measures: (1) using *multiple* interacting experts, (2) asking questions in a *structured* way, and (3) being conducted over *multiple rounds*, so experts can revise their initial judgements in light of feedback and new information. Structured questioning uses clear questions about numerical quantities or probabilities and has minimal linguistic uncertainty and avoidable bias (see Hemming et al. 2018). Conversely, elicitation is "unstructured" when questions are vague, under-specified and with no particular order or design, or when methods to select experts, aggregate elicited judgements, or reach consensus are ad hoc.

Elicitations used in biosecurity commonly use informal or qualitative methods, or the methods are insufficiently documented to ascertain if, and how, they dealt with expert bias and other impediments to achieving reliable judgements. We consider these elicitations to be unstructured. We now discuss three different protocols for eliciting facts (technical judgements)—single-round, Delphi methods, and IDEA (an advanced Delphi protocol)—and give an example of how IDEA might be implemented in a biosecurity setting.

SINGLE-ROUND PROTOCOLS

Elicitation coordinators may choose to conduct elicitation in a single round to balance the goals of capturing expert knowledge reliably while keeping the elicitation burden manageable. Another rationale for single-round elicitation is to avoid introducing dependence between estimates of different experts from, for instance, the influence of exposure to other experts' judgements. However, single-round elicitations assume implicitly that experts share a common, comprehensive understanding of the facts at hand to be able to make well-informed, accurate estimates in one attempt. A comprehensive background document or evidence dossier (scrutinised and complemented by the experts individually) may be distributed prior to the elicitation to ensure all relevant evidence is available to the experts. Single-round elicitations do not accommodate facilitated discussion and feedback, so they typically lack the opportunity to identify and mitigate overconfidence, anchoring and adjustment, dominance effects, and other biasing influences, unless they include calibration questions intended for this purpose. Given only a single round, participants lack the opportunity to share and review each other's evidence and reasoning and update their estimates in light of new information.

DELPHI PROTOCOLS

In response to the shortcomings of single-round elicitations, more sophisticated expert elicitation techniques have been developed to limit error-prone subjective judgements and prevent group judgements from anchoring on dominant individuals. The Delphi method is a structured elicitation method which uses a systematic, collaborative forecasting process to arrive at a group consensus based on the results of two or more rounds of questioning (Mukherjee et al. 2015). After each round, an anonymised summary of the responses from the previous round is provided to the expert panel alongside any reasoning for the judgements. Experts can comment on the responses of others and revise their answers in light of the summary feedback provided by the facilitator. The anonymity of responses helps to avoid the influence of dominant individuals.

Delphi elicitation methods are best suited for estimating well-defined, measurable, and verifiable (but unknown) facts, usually in the form of point estimates. The goal of this iterative process is for the group to converge towards the "true" value of variables that the elicitation aims to predict. An unknown fact may exist in the present (e.g., the number of import shipments contaminated with brown marmorated stink bug in the last month) or be realised in the future (e.g., whether cane toads will establish in Sydney in the next 10 years). Both have an objective truth that can be revealed with perfect detection and/or by waiting long enough to find out. Typically, neither perfect detection nor long decision horizons hold, and most real-world and present-day biosecurity decisions need to be made quickly with imperfect information. This is where structured elicitation methods based on the Delphi technique can help.

A number of attempts at Delphi-type elicitation have been documented in the biosecurity literature, which we review as follows. Kuster et al. (2015) surveyed experts to assess the effectiveness and importance of biosecurity measures on livestock farms by asking experts to rank and rate each measure on a six-point scale. Sawford et al. (2014) applied a similar approach to gather information about Hendra virus, where they elicited ratings on the importance of a range of topics on a five-point scale from a broad selection of "stakeholders" that included scientific experts, vets, wildlife health managers, and representatives from government and the equine industry. All topics in the questionnaire contained a mix of factual questions (e.g. on Hendra virus surveillance and reporting) and value-based questions (e.g. attitudes to vaccination in horses). In both Kuster et al. (2015) and Sawford et al. (2014), the main attribute that was elicited ("importance") is vaguely defined (Regan et al. 2002) and the question is a matter of value judgements rather than facts.

Delphi methods have broad applicability to quantitative elicitations. A Delphi-type approach was used in a study that aimed to predict habitat suitability for feral pigs in northern Australia using a Bayesian network—a probabilistic model that represents dependent relationships between variables using Bayes theorem for quantitative prediction (Froese et al. 2017). Experts were involved in two stages of the study: developing a preliminary model and later, conducting a more detailed review of the model, using a subset of the original group. The initial round of model development was guided by 18 experts who each constructed a conceptual model, defined a set of variables used to predict habitat suitability, and parameterised model states and the relationships between variables. Different steps of the elicitation were performed in break-out groups or individually, although details of the method of elicitation at different steps were omitted. Group consensus was formed from a panel discussion following the elicitation. In the second stage, six experts from the original panel were interviewed to provide judgements about the effects of environmental variables on habitat suitability to inform the conditional probability tables used for inference. The conditional probability tables (the quantitative part of Bayesian networks which specify underlying probabilistic relationships) produced during the initial elicitation were provided and the experts were asked to review and revise them based on their own knowledge. In the study, the authors did not address if and how they dealt with biases arising from group dynamics and individual judgement. They do not address the potential for their elicitation methods to reflect judgements anchored on those of dominant individuals, nor do they acknowledge how to ensure that the subset of experts in the second round provided useful judgements independently from of the values elicited by the initial group.

Anonymity and iteration are essential for Delphi methods to be effective, and Delphi methods that do not adhere to these principles are likely to suffer from the dominance effects that afflict unstructured elicitation methods (Burgman 2015). Other problems that can undermine the quality of estimates are limited opportunities for experts to interact, ignoring or using inappropriate ways to represent uncertainty, and trying to force agreement. Encouraging uniformity in reasoning and in the hope of resolving differences may backfire, leading to hardening of one's position and further polarisation (Lewandowsky et al. 2012).

Delphi elicitations may strive to converge on the "true answer" (where it is verifiable) by taking an average of the group estimates on the assumption that it will provide a reliable and unbiased summary. One interpretation of divergent answers is that differences of opinion represent uncertainty around the truth and extreme values are considered outliers (Hemming et al. 2018). However, they can reflect valid differences of opinion or conceptual understanding, such as different, plausible causal explanations for observed phenomena (Burgman 2015; Hemming et al. 2018). For example, land managers may possess local knowledge about landscape characteristics unknown to ecologists who rely on inferences from spatial data and dispersal rate assumptions to estimate the spread rate of a pest (see Southwell et al. 2017). Simply aggregating estimates across experts will not accommodate differences in causal understanding; however, knowledge shared via feedback and discussion can inform estimates of subsequent rounds.

The IDEA Protocol

Guidelines for Implementing the IDEA Protocol

The IDEA protocol builds on Delphi techniques and provides a structured approach to expert elicitation that has demonstrated success in generating reliable predictions in forecasting (Hanea et al. 2017; Hemming et al. 2018). The IDEA acronym stands for the key steps of the protocol: "Investigate", "Discuss", "Estimate," and "Aggregate". The "Investigate" step includes a first round of estimates (Round 1), and the step "Estimate" consists of the second Delphi round (Round 2). In biosecurity contexts, IDEA can provide the means for quantitative prediction of an event, such as the probability of a foot and mouth disease outbreak in Australia within the next five years. The protocol is flexible and adaptable to different elicitation formats (namely face-to-face meetings, remote workshops, or via other online communication platforms).

Successful implementation of the IDEA protocol requires detailed planning and preparation. Coordinators develop a timeline of tasks and dates for each step of the elicitation and form a project team to manage the elicitation, collect responses, and analyse data. To benefit from the advantages of diverse groups (Page 2007), expert selection aims to include individuals with diverse backgrounds, types of expertise, and scientific and social perspectives. A crucial part of planning an elicitation is designing a set of clear, quantitative questions for which experts can estimate uncertain quantities or probabilities.

The following steps outline the guidelines for using the IDEA protocol.

Investigate: Prior to the elicitation of judgements, experts are asked to clarify the meaning of key terms to minimise linguistic uncertainty and arrive at a common understanding of relevant concepts. To capture uncertainty around a point estimate, for each question for Round 1 of the elicitation, experts may provide private, individual interval estimates (their best estimate; a plausible lower bound; and a plausible upper bound).

Discuss: Feedback is provided to the experts on their estimates and credible intervals together with those of other experts. The feedback can be presented in the form of individual estimates alone, or accompanied by a combined group estimate to give experts the opportunity to see how their estimate compares to the group's. The experts are encouraged to discuss the results from Round 1, resolve different interpretations of the questions, cross-examine one another's reasoning, and share evidence that can be used to inform Round 2 estimates. Contrary to the Delphi method, the discussion may be synchronous and importantly, it is not intended to result in consensus but to resolve residual linguistic uncertainty and encourage critical thinking and sharing of evidence.

Estimate: A second and final confidential round of estimation is undertaken in which experts can revise their interval estimates in light of the feedback and discussion, each providing anonymous judgements.

Aggregate: Final round estimates are standardised and aggregated to produce a combined group interval estimate (see Hemming et al. 2018 for methods). With sharing of information between iteration rounds, estimates tend to improve in both accuracy and range, giving more precise, calibrated, and informative estimates (Hanea et al. 2018). The simplest and most commonly used method for aggregating intervals is to take the arithmetic mean of estimates.

Below, we describe an application of the IDEA protocol in a biosecurity management decision.

Using the IDEA Protocol to Estimate Confidence in the Early Detection of Agricultural Pests and Diseases in Australia

Routine visual inspections of crops are undertaken in the Australian grains industry by agronomists via Crop Safe, a "general surveillance" network of 180 agronomists whose observations of grain crops provide early warning of incursions of a number of exotic grain pests and diseases (Arndt et al. 2022). However, agricultural pests and diseases can be hard to detect at low densities in the early stages of incursion, causing uncertainty around the ability of agronomists to detect pests early enough to effectively manage outbreaks. If issues around early detection could be resolved, the information gathered under the Crop Safe program could also potentially be used to inform area freedom claims (Hammond et al. 2016) at a lower cost than targeted surveillance. To use the programme information in this way, however, requires knowledge about the sensitivity of the surveillance system—the probability of detecting a pest or disease if it is present (see Chapter 6. Detect).

Data on detection probabilities during visual inspections are scarce. The most reliable way of estimating them is via experimentation, but this is resource intensive and infeasible for the wide range of pests and diseases that affect crops (Arndt et al. 2022). In such situations, biosecurity agencies often turn to expert elicitation to help fill critical knowledge gaps.

To estimate threat-specific detection probabilities for visual inspections under Crop Safe, an expert elicitation was conducted to estimate the likelihoods of detecting 14 exotic pests and diseases (Arndt et al. 2022). Prior to undertaking the elicitation, an online survey of agronomists from the Crop Safe network yielded 89 responses and gave an insight into their methods for surveying crop health (e.g. examining plants from good and bad patches, inspecting the crop using a zigzag pattern) in addition to asking how confident they felt (on a 5-point scale) in their ability to detect each of the target pests and diseases. The survey responses were also intended to validate detection probabilities estimated by experts.

A panel of nine experts with a background in biosecurity in the grains industry participated in an expert elicitation workshop using the IDEA protocol. The aim of the workshop was to quantify the experts' estimates of the probabilities of early detection of exotic pests and diseases during the agronomists' visual inspections. In the *Investigate* step, a summary of the results from the online agronomist survey (excluding confidence in early detection, since the exercise sought to elicit experts' judgements without the influence of agronomists' self-rated confidence) was presented to establish a common understanding of how agronomists conduct surveillance. To reduce elicitation burden, experts were tasked with assigning the 15 exotic pests and diseases to 7 groups based on the plant part affected (leaf, stem, root, or grain), whether it was a pest or disease, and plant symptoms. After initial discussion, two subcategories of leaf diseases were defined; the main "leaf diseases" group (barley stripe rust, lentil rust, lupin anthracnose) is known to have distinctive symptoms which are specifically targeted by agronomists during surveys. Lentil anthracnose was considered separately from this larger group and assigned to its own category (type-2 leaf diseases) as its symptoms are less noticeable and frequently misdiagnosed, leading to lower likelihood of detection in the field.

For each of the seven groups, experts were asked "What is the probability of early detection by Crop Safe agronomists according to effort?", with early detection defined as an incursion level of 1% per 100 ha. Four-point estimates (which includes the expert's confidence that the true value lies within their specified bounds; Speirs-Bridge et al. 2010) were elicited to produce credible intervals; the question was posed to workshop participants in terms of frequency (i.e. how often would the pest or disease be detected during routine inspections relative to effort). Estimates were then anonymised, standardised, and aggregated to provide a group interval. In the *Discuss* step, the group interval was presented to the panel to discuss and provide feedback. At this point, the agronomists' self-ratings of confidence in detection were shown to the experts. Round 2 estimates were elicited using the same four-step format as Round 1 (*Estimate* step). Finally, aggregated intervals from Round 2 were used as the final detection probability intervals (*Aggregate* step; Figure 12.1).

These aggregate expert judgements indicate that early detection of leaf pests and diseases (excluding lentil anthracnose) is relatively likely; however, detectability for other groups tends to be low (Figure 12.1). Two takeaway findings from comparing interval estimates between rounds are (i) the slight narrowing of intervals in Round 2 for grain pests, type-1 leaf diseases (leaf diseases excluding lentil anthracnose), leaf pests and stem pests; and (ii) overall lowering of estimated detection probabilities for grain diseases, type-2 leaf diseases (lentil anthracnose), and root diseases. Interval ranges are expected to narrow in Round 2, as this suggests increased confidence and convergence of point (best) estimates after the *Discuss* step. However, this is not always the case where individual confidence remains at the same level or there is continued, honest disagreement between experts. Reduction in the estimates largely reflect findings of the agronomist survey, in particular, lower than expected self-rated confidence in detecting specific pests and diseases during inspections, which indicates systematic overestimation of detectability by experts relative to agronomists' self-assessment.

Quantitative estimates of visual detection probabilities derived from applying the IDEA protocol have been adopted by Crop Safe programme managers. The results can also guide future efforts to improve confidence in the surveillance program such as via targeted training. An important take-home finding from the wide range in estimates observed between experts in this

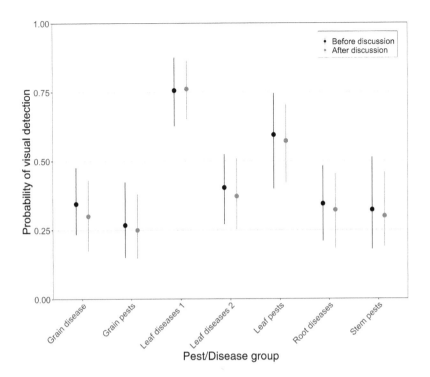

FIGURE 12.1   Probability of early detection during visual inspection for seven groups of exotic pests and diseases affecting grain crops. Aggregated credible intervals (standardised to 80% confidence levels across all experts) are shown for the two rounds of expert elicitation, with circles representing best estimates.

study is to elicit group judgements whenever possible rather than risk relying on a single expert, even the most respected, who may be overconfident or biased (i.e. systematically under- or over-estimates values).

ELICITING VALUES IN BIOSECURITY

Decision making rests not on facts alone, but also on value judgements. Selecting a management option that will deliver the "best" outcome depends on the preferences of decision makers or relevant stakeholders, which are defined by their value judgements. For example, whether food security, economic costs, or biodiversity is most important in a biosecurity context will depend on to whom the question is asked. As far as incorporating science into policy making is concerned, numerous approaches are available, including policy briefs, systematic reviews, science-policy forums, exchange programs, working groups, national funding schemes, and shared governance. All methods are motivated by the need for policy makers to communicate what information is required to make decisions and for scientists to communicate what can be achieved under different courses of action. However, these "technocentric" approaches stop short of providing a framework within which the wider community of stakeholders can evaluate policy options, examine supporting evidence, and provide feedback about their preferences and tolerance for risk in a decision analytic context.

The problem with assuming a technocentric perspective in decision making is that analyses are often undertaken by the analysts alone (sometimes with the help of the problem owner), hastily formulating a few scenarios and asking a group of experts to rank them, or worse, to reach a consensus about the ranking in an unstructured way through informal discussion. Our view is that non-participatory approaches often fail to consider all the social and cultural aspects of a problem, leading analysts to formulate a formal model that may be analytically impeccable but that omits stakeholder concerns, holds hidden assumptions, and is difficult to challenge. When the implications of a bad decision are significant, then the consequences of failing to define the context, engage with stakeholders, formalise the objectives and measures, and identify the full set of options may lead to double counting some values, omitting others, alienating stakeholders, and monitoring redundant measures.

Unlike eliciting technical judgements (facts), there is no rationale for eliciting value judgements specifically from experts. Nor is there an objective truth against which the correctness of value judgements can be gauged. One's feelings, opinions and preferences simply do not lend themselves to rational analysis. Capturing value judgements in a transparent, fair, and defensible way to incorporate them into decision making is no straightforward task, given the competing priorities and strategic interests of biosecurity stakeholders (see Chapter 2. Biosecurity Systems and International Regulations). Stakeholder values might only be considered as an afterthought or a post-hoc justification of a pre-existing choice. Conventional approaches to decision making such as BOGSAT (a "bunch of guys/people sitting around a table") have historically been bureaucratic, hierarchical, opaque, and exclusive. They are susceptible to groupthink, leading to conformity in values, preferences, and priorities (Janis 1982). Decisions resulting from such processes may fail to serve the interests of the broader community or impact disproportionately on underrepresented stakeholder groups. Value judgements can masquerade as technical judgements in scientific advice, allowing covert intrusion of values into decision making.

Equitable decision making should acknowledge the range of values in a decision, and integrate stakeholder value judgements. This requires elicitation techniques that capture participant value judgements in a way that is technically sound, encourages dialogue, contributes insights in a deliberative environment, and is easily understood by participants irrespective of technical expertise (Gregory et al. 2012). If these principles are neglected, decisions may not deliver acceptable outcomes or reflect the needs of stakeholders.

There are many ways to ask someone about what they value. Popular techniques include: direct rating (e.g. by scoring importance out of 100), point allocation (i.e. distributing a set total number of points based on importance), and paired comparisons such as the analytic hierarchy process (AHP; Hajkowicz, McDonald, and Smith 2000). Criteria for decision making measure the satisfaction of objectives. For value elicitation to provide valuable input to a decision analysis, measurement scales for the desired objectives must be clearly defined so that the difference between consequences is quantifiable. Naïve weighting methods commonly elicit weights independently of the measurement scales used, leading to range insensitivity bias (Montibeller and von Winterfeldt 2015). Weight elicitation approaches that attach weights to try to quantify individual preferences within multiple-criteria decision analysis frameworks have been used widely in natural resource management (Mendoza and Martins 2006) and environmental science (Huang, Keisler, and Linkov 2011); however, examples relevant to biosecurity are scarce.

Several approaches to incorporating preferences using value function methods are used for multicriteria decision support. Value function methods combine consequences (Chapter 3. Anticipate), translated into performance measures with elicited weights to give an evaluation of how well each alternative will perform on fundamental objectives (Belton and Stewart 2002). In most cases, simple weighted summation is sufficient for calculating decision scores—numerical values that measure how well alternatives perform. Essentially, the weight for a given objective (e.g. minimise the amount of biosecurity risk material reaching the border) is multiplied by the normalised performance score measuring the expected consequence of the given alternative towards that objective. Then, these products are summed to arrive at the decision score for each alternative. Conceptually, the alternative that maximises the decision score is the "best" option. Breaking down a complex decision into facts and values and combining them using the logic described here is a systematic and transparent way to evaluate alternatives.

As real-world decisions in the biosecurity continuum vary in scope, complexity, and data availability, different value elicitation approaches may be appropriate depending on the specific context. Formal methods for eliciting preferences should be transparent, encourage buy-in in multi-stakeholder contexts and result in more publicly defensible decisions. Next, we discuss some common approaches to eliciting value judgements and their limitations. In discussing these approaches, we assume that analysts have undertaken the preliminary steps of decision structuring, and identifying aspects of the decision that are value-based (see Von Winterfeldt and Edwards 2007, and Chapter 5 of Gregory et al. 2012 for guidance on decision structuring).

ANALYTIC HIERARCHY PROCESS

The AHP (Saaty 1977) is one of the most popular methods for weighting alternative options in decision analysis for environmental management (Huang et al. 2011). This method decomposes a complex problem into a hierarchy of clusters representing the elements of importance to the decision (i.e. the criteria for meeting objectives; Forman and Gass 2001). Eliciting individual preferences is achieved using pairwise comparisons between all possible combinations. For each pair, individuals determine the extent to which they prefer one element over the other according to a semantic scale using verbal descriptors equating to levels on a numeric scale. The standard scale lies on a range from 1–9, where 1 means that both elements are equally preferred and 9 means that one element is favoured at the highest possible level. Pairwise combinations are converted into scaled weights that are combined to calculate the performance of alternatives (see Saaty and Vargas 2012).

As simple as elicitation using the AHP may seem, a number of critiques have been raised relating to transitivity (the internal consistency of preference judgements, i.e. if A > B > C, then A must be preferred over C) and changes in rank that can occur when changing the scope of alternatives (discussed in Forman and Gass 2001). It is possible for rank reversals to occur if alternatives are added or removed, which violates rationality principles. Consistency checks are built into the method, but some degree of inconsistency is common. The analysis does not explicitly compare gains on one criterion against losses on another. Questions of relative importance have been argued to be meaningless as they are open to interpretation and responses are frequently riddled with inconsistencies, resulting in a disconnect between responses and the weights used to arrive at a management recommendation (Belton and Stewart 2002).

POINT SCORING

A more direct way of eliciting weights is to allocate points, where a fixed number of points (e.g. 100) are distributed among criteria. This approach offers the convenience of requiring the fewest transformations of responses into weights for use in a value function. For example, van Poorten and Beck (2021) detail a step-by-step application of structured decision making for the control of the invasive smallmouth bass (*Micropterus dolumieui*) in British Columbia, Canada. The decision process set out to achieve three fundamental objectives for management: (1) low final abundance of smallmouth bass, (2) improved public participation, and (3) avoiding impacts on non-target species. A set of means objectives were also nested within these fundamental objectives. During consultation, stakeholder representatives were asked to allocate 100 points to the three fundamental objectives and to further subdivide the points among the means objectives. Normalised scores from each participant were averaged to converge on a single set of criterion weights to input into a value function used to determine the performance of management options.

A number of improvements could be made on this approach:

- **Weights should be elicited once the expected consequences have been clearly laid out** (not prior to listing the management options) so that all relevant information on the predicted outcome of each option is clear. Any dominated alternatives (i.e. performing worse than all other options on all criteria) or redundant criteria (i.e. that are invariant across the set of options) can also be eliminated from analysis.

- **Considering the measurement of trade-offs among objectives**, specifically the magnitude and balance of gains versus losses on different objectives under different courses of action should be made explicit.
- **Using direct ratings instead of point allocation**. The vague framing of objectives in conjunction with the unintuitive task of dividing up points among multiple objectives proportionally to their importance makes obtaining meaningful weights difficult and may be dissatisfying to participants. In comparison to other weighting methods, weights elicited using point allocation tend to be top-heavy (i.e. proportionally higher for higher ranked items and lower for lower ranked items). Responses of individuals asked the same question on different occasions are often inconsistent, suggesting that weights are poorly calibrated with underlying value judgements (Bottomley, Doyle, and Green 2000).
- **Incorporate the diversity of stakeholder values in decision making**. Any attempt to combine (e.g., through averaging) the weights of multiple decision makers or stakeholders into a single set of weights is problematic. In complex decisions, stakeholders are unlikely to agree on matters of importance. Structured decision-making techniques can allow analysts to explore different sets of attribute weights. For example, they can calculate decision scores (the overall value of an option) using various combinations of weights to determine how the recommended option(s) might change depending on whose values inform the decision, or what magnitude of a change in weight would change the recommendation. Additionally, a sensitivity analysis could be conducted to explore the relationship between attribute weights and management recommendations. We also recommend undertaking sensitivity analysis as part of a deliberative process so that stakeholders can work together on deciding the preferred option.

SWING WEIGHTING

Guidelines for Implementing Swing Weights

When formally quantifying stakeholder preferences as weights to be meaningfully incorporated into decision analysis, the weight assigned to any single fundamental objective needs to account for two elements: (1) the inherent importance of the objective, and (2) the range of the consequences estimated for the attribute across all management options (Keeney 2002; Steele et al. 2009). Range is pivotal to judgements as (all else being equal) consequences that vary substantially within the decision frame should be judged to be more important than those with a narrower range. In other words, measures with larger potential gains or losses under different alternatives have a greater bearing on the outcome of a decision. Swing weighting lays out the range of expected consequences for the decision and can help make weight elicitation more intuitive, consistent, transparent, and well-calibrated to a person's true value judgements (Fischer 1995).

Swing weighting relies on considering the best and worst possible consequences for each objective—objectives are first ranked by asking how important it would be to "swing" one objective from its worst to best consequence. To identify the top-ranked objective, participants are prompted think about which objective they consider most important to swing from its worst to its best outcome under the set of management options. Then, their next most important objective to swing from worst to best is ranked second, and so on until all objectives are ranked.

The second step is to assign a numerical weight (typically 100) to the top ranked objective, and then assigning weights moving down the ranks relative to swinging the top-ranked objective. For example, if an objective's swing is judged to be half as important as that of the first-ranked objective, it should be assigned a weight of 50. Raw swing weights are normalised to sum to one by dividing the raw weight by the sum of the weights. The normalised weights can then be used as the weight input, to be multiplied with normalised performance scores to calculate decision scores. In most contexts (especially those with high-stakes and multiple stakeholders), weighting should be performed individually to minimise potential interference of personalities, social friction, and communication problems (Gregory et al. 2012).

Under a swing weighting approach, preferences and the weights applied to the different criteria are an emergent property of the deliberative process. Calculating weights this way relieves decision makers of the complicated mental computations required by techniques such as point allocation, and theoretical and practical concerns with translating value judgements into pairwise comparison methods (e.g. AHP; Belton and Stewart 2002). This method strikes a balance between feasibility, owing to its simple implementation, and transparency and consistency of judgements (Németh et al. 2019). Capturing both personal judgements of importance and the extent to which the measurement scale discriminates between alternatives is central to the concept of swing weights (Belton and Stewart 2002). Explicitly focusing the elicitation on the net change in performance measures results in weights that have the property of range sensitivity, meaning that the magnitude of the "swing" from worst to best (i.e. how much is gained or lost from the choice of alternative) directly factors into the elicited weights (Fischer 1995). When grounded in structured decision making, focusing on specific trade-offs rather than vague statements of priority provides important insight and learning opportunities to foster meaningful stakeholder engagement (Gregory et al. 2012).

We emphasise again that stakeholders are unlikely to agree on matters of importance. Each stakeholder will apply different weights to each of the criteria. Reconciling the different resulting decision scores may not be possible—or even necessary. Gregory et al. (2012) advise that the process of negotiation begins at this point, by facilitating interactions between stakeholders that will help people reconcile differences and arrive at an alternative that is "good enough"—that "satisfices" their objectives rather than attempting to maximise them.

Worked Example for Swing Weighting Elicitation

Here, we provide a hypothetical example of how stakeholder value judgements can be elicited via swing weighting and combined with consequences in a biosecurity decision problem. In this scenario, stakeholders were asked to provide judgements about which of three exotic pest species of concern (pests x, y, and z) should be prioritised for increased pre-border and border management to prevent entry, establishment, and spread (EES) based on their economic, social, and environmental impacts. The three fundamental objectives for which preferences will be elicited relate to consequences of each pest on (i) economic damages, (ii) loss of social amenity, and (iii) impacts to biodiversity, with respective performance measures of cost (millions of dollars), social impact assessed on a 1–5 categorical scale, and number of native species threatened with extinction by pest EES. Cost is measured on a natural (direct) scale (dollars); social impact is rated on a constructed scale defined specifically for the problem at hand, and environmental impact uses a proxy (indirect) measure (or indicator; Gregory et al. 2012).

Expected consequences—fabricated for demonstration in this example—are given in Table 12.1. Notice the trade-offs inherent to the decision. Pest z will incur the greatest economic damages, but

---

TABLE 12.1

Consequence Table for the Decision Scenario

Objective	Performance Measure	Consequence Estimate			Normalised Consequence		
		Pest x	Pest y	Pest z	Pest x	Pest y	Pest z
Economic damages	Million $	1.2	0.5	4	0.200	0.000	1.000
Social amenity impact	Constructed scale (1 to 5)	4	3	1	1.000	0.667	0.000
Environmental impact	Number of susceptible native species	5	60	37	0.000	1.000	0.582

the least impact to social amenity. Pest y potentially threatens many native species but has minimal impact on economic assets. For simplicity, no uncertainty bounds are included in this analysis, but intervals should accompany point estimates in a real-world application of decision analysis. Since the decision is framed in terms of impact in prioritising species for management, the direction of "preference" is always in favour of maximising consequence values, as species that are more damaging across economic, social, and environmental objectives have the greatest mandate for targeting risk management activities. Dealing with the different scales of consequence estimates requires normalisation using the equation:

$$Normalised\ value = \frac{Data\ point\ value - Worst\ value}{Best\ value - Worst\ value},$$

bearing in mind that "best" and "worst" correspond to the maximum and minimum, respectively across all pests in this context—in other decision contexts "best" and "worst" will depend on the direction of preference, such as minimising cost or maximising crop yield. This results in consequence values that are scaled between 0 and 1 so that they can be combined with stakeholder elicited weights to calculate decision scores, which measure the value or "satisfaction" each option provides the assessor.

During a workshop, swing weighting was used to elicit preferences from three stakeholder participants. First, they considered the three objectives in terms of their "swing" in performance from worst to best. They were asked to rank them in order of this swing. Then, they weighted each objective, assigning a weight of 100 to the top ranked objective, and proportional weights (between 0 and 100) sequentially to their perceived importance of the "swing" from worst to best for the remaining objectives relative to the "swing" from worst to best for their top ranked objective. These weights were normalised to sum to one; both raw and normalised swing weights are shown in Table 12.2. For example, Participant 3 considered a swing from $0.5 million to $4 million in damages to be 20% as important as a swing from 5 to 60 threatened species in the event of EES.

Combining weights with pest performance scores on each objective produces a rank-order priority recommendation for each participant. Normalised swing weights are multiplied with the normalised consequence scores for pest i on objective j and summed to calculate the participant's aggregate decision score for each pest (Figure 12.2). For Participant 1 who was most concerned about economic damages and somewhat less so with amenity, and less so again with environmental impacts, the recommendation generated by this analysis is to prioritise Pest z, followed by Pest y and then Pest x. In contrast, decision scores based on swing weights elicited from Participant 3 who is primarily concerned about environmental impacts suggest prioritising Pest y over Pest z and Pest z over Pest x. The relative contribution of the "environment" objective (shaded in light grey)

TABLE 12.2
Participant Elicited Swing Weights.

Objective	Impact		Participant 1		Participant 2		Participant 3	
	Lowest	Highest	Weight	Weight (norm)	Weight	Weight (norm)	Weight	Weight (norm)
Damages	0.5	4	100	1.471	70	0.333	20	0.131
Amenity	1	4	50	0.735	100	0.476	33	0.216
Environment	5	60	40	0.588	40	0.190	100	0.654

Note: Raw weights are scaled relative to the "swing" from worst to best for the top-ranked objective, assigned a weight of 100. Values under "weight (norm)" are raw swing weights normalised to sum to 1.

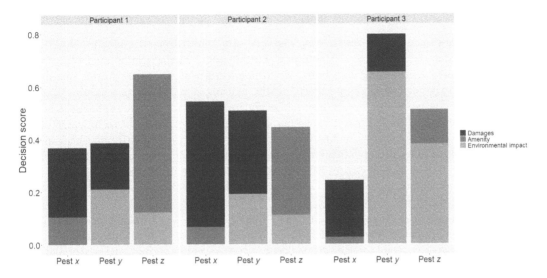

FIGURE 12.2 Decision scores aggregating the weighted sum of normalised stakeholder elicited swing weights and the normalised performance of alternatives (pests) on objectives.

is substantially higher for Participant 3 than Participant 1, reflecting their differing preferences captured by the weights. Note that only two objectives contribute to decision scores for each pest. This is because performance scores for the pest species with the lowest impact across objectives are normalised to zero in this simplified example. If uncertainty around consequences were appropriately captured by intervals, the performance in the best estimate scenario shown here would be scaled relative to the "worst" outcome in the worst-case scenario and would be some value greater than zero.

The choice of how to prioritise the pests has no clear answer, given the range of elicited stakeholder preferences and trade-offs between alternatives. Ultimately, a decision must be made which may or may not satisfy everyone. However, eliciting stakeholder values using this approach helps foster an understanding of the decision context, the relative performance of alternatives and how these trade off against one another, and invites stakeholders to engage in a participatory process. The results provide valuable insight to decision makers who can consider the performance of alternatives in light of the range of value judgements within the scope of the specific problem context. In facilitated negotiations, it is often possible for stakeholders to discuss differences and to consider alternative actions that may satisfy their fundamental objectives. Thus, participants are encouraged to seek satisfactory rather than optimal solutions (Gregory et al. 2012).

This highly simplified hypothetical case study is limited to only three alternatives and three objectives and does not consider uncertainty in the consequences for the sake of clarity. In practice, biosecurity agencies deal with complex problems in which large numbers of exotic pests and diseases need to be prioritised. It would be impossible to predict precisely the economic damages a potential pest will incur if it establishes and spreads or the number of native species it will impact to a single value, so uncertainty in consequences should be clearly represented. However, the application of preference elicitation demonstrated here can be (and regularly is) scaled up to the level of real-world problems.

Introducing uncertainty to the decision framework provides a new dimension in which to seek acceptable compromises. When uncertainty is explicit, stakeholders have the option to trade-off expectation against reliability. A stakeholder may propose an alternative for which the expected value of their decision score is lower, but the reliability of achieving a specified, minimally acceptable outcome is higher. Different stakeholders typically have different attitudes towards risk, and dealing with uncertainty in the estimates of a decision problem provides a framework in which different risk tolerances provide new opportunities for negotiating acceptable trade-offs.

DISCUSSION

Decision science provides tools for framing and analysing multivariate decision problems and to engage with stakeholders to identify solutions. Formulating management options, objectives, and trade-offs involves working with both facts (the uncertain state of the "world" one models) and values (what one deems important). Ideally, in decision making, the measures used to describe the uncertain states of the world should be independent of the decisions or actions being evaluated. Accordingly, these aspects carve out separate roles for groups of technical experts and stakeholders; the former group focuses on estimating facts and quantifying uncertainty, and members of the latter group indicate their personal (subjective) preferences for the decision criteria and negotiate acceptable trade-offs. In this chapter, we highlight the importance of, and the approaches to dealing appropriately with expert judgements and stakeholder values in biosecurity decision making.

Ideally facts and values should not be conflated, as far as possible, even when it looks as though they are entangled. Decision analyses undertaken by analysts working alone, that engage experts in naïve ways, or fail to adequately consider stakeholder concerns are unlikely to have the methodological rigour to clearly lay out the components of a decision and should be treated with caution. These informal but widespread practices lead to narrowly constructed solutions, that can result in imposing values held by a small group of experts O'Brien (2000). Elicitations that muddle factual and value-based judgements push experts to overstep their remit as forecasters or "honest brokers" of policy advice (Pielke 2007) and (often unwittingly) embed the personal value judgements of individual experts into elicited judgements. When properly applied, structured expert elicitation approaches such as IDEA generate reliable and useful technical estimates to inform decisions, with features in place to minimise the influence of value judgements.

IDEA and similar structured elicitation protocols are designed to estimate facts or predict the outcome of future events, and not to choose courses of action or decide the best way to manage a problem (these latter choices depend on stakeholders and their preferences). They should have design features in place to insulate against cognitive biases and influencing effects of group social dynamics, and importantly, to treat expert elicited data with the same scientific rigour as is applied to data collected conventionally (Burgman 2015). Examples described in this chapter suggest that *ad hoc* methods and incomplete reporting are the norm for expert elicitation in biosecurity and related fields, limiting their credibility and reproducibility.

Problem owners should abandon the idea that a single expert, even the most highly regarded, can reliably estimate unknown facts. The convergence of independent interval judgements of multiple experts parallels the concept of replication in science. Individual accuracy can improve by testing and validating estimates with data and providing experts with specific feedback on performance (Burgman et al. 2011). Performance on similar, previous questions can distinguish experts based on their ability to make accurate judgements and any attempt to weight individual expert judgements when aggregating estimates should be performance based (Hanea et al. 2018) and not based on expert status, years of experience, self- or peer-rated confidence, or any other unproven proxy for accuracy.

Part of the solution to mitigate bias-prone individual judgement is to use groups effectively. Groups consistently perform better than individuals, but not just any groups (Burgman 2015; Hong and Page 2004). Evidence from forecasting demonstrates that groups that are diverse in gender and culture do well, and the reason they do well is that they bring diverse experiences and backgrounds that pool together different sources of information to come up with independent estimates (Hong and Page 2004; Page 2007). Diversity helps buffer against biasing influences of shared professional backgrounds, anchoring on shared prior knowledge, or common motivations. Facilitators should create opportunities and encourage people to constructively engage, actively consider information and reasoning from a wide range of sources and explore counterarguments and alternative explanations.

Discussions among experts should aim to clarify the logic of arguments, provide opportunities to present and share data, and examine the evidence for competing explanations underlying an estimate (Burgman 2015; Hemming et al. 2018). Tools for capturing the logic of these complex, competing narratives (e.g. argument maps; van Gelder 2007) may be employed with the intention of promoting clarity and insight, improving communication, resolving disagreements, and seeking consensus (Burgman 2015). However, an underlying principle of the IDEA protocol is that consensus is not a necessary outcome of discussion; rather, discussion aims to minimise linguistic uncertainty, promote critical and counterfactual thinking, and allow relevant evidence to be shared between participants (Hemming et al. 2018). In some situations, consensus might be desirable, such as decisions that concern organisations in which different decision makers are unified under common goals. In any case, we encourage analysts to use more thoughtful aggregation methods than simply averaging answers. Analysts should transparently communicate individual expert judgements and how they are combined rather than feeling obligated to achieve consensus in order to act. Even if aggregated uncertainty bounds remain large or experts disagree over causal models after discussion, ongoing uncertainty may not be critical to decision making if the uncertainty falls within a tolerable range or differences are inconsequential to stakeholders, which comes down to value judgements.

This leads to the second type of elicitation often conducted in environmental management—incorporating preferences (value judgements) into decision making—an activity which is limited by our ability for legitimate measurement. There has been a tendency in the past for analysts to take charge of decision-making processes, marginalising stakeholders and thereby ignoring plausible options or important decision criteria. However, when the decision context remains underspecified and not all relevant stakeholders come to the table, relevant objectives are likely to be omitted or not appropriately measured and analysts are unlikely to fully understand the stakeholders' values. Ensuring that decision making is truly participatory takes effort, but when stakes are high, structured decision making provides a means to reconcile different agendas transparently and equitably (Gregory et al. 2012).

Those working in biosecurity should not shy away from frank and open conversations about values. Decision makers need to be honest to themselves and others about whether elicitations aim to predict facts or capture preferences, as these inputs must not be conflated in decision processes. Likewise, experts need to be honest about their own value-based positions and how their motivations or concerns might skew their opinions about factual matters.

Considering that value judgements define and seek to fulfil social goals, the elicitation of values is as important to biosecurity decision making as technical judgements. Accordingly, techniques to capture and translate stakeholder values into weights for use in decision analysis should also be rigorous and defensible. Used in this way—as illustrated by the worked example in this chapter—swing weighting demonstrates that employing technically sound, systematic, and transparent procedures to capture value judgements does not have to impose unreasonable cognitive burden on participants nor be technically onerous on analysts. Elicitations of facts and values can be complicated, but the right tools and techniques can help decompose and work through that complexity to make biosecurity decision making more tractable and lead to better management outcomes.

IN A NUTSHELL

- Decision science provides tools for analysts to frame and analyse complex decision problems and to engage with stakeholders to identify solutions.
- In this process, both facts (e.g. about future events) and values (preferences) may be evaluated, but they should not be conflated.
- Structured expert elicitation protocols are designed to ensure expert-derived data are subject to the same standards as empirical or modelled data.

- Rigorous methods (e.g. swing weights) are also available to transparently incorporate stakeholder values in decisions.
- Modes of group interaction, task complexity, and elicitation burden should be factored in when planning an elicitation.

REFERENCES

Arndt, Edith, Libby Rumpff, Stephen Lane, Sana Bau, Martin Mebalds, and Tom Kompas. 2022. "Estimating Probability of Visual Detection of Exotic Pests and Diseases in the Grains Industry— An Expert Elicitation Approach." *Frontiers in Ecology and Evolution*, 10. https://doi.org/10.3389/fevo.2022.968436.

Belton, Valerie, and Theodor J. Stewart. 2002. "Value Function Methods: Practical Basics." In *Multiple Criteria Decision Analysis: An Integrated Approach*, 119–161. Boston, MA: Springer US.

Böhm, Gisela, and Carmen Tanner. 2018. "Environmental risk perception." In *Environmental Psychology*, 13–25. https://doi.org/10.1002/9781119241072.ch2

Bottomley, Paul A., John R. Doyle, and Rodney H. Green. 2000. "Testing the Reliability of Weight Elicitation Methods: Direct Rating Versus Point Allocation." *Journal of Marketing Research* 37 (4):508–513. https://doi.org/10.1509/jmkr.37.4.508.18794.

Burgman, Mark. 2004. "Expert Frailties in Conservation Risk Assessment and Listing Decisions." In *Threatened Species Legislation: Is It Just an Act?*, edited by Pat Hutchings, Daniel Lunney and Chris Dickman. Royal Zoological Society of New South Wales.

Burgman, Mark A. 2015. *Trusting Judgements: How to Get the Best Out of Experts*. Cambridge: Cambridge University Press.

Burgman, Mark, Anna Carr, Lee Godden, Robin Gregory, Marissa McBride, Louisa Flander, and Lynn Maguire. 2011. "Redefining Expertise and Improving Ecological Judgment." *Conversation Letters*, 4 (2):81–87. https://doi.org/10.1111/j.1755-263X.2011.00165.x.

Cooke, R. M. 2008. "Special Issue on Expert Judgment." *Reliability Engineering & System Safety* 93 (5):655–656. https://doi.org/10.1016/j.ress.2007.03.001.

Fischer, Gregory W. 1995. "Range Sensitivity of Attribute Weights in Multiattribute Value Models." *Organizational Behavior and Human Decision Processes* 62 (3):252–266. https://doi.org/10.1006/obhd.1995.1048.

Forman, Ernest H., and Saul I. Gass. 2001. "The Analytic Hierarchy Process—an Exposition." *Operations Research* 49 (4):469–486. https://doi.org/10.1287/opre.49.4.469.11231.

Fraser, Robert W, David C Cook, and Janet Haddock-Fraser. 2019. *The WTO and Environment-Related International Trade Disputes*. London: World Scientific.

Froese, Jens G., Carl S. Smith, Peter A. Durr, Clive A. McAlpine, and Rieks D. van Klinken. 2017. "Modelling Seasonal Habitat Suitability for Wide-Ranging Species: Invasive Wild Pigs in Northern Australia." *PLOS ONE* 12 (5):e0177018. https://doi.org/10.1371/journal.pone.0177018.

Gregory, Robin, Lee Failing, Michael Harstone, Graham Long, Tim McDaniels, and Dan Ohlson. 2012. *Structured Decision Making: A Practical Guide to Environmental Management Choices*. Chichester, West Sussex; Hoboken, NJ: Wiley-Blackwell.

Hajkowicz, Stefan A., Geoff T. McDonald, and Phil N. Smith. 2000. "An Evaluation of Multiple Objective Decision Support Weighting Techniques in Natural Resource Management." *Journal of Environmental Planning and Management* 43 (4):505–518. https://doi.org/10.1080/713676575.

Hammond, N. E. B., D. Hardie, C. E. Hauser, and S. A. Reid. 2016. "Can General Surveillance Detect High Priority Pests in the Western Australian Grains Industry?" *Crop Protection* 79:8–14. https://doi.org/10.1016/j.cropro.2015.10.004.

Hanea, A. M., M. F. McBride, M. A. Burgman, B. C. Wintle, F. Fidler, L. Flander, C. R. Twardy, B. Manning, and S. Mascaro. 2017. "Investigate Discuss Estimate Aggregate for Structured Expert Judgement." *International Journal of Forecasting* 33 (1):267–279. https://doi.org/10.1016/j.ijforecast.2016.02.008.

Hanea, Anca M., Marissa F. McBride, Mark A. Burgman, and Bonnie C. Wintle. 2018. "The Value of Performance Weights and Discussion in Aggregated Expert Judgments." *Risk Analysis* 38 (9):1781–1794. https://doi.org/10.1111/risa.12992.

Hanea, Anca M., and Andrew Robinson. 2020. Comparability of Elicitation Approaches: Literature Review & Experimental Design. Centre of Excellence for Biosecurity Risk Analysis.

Hemming, Victoria, Mark A. Burgman, Anca M. Hanea, Marissa F. McBride, and Bonnie C. Wintle. 2018. "A Practical Guide to Structured Expert Elicitation Using the IDEA Protocol." *Methods in Ecology and Evolution* 9 (1):169–180. https://doi.org/10.1111/2041-210X.12857.

Higgins, Vaughan, and Jacqui Dibden. 2011. "Biosecurity, Trade Liberalisation, and the (anti)politics of Risk Analysis: The Australia-New Zealand Apples Dispute." *Environment and Planning A: Economy and Space* 43 (2):393–409. https://doi.org/10.1068/a43289.

Hong, Lu, and Scott E. Page. 2004. "Groups of Diverse Problem Solvers can Outperform Groups of High-Ability Problem Solvers." *PNAS* 101 (46):16385–16389. https://doi.org/10.1073/pnas.0403723101.

Howard, Ronald A.. 1988. "Decision Analysis: Practice and Promise." *Management Science* 34 (6):679–695. https://doi.org/10.1287/mnsc.34.6.679.

Huang, Ivy B., Jeffrey Keisler, and Igor Linkov. 2011. "Multi-Criteria Decision Analysis in Environmental Sciences: Ten Years of Applications and Trends." *Science of the Total Environment* 409 (19):3578–3594. https://doi.org/10.1016/j.scitotenv.2011.06.022.

Janis, Irving Lester. 1982. *Groupthink: Psychological Studies of Policy Decisions and Fiascoes*. 2nd ed. Boston, MA: Houghton Mifflin.

Keeney, Ralph L. 2002. "Common Mistakes in Making Value Trade-Offs." *Operations Research* 50 (6):935–945. https://doi.org/10.1287/opre.50.6.935.357.

Keeney, Ralph L. 2007. "Developing Objectives and Attributes." In *Advances in Decision Analysis: From Foundations to Applications*, edited by Detlof von Winterfeldt, Ralph F. Miles Jr and Ward Edwards, 104–128. Cambridge: Cambridge University Press.

Kuster, Karin, Marie-Eve Cousin, Thomas Jemmi, Gertraud Schüpbach-Regula, and Ioannis Magouras. 2015. "Expert Opinion on the Perceived Effectiveness and Importance of on-Farm Biosecurity Measures for Cattle and Swine Farms in Switzerland." *PLOS ONE* 10 (12):e0144533. https://doi.org/10.1371/journal.pone.0144533.

Lewandowsky, Stephan, Ullrich K. H. Ecker, Colleen M. Seifert, Norbert Schwarz, and John Cook. 2012. "Misinformation and Its Correction: Continued Influence and Successful Debiasing." *Psychological Science in the Public Interest*, 13 (3):106–131. https://doi.org/10.1177/1529100612451018.

Longino, Helen E. 1990. *Science as Social Knowledge Values and Objectivity in Scientific Inquiry*. Princeton, NJ: Princeton University Press.

Mendoza, G. A., and H. Martins. 2006. "Multi-Criteria Decision Analysis in Natural Resource Management: A Critical Review of Methods and New Modelling Paradigms." *Forest Ecology and Management* 230 (1):1–22. https://doi.org/10.1016/j.foreco.2006.03.023.

Montibeller, Gilberto, and Detlof von Winterfeldt. 2015. "Cognitive and Motivational Biases in Decision and Risk Analysis." *Risk Analysis*, 35 (7):1230–1251. https://doi.org/10.1111/risa.12360.

Mukherjee, Nibedita, Jean Hugé, William J. Sutherland, Jeffrey McNeill, Maarten Van Opstal, Farid Dahdouh-Guebas, and Nico Koedam. 2015. "The Delphi Technique in Ecology and Biological Conservation: Applications and Guidelines." *Methods in Ecology and Evolution* 6 (9):1097–1109. https://doi.org/10.1111/2041-210X.12387.

Németh, Bertalan, Anett Molnár, Sándor Bozóki, Kalman Wijaya, András Inotai, Jonathan D Campbell, and Zoltán Kaló. 2019. "Comparison of Weighting Methods Used in Multicriteria Decision Analysis Frameworks in Healthcare With Focus on Low- and Middle-Income Countries." 8 (4):195–204. *Journal of Comparative Effectiveness Research*, 8(4). https://doi.org/10.2217/cer-2018-0102.

O'Brien, Mary. 2000. *Making Better Environmental Decisions: An Alternative to Risk Assessment*. Cambridge, London: MIT Press. https://mitpress.mit.edu/9780262650533/making-better-environmental-decisions/

Page, Scott E. 2007. "Making the Difference: Applying a Logic of Diversity." *The Academy of Management Perspectives* 21 (4):6–20. https://doi.org/10.5465/AMP.2007.27895335.

Pielke, Roger A Jr. 2007. *The Honest Broker: Making Sense of Science in Policy and Politics*. Cambridge: Cambridge University Press.

Regan, Helen M., Mark Colyvan, and Mark A. Burgman. 2002. "A Taxonomy and Treatment of Uncertainty for Ecology and Conservation Biology." *Ecological Applications* 12 (2):618–628. https://doi.org/10.1890/1051-0761(2002)012[0618:ATATOU]2.0.CO;2.

Saaty, Thomas L. 1977. "A Scaling Method for Priorities in Hierarchical Structures." *Journal of Mathematical Psychology* 15 (3):234–281. https://doi.org/10.1016/0022-2496(77)90033-5.

Saaty, Thomas L., and Luis G. Vargas. 2012. "The Seven Pillars of the Analytic Hierarchy Process." In *Models, Methods, Concepts & Applications of the Analytic Hierarchy Process*, edited by Thomas L. Saaty and Luis G. Vargas, 23–40. Boston, MA: Springer US.

Sawford, Kate, Navneet K. Dhand, Jenny-Ann L. M. L. Toribio, and Melanie R. Taylor. 2014. "The Use of a Modified Delphi Approach to Engage Stakeholders in Zoonotic Disease Research Priority Setting." *BMC Public Health* 14 (1):182. https://doi.org/10.1186/1471-2458-14-182.

Southwell, Darren, Reid Tingley, Michael Bode, Emily Nicholson, and Ben L. Phillips. 2017. "Cost and Feasibility of a Barrier to Halt the Spread of Invasive Cane Toads in Arid Australia: Incorporating Expert Knowledge into Model-Based Decision-Making." *Journal of Applied Ecology*, 54 (1):216–224. https://doi.org/10.1111/1365-2664.12744.

Speirs-Bridge, Andrew, Fiona Fidler, Marissa McBride, Louisa Flander, Geoff Cumming, and Mark Burgman. 2010. "Reducing Overconfidence in the Interval Judgments of Experts." *Risk Analysis* 30 (3):512–523. https://doi.org/10.1111/j.1539-6924.2009.01337.x.

Steele, Katie, Yohay Carmel, Jean Cross, and Chris Wilcox. 2009. "Uses and Misuses of Multicriteria Decision Analysis (MCDA) in Environmental Decision Making." *Risk Analysis* 29 (1):26–33. https://doi.org/10.1111/j.1539-6924.2008.01130.x.

Steel, Daniel, Sina Fazelpour, Bianca Crewe, and Kinley Gillette. 2021. "Information Elaboration and Epistemic Effects of Diversity." *Synthese* 198 (2):1287–1307. https://doi.org/10.1007/s11229-019-02108-w.

van Gelder, Tim. 2007. "The Rationale for Rationale™." *Law, Probability and Risk* 6 (1-4):23–42. https://doi.org/10.1093/lpr/mgm032.

van Poorten, Brett, and Martina Beck. 2021. "Getting to a Decision: Using Structured Decision-Making to Gain Consensus on Approaches to Invasive Species Control." *Management of Biological Invasions* 12 (1):25–48. https://doi.org/10.3391/mbi.2021.12.1.03.

Von Winterfeldt, Detlof, and Ward Edwards. 2007. "Defining a Decision Analytic Structure." In *Advances in Decision Analysis: From Foundations to Applications*, edited by Detlof von Winterfeldt, Ralph F. Miles and Edwards Ward, 81–103. Cambridge: Cambridge University Press.

13 Profiling and Automation

Evelyn Mannix, John B. Baumgartner, Sana Bau,
Lucie M. Bland, Andrew P. Robinson, and Natasha Page

SUMMARY

High-volume pathways with low rates of biosecurity non-compliance pose a significant challenge to regulators because of the intense effort required to detect biosecurity risk material. As trade and passenger volumes increase, it is unlikely that budgets will be available to maintain current residual risk levels. Profiling allows regulators to achieve a greater reduction in risk with the same resources by targeting interventions (e.g. screening or inspection) towards entities that are more likely to be non-compliant. This improvement in efficiency is enabled by recent advances in automation using statistical and machine learning methods, particularly on high-volume pathways. This chapter introduces profiling and automation methodologies via several examples and discusses how profiling can be used as part of a strategy to drive behavioural change within a pathway. We conclude with a discussion of some common challenges in applying profiling and automation in real-world biosecurity settings.

GLOSSARY

Biosecurity risk material (BRM) Living organisms exotic to a country or biological material carrying an exotic pest or disease. BRM includes live animals and plants, animal and plant material (e.g. fur, feathers, pollen, and flowers), food scraps, and soil.

Non-compliance The failure of participants in the biosecurity system to act in accordance with biosecurity requirements, such as import conditions.

Profiling Profiling relies on prior information or beliefs about what makes an entity risky (non-compliant) in order to prioritise intervention activities. The goal of profiling is to maximise the reduction in risk that can be obtained with limited intervention resources.

Border interventions Include measures and actions implemented by a biosecurity agency or requested of importers to reduce the biosecurity risks of the imported products to which they are applied (e.g. screening or inspection).

Intervention effort The number of interventions undertaken on given items.

Endpoint (also called slippage or leakage) survey Inspections undertaken behind screening tools (e.g. manual inspections, detector dogs, and X-ray units) to determine the amount of BRM passing undetected through an intervention point.

Risk model A set of rules or a statistical or machine learning construct that describes risk within a pathway. For example, a risk model might predict the probability a passenger is non-compliant based on their characteristics.

Machine learning model Machine learning is a field of artificial intelligence focused on designing algorithms that improve themselves using data or experience (Nguyen et al. 2019).

DOI: 10.1201/9781003253204-18

INTRODUCTION

From manually checking documents during import clearance processes to X-raying luggage, many activities that are used to manage biosecurity risks can be labour intensive and expensive (Stoneham et al. 2021). While it is comforting to think that the purpose of these interventions is to "stop things at the border", in reality, not all instances of biosecurity risk material (BRM) will be found and, in some cases, a large proportion of BRM may be missed (Turner et al. 2020). This is particularly the case for large-volume pathways with diverse participants (e.g. international mail and travellers).

When inspection activities that are undertaken to manage or identify risks are expensive and time consuming, it can become impractical and inefficient to examine every single item or to inspect at frequencies that were once realistic for a given budget. Given projected increases in pathway volumes (BITRE 2012, 2014), responding by simply scaling up inspection activities is unlikely to be a realistic solution, as resources available for biosecurity activities would need to increase at a higher rate to ensure the absolute risk of the pathway remains identical (see Craik, Palmer, and Sheldrake 2017 for an illustrative example).

Rather than reduce the rate of intervention or discontinue risk management activities altogether, regulators have the opportunity to use techniques such as automation and risk profiling to cope with the increasing scale of biosecurity risk management. Profiling refers to the use of prior information or beliefs about what makes an entity risky in order to prioritise intervention activities, with the goal of maximising the reduction in risk that can be obtained with limited intervention resources. To fix our ideas, if BRM were assumed to be randomly distributed, then it could be expected that if 10% of the pathway were randomly selected for intervention, at least 90% of the BRM present within the pathway would evade detection. By using a profiling approach, a specific 10% of the pathway that captures a much greater proportion of the BRM present can be selected for intervention.

Profiling is used in a variety of industries where intervention efforts need to be targeted to save costs, including airline passenger inspections for security risks (Barnett 2004; Cavusoglu et al. 2013; Zheng et al. 2017), food safety compliance inspections (Oldroyd, Morris, and Birkin 2021), taxation audits (Lahann, Scheid, and Fettke 2019), and manufacturing quality assurance inspections (Sankhye 2020). Within the field of biosecurity, profiling has been used to prioritise interventions in many contexts, from vessel inspections to airline passenger screening (Clarke et al. 2017; Lane et al. 2017).

Similarly, automation using statistical and machine learning models can enable more efficient and targeted biosecurity interventions, especially in high-volume pathways. These models learn about a system from data or experience (Nguyen et al. 2019), thus potentially removing much of the reliance on humans in biosecurity screening and interventions. In particular, machine learning models (unlike many statistical models) are capable of making findings from data inputs of varied formats (e.g. text, images, and sounds), expanding their scope of application in biosecurity. For example, machine learning methods are being used to automate document checks and to detect items with a high likelihood of posing a biosecurity risk in X-ray images (IGB 2019).

As the number of passengers and cargo consignments increase, biosecurity agencies will be under pressure to increase the efficiency of intervention efforts on more pathways (BITRE 2012, 2014). This chapter provides a step-by-step introduction to profiling and automation as methods used to better target inspection activities. The final section explores how profiling and automation strategies can be put into practice in biosecurity.

PROFILING BASICS

When pathway volumes become sufficiently large, it is no longer practical or cost-effective to maintain long-standing complete inspection practices. In this context, a decision needs to be made on how individuals or consignments are to be selected for inspection. This could be done by (1) randomly choosing passengers or consignments, (2) only intervening in cases where other issues are identified, such as paperwork deficiencies (e.g. containerised cargo pathway in IGB 2019), or (3) intervening based on prior experience or beliefs about what makes an entity risky—known as *profiling* (e.g. passengers within a certain age group and with a passport from a specified list of countries are to be directed for screening).

The effectiveness of profiling will depend on the quality of information available to understand risk within a pathway, as well as how risk is distributed. There are several reasons why non-compliance could be more common within specific passenger groups, containers, or commodity types—from different levels of knowledge about, and desire to follow, biosecurity regulations, social norms, values, expectations, or business processes (Sherring 2021) to additional risk factors that might make non-compliance more common, such as the transit of goods through locations where there is a high risk of introducing contamination (Ashcroft et al. 2008). The insights gained from identifying these patterns enable regulators to prioritise particular groups for intervention to maximise the amount of BRM found.

In terms of managing biosecurity risk at the border, focus is generally placed on the *proportion of non-compliant consignments* rather than the *raw number of interventions found to be non-compliant*. This is because counts of raw non-compliance generally conflate two signals: the number of consignments or passengers coming from a particular pathway segment and the proportion that are non-compliant. If pathway segments with the highest proportion of non-compliance are targeted, then the largest amount of non-compliance given the number of inspections conducted will be found. Furthermore, this provides an evidence-based method to rank pathway segments into high- and low-risk entities provided sufficient data are available for each segment. Pathways that carry high raw quantities of BRM due to their high pathway volume are arguably more efficiently risk-managed by making changes to policy than by increased border intervention.

For less complex pathways, targeting segments that have the highest proportion of non-compliance may be sufficient and reasonably straightforward. However, for more complex pathways, it can be difficult to navigate the additional information available for incoming arrivals and cargo in order to understand which segments should be targeted. For example, agencies have information about which flight a passenger took and which country their passport is from, or the supplier of a consignment and where it was manufactured. However, as more variables are added, assessing the rate of non-compliance becomes challenging because some variable combinations are likely to be infrequently inspected and therefore lacking in data to make reliable inferences. For example, if we are interested in inspections of passengers who are New Zealand nationals, female, between 30 and 40 years old, and flying to Australia from Tahiti, we may find that only a small number of inspections match these criteria. Inference about the non-compliance rate of these passengers based solely on a few instances will have high uncertainty and may not be reliable. In order to solve this problem, statistical or machine-learning models can provide more robust predictions about the likelihood of non-compliance for combinations of categories with little data. They do this by "borrowing" information from closely related combinations of categories—for example, the combination "male and aged 45–50" could be closely related to either of "female and aged 45–50" and "male and aged 50–55".

PROFILING CARGO CONSIGNMENTS—A SIMULATED EXAMPLE

Imported goods vary in terms of commodity type, exporting country, trade volumes, and many other factors. As a result, risks of BRM introduction also vary, making the selection of cargo consignments for inspection to minimise leakage of non-compliant consignments a challenging task. In the following example, we demonstrate how prioritisation of inspection effort to target consignments by country of origin or traded good might typically occur, using a simple profiling approach based on the proportion of non-compliant items within different groups.

We consider a simulated dataset containing data on imports from five countries (A, B, C, D, and E) and four types, classified by tariff codes (#1010, #2020, #1050, and #3000). For simplicity, each consignment has only one country of origin and tariff code. The dataset also includes import quantities (number of containers) for each consignment.

The level of non-compliance differs greatly depending on country of origin (Figure 13.1a). At first glance, it would seem that Country A presents the greatest risk of non-compliance based on the number of non-compliant inspections. However, as most inspections were carried out on consignments originating from Country A (Figure 13.1b), the higher level of non-compliance is more reflective of inspection effort than inherent risk—more inspections will generally result in finding more non-compliance. Rather than basing risk judgements on absolute interception numbers, profiling uses information about the proportion of inspections that detect non-compliance (the non-compliance rate; Figure 13.1c). For Country A, the non-compliance rate is only around 5% of inspected consignments, compared with around 9% for Country B. Thus, targeting goods from Country A purely based on the number of non-compliant inspections is naïve to the underlying non-compliance rate. Note that in Figure 13.1c, the 95% confidence interval (CI) width reflects the

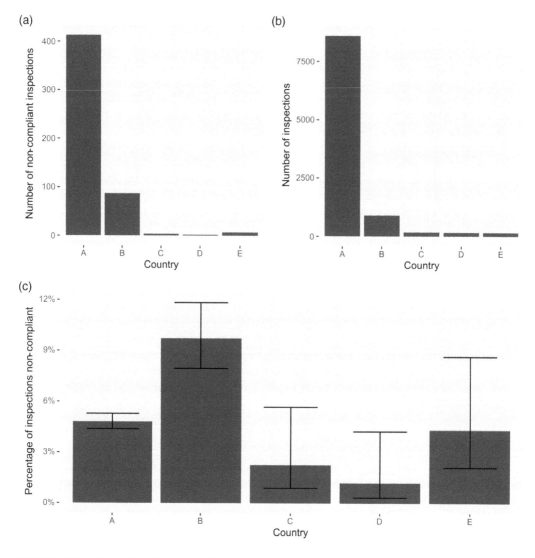

FIGURE 13.1   (a) Number of non-compliant inspections by country, (b) number of inspections by country, and (c) percentage of non-compliant inspections by country. Error bars show 95% confidence intervals.

TABLE 13.1

Highest Risk Consignment Types Based on Country, Tariff, and Number of Containers

Country	Tariff	Containers	Total Inspections	Non-Compliant Inspections	Proportion Non-Compliant (95% CI)
E	2020	1	6	3	0.50 (0.19–0.81)
B	2020	1	54	21	0.39 (0.27–0.52)
E	1010	1	7	2	0.29 (0.08–0.64)
B	1010	1	236	46	0.19 (0.15–0.25)
A	2020	1	996	179	0.18 (0.16–0.20)
B	3000	2	6	1	0.17 (0.01–0.56)
D	2020	2	7	1	0.14 (0.01–0.51)
C	2020	1	24	3	0.12 (0.04–0.31)
D	1010	2	9	1	0.11 (0.01–0.43)
A	1010	1	1015	87	0.09 (0.07–0.10)

number of inspections undertaken (i.e. the inspection effort) in line with general sample size principles. We have greater confidence in our estimate for Country A because the sample size is much greater than for the other countries.

With further information, inspection effort could be better targeted to improve effectiveness. Here, tariff codes and the number of containers within each consignment provide additional detail for profiling. Table 13.1 provides the ten "riskiest" entry types (defined as a specific combination of country, tariff, and number of containers) ranked in descending order by the proportion of non-compliant inspections. Of these, half have fewer than ten inspections, so confidence in the riskiness of these entries is low (wide CIs indicate lower confidence or greater uncertainty). Again, small sample sizes reduce the confidence we can have in non-compliance rates—their apparent high risk could be a result of chance. Take for instance the sixth row of Table 13.1; one out of six inspections were non-compliant, resulting in a best estimate proportion of non-compliant consignment types of 0.17. However, based on the 95% CI, it could be as low as 0.01 and as high as 0.56. By contrast, the final entry shows a consignment type that was inspected many more times and with a much more precise CI of 0.07–0.10. Targeting inspections based on best estimates that neglect to consider sample size could overestimate risk and impose unfair burden on smaller importers.

Methods for addressing this issue include grouping consignments with a small number of inspections and similar features together (Huang and Ng 1999), and an approach known as Empirical Bayes, which anchors the predicted rate of non-compliance closer to the mean for groups with a small number of inspections (Casella 1985). Although they can be useful, such techniques lose information that could be used to obtain better profiling results. Instead, using statistical or machine learning models can provide more robust predictions about the likelihood of non-compliance for combinations of categories with little data by considering how each of these categories relate to risk, independently and in combination with other variables throughout the dataset.

PROFILING WITH MODELS

As noted in the previous example, statistical or machine learning modelling of compliance data can help construct profiles of consignments for targeting inspection effort where information is sparse. For situations with greater information availability, statistical and machine learning models can be employed to create sophisticated representations of risk with high predictive performance. The process of training these models is technically complex but conceptually simple. Building profiling models starts with splitting data containing relevant attributes into two sets: training (used to train the model) and testing data (withheld data used to test the model's performance). Given the training set, models learn patterns within the data that are associated with an outcome (non-compliance), which allows them to make a prediction about whether non-compliance is likely. Profiling models may be able to achieve perfect accuracy on training data because they are optimised to fit the answers as closely as possible in these cases, but they may not necessarily work well when tested against new (i.e. unseen) data.

These training and testing datasets should be selected in a way that reflects how the model will be used. Generally, training data should be randomly selected, ensuring that data cover the range of important categories (e.g. countries and tariff codes) so that the model algorithm is trained to recognise variants within the scope of the profiling problem. For example, if a model is going to be used to predict the biosecurity risk posed by airline passengers, it makes sense to select a portion of the most recent data for testing and use the rest for training. This ensures that the test data measures how well the model's understanding of risk generalises into the future.

The following explanation assumes some knowledge of statistical modelling and may be beyond the interest of some readers. In the current example, we randomly allocate 80% of the dataset to provide the training data. We fit a logistic regression model to this training set with the binary outcome response variable: compliance versus non-compliance (Gelman and Hill 2006) and the main effects of country, tariff code, and number of inspections. The output describes the direction of correlation (positive or negative) and the strength of effects on compliance. The outputs predict the probability of non-compliance for the different classes of consignments (where the classes are defined as combinations of the main effects listed earlier). To predict how well the model performs in predicting non-compliance in new data, it is then applied to the remaining 20% of the data assigned to the testing set. For further details on the process of fitting logistic models and machine learning models, see Gelman and Hill (2006), Harrell (2015), and Lindholm et al. (2022).

Model performance can be measured by comparing the percentage of non-compliant inspections found versus the total percentage of inspections undertaken. Plotting this relationship for the current example produces a curve showing the proportion of risky consignments captured with increasing inspections (Figure 13.2). The performance of the model is compared relative to the diagonal line starting from the bottom left-hand corner where no inspections are conducted and therefore none of the non-compliant consignments can be found, moving towards the top right-hand corner, where

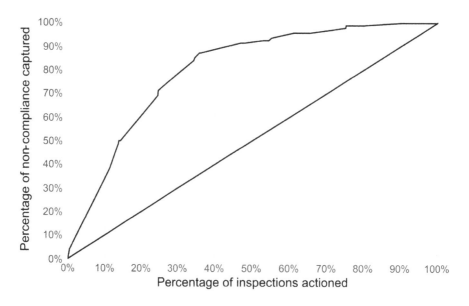

FIGURE 13.2   Percentage of non-compliance captured versus percentage of inspections actioned, using a logistic model to predict non-compliance in the simulated dataset to sample from the riskiest items in the first instance.

all consignments are inspected, thus finding all the non-compliant consignments (assuming perfect detection). Curves that represent better models are in the region towards the top left corner, which capture more non-compliant consignments while having to rely on fewer inspections. The diagonal represents parity between the proportion of consignments inspected and the proportion of non-compliant consignments found, meaning that if a model is plotted along this line, it performs no better than expected if inspections are randomly conducted (e.g. a 50% non-compliance rate from inspecting 50% of consignments). For the example dataset, the figure demonstrates that inspections guided by the profiling model can capture 90% of non-compliant consignments by searching 50% of the consignments. (In practice, such a model would be considered extremely good!) Thus, targeting high-risk entities identified by the profiling model would substantially improve the efficiency of cargo inspections.

INTERPRETING MODELS

In cases like this, profiling is worthwhile. However, a regulator may be uncomfortable singling out consignments for inspection based on a model without understanding who is being targeted (Rudin 2019). This is a sound concern, as without this information, it would also be impossible to check if the model confirms what is generally accepted as being the riskiest types of consignments or whether it tells a different story. This "truth-checking" process is important in profiling, as there may be issues within the data or modelling process. For logistic models, a plot of effect size can be produced to show which factors are found to be riskier by the model (Figure 13.3).

For factors listed in Figure 13.3, effect sizes on either side of 0 indicate a positive or negative effect on non-compliance relative to the reference class: Country A and Tariff #1010. (NB: The software assigns variables to this reference class alphabetically by default.) Positive effect sizes (e.g. Tariff #2020 and Country B) indicate a higher likelihood and negative effect sizes (e.g. Tariffs #1050 and #3000) indicate a lower likelihood of non-compliance. In the case of numeric variables (in this instance, number of containers), interpretation is different. The negative effect size means that the relationship between consignment size and non-compliance is inverse; that is, larger consignments (with many containers) have lower risk than small consignments (and vice versa). Alternatively, if the effect size had been positive, it would mean the larger the consignment the greater the risk.

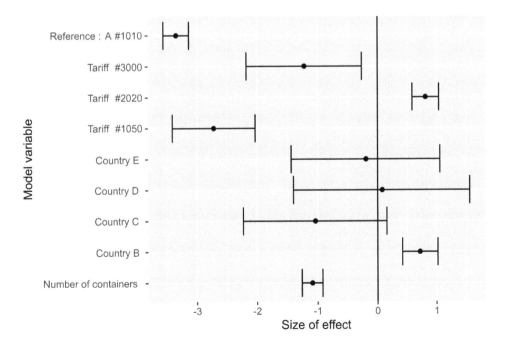

FIGURE 13.3   Plot of effect size from the generalised linear model. Error bars indicate 95% confidence intervals.

Note that the model has predicted a similar risk level for Country D as the reference Country A, in contrast to its much lower non-compliance rate suggested by Figure 13.1c. This is not contradictory, but rather it shows that the apparent low risk of consignments from Country D is explained by other effects, namely the tariff codes and container counts.

In totality, Figure 13.3 shows that this profiling model makes a similar recommendation as in the previous section, namely that consignments from Country B are high risk and should be prioritised. However, we also obtain richer information, as we can see that consignments that are smaller *or* have Tariff #2020 are also higher risk. Using a modelling approach generally provides a much more precise and informative insight into how the patterns of risk within a pathway are composed, especially when there is detailed information available about items on the pathway.

Important considerations for applying the example profiling model:

- *The reference class provides a "baseline" for estimating effect sizes for categorical variables* (i.e. the magnitude and direction of effects are relative to this reference class). In practice, reference categories (labelled "Reference: A #1010" in Figure 13.3) should be selected based on the application context. Choice of reference categories does not affect model predictions but will affect uncertainty estimates (indicated by the error bars in Figure 13.3) as the model will be comparing different combinations of variables.
- *Real-world observations may not match model predictions.* Inconsistencies between model explanations and on-the-ground experience can arise from biases in, or misinterpretation of the data, or errors in the models (Gilpin et al. 2018). Alternatively, prior beliefs may prove to be inaccurate in light of the data.
- *The real world is more complex than this example.* Consignments can contain goods with multiple tariff codes and exporting countries, and variable stated quantities. Defining non-compliance is a non-trivial task. For the container pathway, it could take a variety of different forms, from procedural issues such as container seals being broken before an inspector arrives, to genuine biosecurity risks being present, like plant material or live insects (Clarke et al. 2017; Hester et al. 2020). Different inspection methods may have also been applied in different cases, preventing direct comparisons between different sets of inspection data (Clarke et al. 2017).

However, the key ideas presented here are readily extensible and a more complex example is explored in the following section.

PROFILING AIRLINE PASSENGERS

Managing the biosecurity risk posed by international airline passengers is a complex problem. Depending on a person's understanding of biosecurity regulations and their desire to follow them (Chapter 8. Incentives), non-compliance can be expected to vary from passenger to passenger (Sherring 2021). Within an airport environment, biosecurity officers need to process incoming passengers efficiently through exit channels, particularly during peak arrival times as delays place a significant burden on incoming passengers and airport infrastructure.

In addition to processing passengers that have declared BRM, officers need to conduct inspections to find undeclared non-compliance. Biosecurity officers may manually inspect a passenger's belongings, or use screening methods (e.g. X-raying luggage and detector dogs) to assist in identifying passengers that are more likely to be carrying BRM. To keep passengers moving through the airport smoothly, a direct exit channel may be used to allow travellers to exit the airport without screening or inspection. During peak times, more passengers may be directed to this direct exit channel to ensure wait times remain acceptable.

Collecting intervention data in this environment is challenging in the absence of automated systems. In many airports, it may be the case that records are only made when BRM is seized from a

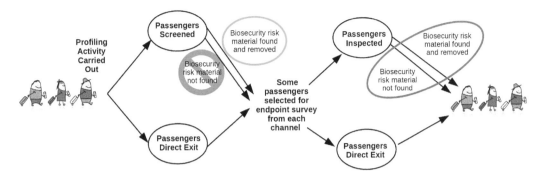

FIGURE 13.4   Data collection to inform profiling for the international airline passenger pathway. The green circle highlights the dataset of positive finds and the red circle shows negative finds missing from typical interception data. The blue circle shows data collected in the endpoint survey.

traveller after it is declared or detected through screening and manual searches. This means inspection data will be incomplete—instances of compliance, often called nil-finds, are not recorded. These data cannot directly be used for profiling as there is incomplete information about the intervention effort (i.e. how many interventions were done and of what items) spent on uncovering instances of non-compliance. This means that estimating the non-compliance rate of cohorts is impossible with the readily available data.

To determine the non-compliance rate, intervention effort can be estimated using an "endpoint" or leakage survey (Figure 13.4). Endpoint surveys are commonly used to report on the effectiveness of screening methods and to derive a measure of the amount of BRM leaving the airport, and they may be a requirement for reporting against key performance indicators (see Chapter 10. Monitoring, Evaluation, and Reporting). They involve inspecting a subset of passengers after they have exited and recording compliant as well as non-compliant inspections, which can help determine the effectiveness of the different screening channels. Here we show how the endpoint survey can also be used to estimate the number of nil-finds in each cohort, which in turn enables estimation of the non-compliance rate.

Passengers selected for the endpoint survey are fully inspected for BRM after they have either been screened or have left through the direct exit channel. Ideally, this survey is a random sample of passengers. If data on the screening channels that the passengers went through are also collected in the endpoint survey, then these data can be used to predict the likelihood a passenger with particular characteristics was screened. This allows inspection effort to be accounted for during profiling.

If the endpoint survey is sufficiently large, then the data collected could be used to predict the likelihood of passengers being non-compliant. However, screening procedures will likely differ between the endpoint survey and regular screening. For instance, passenger screening could be done using X-rays, where positive finds from the X-rays result in a manual inspection, while the endpoint survey could involve selecting one bag the passenger carries and manually inspecting its contents. These two methods would likely have different probabilities of finding an item of biosecurity risk if it were present.

Instead, we can combine the information from both the positive finds through screening, and the endpoint survey, to create a profiling model that incorporates all of the information available about the potential BRM carried by a passenger.

AIRLINE PASSENGER PROFILING EXAMPLE

In the context of the airline passenger pathway, the aim of profiling is to maximise interception of undeclared BRM (i.e. passenger non-compliance) by targeting inspections based on passenger characteristics. However, in this setting, we have incomplete information about the inspection effort. Again, we use simulated data to illustrate, and the technical detail may exceed the interest of some readers. The simulated dataset contains details of passengers with citizenship of one of five countries (BRI, GLD, LON, MAN, and VAL) arriving on one of three flights (QE101, QE314, and QE905). These characteristics and gender (F or M) are sampled with equal probability. These three categorical characteristics are used to form cohorts of seemingly alike passengers. In general, determining the best characteristics to form cohorts is an exercise in statistical modelling.

Some attributes within our simulated data have been set to be of higher risk than others, as follows. Citizens from countries BRI and GLD were more likely to be non-compliant than those from LON, MAN, or VAL. The likelihood of being selected for screening also differed by citizenship; citizens of GLD and VAL were more likely to be screened than BRI, LON, or MAN. As expected, raw non-compliance was highest for GLD in the simulated data (these citizens are more likely to be screened and more likely to be non-compliant). Sampling of citizenship country was deliberately biased; oversampling of VAL and undersampling of BRI skew the apparent risk between these countries (Figure 13.5).

In this instance, we cannot calculate the non-compliance rate for each country directly, because we do not have records for which passengers were screened. We only know if they were screened and non-compliant. However, we can recover this information by using a generalised mixed effects model (Stroup 2012) to predict the likelihood a passenger was screened using their characteristics (country, gender, and flight) using the data from the endpoint survey.

This provides us with a probability that a passenger was screened given their citizenship country, gender, and flight. As we also have data describing the total number of passengers from a given cohort, we can combine these datasets to end up with data in the form shown in Table 13.2.

From here, we can estimate the total number of passengers who were screened by multiplying the total number of passengers by the predicted screened proportion. We can then either estimate the proportion of passengers in each cohort that had a non-compliant inspection directly, as per the table, or in a model-based way like the previous example to allow for sparse data. This allows us to predict the non-compliance rate as we did previously.

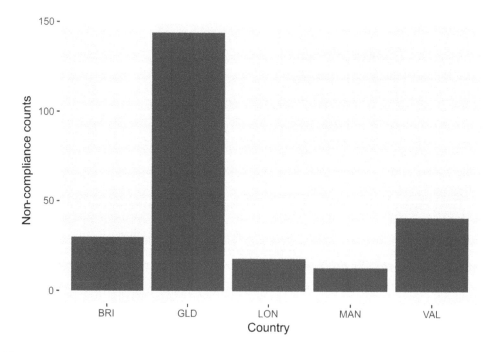

FIGURE 13.5 Non-compliance interception counts by citizenship country in a simulated airline passenger example.

TABLE 13.2

Passenger Cohort Data and the Proportion of Travellers Screened, Predicted from a Generalised Mixed Effects Model

Country	Flight	Gender	Total in Cohort	Non-Compliance Found in Screening	Predicted Proportion Screened	Total Passengers Screened	Proportion of Inspections Non-Compliant
BRI	QE101	F	27	2	0.111	3	0.667
LON	QE314	F	23	0	0.000	0	0.000
MAN	QE101	M	35	1	0.086	3	0.333
GLD	QE905	F	31	0	0.097	3	0.000
BRI	QE314	M	17	0	0.118	2	0.000
VAL	QE905	M	19	0	0.105	2	0.000
BRI	QE905	M	23	0	0.087	2	0.000
GLD	QE101	F	36	2	0.139	5	0.400

In Figure 13.6, the true (known, these are simulated data!) versus predicted non-compliance rate is plotted for each unique cohort in the test dataset for this example. The positive correlation indicates that the model is correctly identifying cohorts based on their risk of non-compliance. At higher levels of risk, the model tends to underestimate non-compliance rates; however, this is of little consequence to the real-world application of passenger profiling. In practice, ranking cohorts on their non-compliance likelihood predicted by the model to prioritise screening of passengers (i.e. selecting the top $x\%$ of passengers for screening) means that the absolute value of the risk estimate is not important.

Figure 13.7 plots the predicted percentage of non-compliant passengers found by screening against the total percentage of passengers screened in the same way as shown in the previous example. The model performance tracks significantly above the $x = y$ line demonstrating its effectiveness in finding a large proportion of the non-compliance in the population when targeting a small

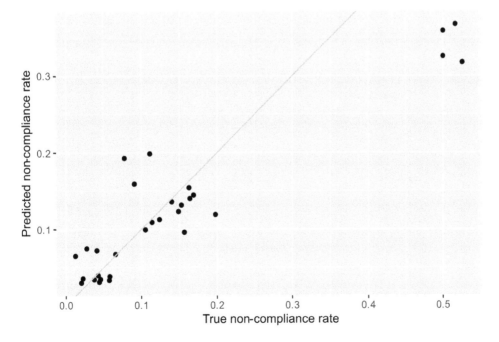

FIGURE 13.6   True versus predicted non-compliance rate for each cohort. Points closer to the y = x line are more highly correlated.

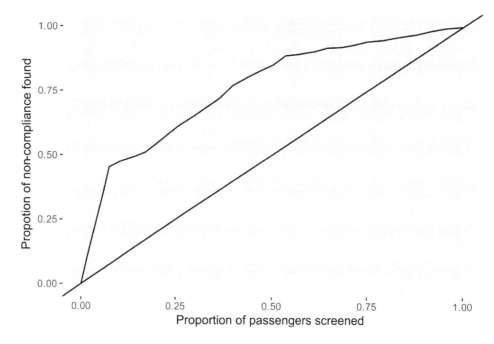

FIGURE 13.7 Comparison of non-compliance found through screening relative to the proportion of passengers screened.

proportion of passengers. Profiling substantially improves on random sampling in this confected example. For example, screening 25% of passengers at random would result in finding an expected 25% non-compliance in the population, whereas targeting passengers based on model predictions would find closer to 63% of non-compliance.

Effects of each predictor characteristic are plotted in Figure 13.8. In this example, flight and citizenship country are important predictors of the likelihood that a passenger is non-compliant

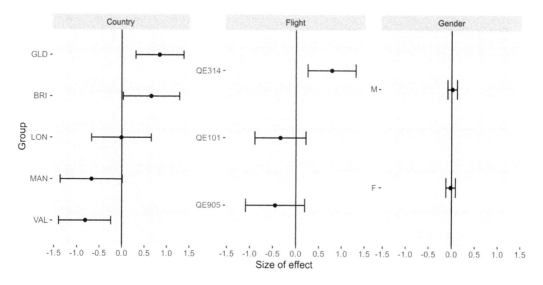

FIGURE 13.8 Plots of generalised mixed effects model effects of passenger characteristics on the likelihood of non-compliance.

but gender is not. The model also correctly identifies BRI as being higher risk than VAL, which is consistent with how the data were sampled in this simulated example.

AUTOMATION BASICS

Profiling can be seen to be a simple form of automation, where the decision of who or what to inspect is informed by a profiling model, using a defensible representation of risk based on data. Machine learning methods can not only be used to create profiling models, but they can also automate a much greater range of activities including reviewing documents, imagery and video, making large-scale undertakings cheaper, and allowing resource-intensive human intervention to be targeted to where it is most useful (e.g. automatically flagging baggage likely to contain BRM for review by a human operator). Machine learning methods are currently being applied to a wide range of biosecurity challenges, from automating document-checking processes to detecting items with high biosecurity risk in X-ray images (IGB 2019). Machine learning is also being used in the post-border and surveillance spaces. Automated methods are being developed to identify infrastructure in satellite imagery associated with particular agricultural activities, to assist future biosecurity responses that will need to quickly identify where particular types of livestock are located (Sheffield et al. 2018). Machine learning is also being used to conduct surveillance for plant diseases using aerial imagery, allowing much larger areas to be surveyed compared to traditional approaches (Miranda et al. 2020; Zarco-Tejada et al. 2018; see Chapter 6. Detect).

Machine learning models are algorithms that are designed to improve themselves through the use of data or experience (Nguyen et al. 2019), allowing them to make predictions or decisions without being explicitly programmed to do so. This allows them to perform tasks that imitate intelligent human behaviour. Some common machine learning techniques include deep neural networks and ensembled decision trees (Russell and Norvig 2021). They are often trained using a supervised approach, where a set of data is categorised and the model is trained to assign information to the correct category. For example, X-ray images could be tagged as containing biosecurity risk, or not containing biosecurity risk, and the model is optimised to return these categories when shown a particular image.

Machine learning methods have great potential to improve the efficiency and scale at which biosecurity activities can be conducted, especially those involving complex data inputs. Although they have the potential to be used to create more efficient workflows, prioritise effort towards more valuable activities and be used to create insights and improve decision making, they can be challenging to develop and implement. There are three key steps to the development process:

1. *Identify the challenge*: Machine learning is best suited to cases where there is a well-defined input from which a decision needs to be made. This could be an X-ray image of a passenger's bag, for which operators need to decide whether BRM may be present, to a document where border staff have to check that certain claims are made. In both cases, machine learning may be able to carry most of the burden, and only refer images or documents to a person where the model is uncertain, or an issue is detected. This will allow staff to process a much greater volume of activity, with much less effort.

2. *Build a prototype dataset*: To determine whether machine learning is actually feasible to solve this challenge, a prototype dataset will need to be put together and manually labelled. These labels should correspond to the decision the machine learning model would need to make.

3. *Train and evaluate the machine learning models*: If the model performs poorly, it may be the case that the problem was not sufficiently well specified, or that the categories the models need to learn need to be simplified. The threshold for a "useful" model will always depend on the application—a poorly performing model may still be worth implementing if it can reduce the amount of time taken to complete a high-volume task.

Implementation requires collaboration between data-scientists, problem owners, front-line staff, and software specialists to deploy the finalised machine learning model. For example, documents submitted by brokers for import clearance could be uploaded directly into a web portal, through which a model that checks compliance could automatically be run. Conversely, software which automatically identifies BRM could be deployed within the X-ray machines themselves. When considering what deployment might look like for a particular problem, it is important to consider the costs and ethics of making a wrong prediction and adjust the thresholds used for decision making accordingly. Currently, machine learning methods are most effective in this fashion if they are applied as screening tools, where issues are flagged and returned to a human reviewer. It is also important to develop the infrastructure (e.g. continual collection of new data) that will be required to support the model into the future, to ensure predictive performance is maintained over time.

PROFILING AND AUTOMATION IN PRACTICE

Profiling and automation are tools within the broader scope of formalised risk models that can be used to inform decisions (Box 13.1).

BOX 13.1 AUTOMATING BIOFOULING ASSESSMENTS

Biofouling, the accumulation of organisms on surfaces immersed in water, is of concern to the international shipping industry because it impacts fuel costs (Davidson et al. 2010). Furthermore, biofouling poses a biosecurity risk by providing an invasion pathway for non-indigenous marine species. There is growing interest by regulators to strengthen biofouling risk management (Scianni et al. 2021) but conducting in-water inspections by dive teams and assessing data to determine the degree of biofouling on a vessel can be prohibitively expensive.

Currently, inspections require the presence of biofouling experts to undertake a dive and/or review inspection footage. These experts may also be responsible for determining whether biofouling risk is acceptable, often based on subjective judgement, which is susceptible to a range of cognitive biases (Bloomfield et al. 2021). Developing an automated method to identify the degree of biofouling from in-water inspections could expedite such assessments and improve their robustness and consistency.

The Centre of Excellence for Biosecurity Risk Analysis (CEBRA) collaborated with Department of Agriculture, Fisheries and Forestry (DAFF), the New Zealand Ministry of Primary Industries (MPI), and the California State Lands Commission, to collate their respective in-water vessel survey datasets to train deep neural networks to assess biofouling (Bloomfield et al. 2021). Deep neural networks are a form of machine learning model that have been applied to similar machine vision problems, ranging from the identification of wild animals in camera-trap images (Norouzzadeh et al. 2018) to the identification of coral species (Gómez-Ríos et al. 2019).

The combined biofouling dataset contained 10,263 images, labelled using a simple three class level of fouling scheme: (1) no fouling present, (2) lightly fouled, and (3) heavily fouled (Figure 13.9). A supercomputer with access to graphics-processing units was used to train the models and test different sets of neural network architectures and hyper-parameters. The final model was an ensemble of the best-performing neural networks, where predictions for each image were averaged to obtain a final prediction.

To evaluate model performance, the three-class problem was split into two binary classification problems. Although there are three classes, there are only two binary problems

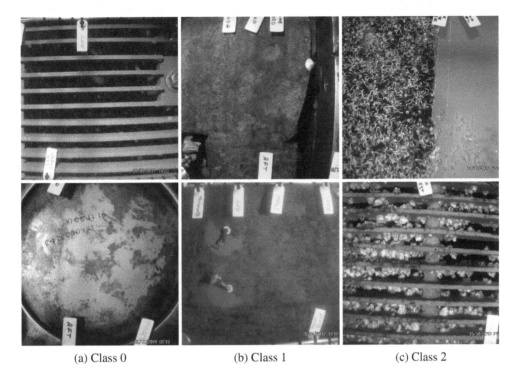

(a) Class 0   (b) Class 1   (c) Class 2

FIGURE 13.9   Example photographs of the simplified level of fouling scheme. Classes are defined as: (a) class 0: fouling organisms not present, but there may be slime or biofilm; (b) class 1: some fouling organisms are present (e.g. barnacles, mussels, seaweed, tube worms), but coverage is patchy (1–15% surface covered); and (c) class 2: large amount of fouling organisms present (16–100% surface covered).

to consider as classes are ordered. The first problem was to identify class 2 (heavily fouled vs. light/no fouling) and class >0 (fouling present vs. absent). The third problem, identifying class 1 (light fouling), was defined by the thresholds obtained for the previous two problems.

The neural networks achieved high consistency (70–87%) in classifying images into fouling classes with a group of experts, similar to the consistency between experts (62–86%). This suggests that the machine learning algorithm performed as well as experts in accurately categorising images by their level of biofouling within this dataset.

Real world complexities can complicate the use of automation in image analysis. In this study, the imagery that was used to train the neural networks was taken using a camera frame facing the vessel hull. However, this may not be common practice for all vessel surveys as lighting conditions, water turbidity, or camera angles can vary widely, which may impact predictive performance. Training neural networks on more challenging examples such as with footage taken at different angles could help them adapt to and improve predictions in a wide range of survey contexts.

These types of risk model incorporate a set of rules, characteristics, or a statistical or machine learning construct that describes our understanding of risk within a pathway or task (Copas 1999). However, putting profiling and automation into practice requires more than just an understanding of how risk is distributed across a pathway or set of inputs. An effective implementation strategy comprises three key parts:

- A defensible risk model that describes the high-risk and low-risk segments of a pathway.
- A mechanism to enable biosecurity officers to prioritise high-risk items.
- An intelligence-gathering strategy to update the risk model.

DEFENSIBLE RISK MODELLING

Prioritising particular areas of a pathway for inspection can have negative consequences (e.g. delays and financial losses) for the people impacted. Profiling activities (e.g. airport security screening) can also be viewed as unethical, discriminatory, and unfair (Wilson and Weber 2008), particularly in cases where officers have discretion in who is selected (Lum et al. 2015). Adding barriers to travel or the importation of goods from particular groups without reliable evidence could potentially breach international expectations and be challenged by other countries as non-tariff trade barriers (Higgins and Dibden 2011). As such, risk models need to be transparent and defensible (see Chapter 3. Anticipate).

To minimise bias, profiling is best conducted using intervention data as an evidence base. If data analysis and statistical modelling are used to build a risk model, then care should be taken to understand how the data were collected. It may be that intervention data are incomplete (e.g. only covering a portion of the pathway, only recording non-compliant inspections, or having inspections contingent on some other issue being identified). In those cases, conclusions drawn from the data will not be applicable to the whole pathway and will provide an unreliable basis for profiling. This is an issue in many profiling contexts (e.g. when crime reporting is geographically biased; Brantingham 2018). Over-sampling or under-sampling a particular group can reduce the accuracy of profiling approaches if the variables most correlated with non-compliance differ among groups (Larrazabal et al. 2020). Under-sampling a group also increases the uncertainty in the predicted biosecurity risk for this group (Altman and Bland 2014; Dakin et al. 2020).

These issues are also particularly important in the context of machine learning models. These models can readily pick up human biases, as they are trained using datasets curated and categorised by humans. Care needs to be taken to ensure the approaches used are defensible and transparent (Sveen, Dewan, and Dexheimer 2022).

PRIORITISING INTERVENTIONS

Models of profiling risk are used to create a simple set of rules targeting high-priority intervention groups. More sophisticated profiling and machine learning approaches could involve directly integrating these risk models into IT systems, to automate the selection of passengers, consignments for intervention, or screening of documents and images. This would allow a more consistent application of profiling rules, finer control of intervention strategies, and greater efficiency gains. In Australia, the "SmartGate" system performs this role (ANAO 2012). Integrating profiles into border activities can also require collaboration between biosecurity regulators and the agencies responsible for border security (Magarey, Colunga-Garcia, and Fieselmann 2009).

When using a risk model, care needs to be taken to consider the resources required. In the context of airline passenger profiling, the arrival of high-risk passengers is likely to be clustered, particularly during times when multiple flights from high-risk countries arrive. Similarly, the arrival of high-risk luggage to be reviewed by a machine learning model using X-ray images is also likely to be clustered. This has implications for how inspection and screening resources are structured, and either compromises will need to be made in regard to the number of high-risk passengers sent through the direct exit channel, or the structure of officers' timetables and tasks will need to be changed to reflect risk patterns (ANAO 2012). While potentially difficult from an operational perspective, the latter choice will likely have significant advantages to effectively managing biosecurity risk.

UPDATING THE RISK MODEL

As new global threats emerge, biosecurity risks change over time (Cheke 2018; Grosholz 2018; Ristaino et al. 2021; Robinson, Cannon, and Sonia 2012). Patterns of trade can shift between countries as consumer preferences change and new industries emerge. Importers targeted by more frequent inspections might become more compliant to avoid delays and fines. Using a profiling strategy to prioritise intervention activities provides little information on low-risk groups in the pathway. This means if conditions change, nothing will be known about the biosecurity risk of these goods in comparison to the current pathway participants. A profiling strategy based on outdated data or experience will become less effective over time (Sedee 2019).

The best way for regulators to stay informed of these changes—so that they can know when the high-risk areas of a pathway shift, and alter their management strategies accordingly—is to split inspection effort between high-risk areas and the remainder of the pathway (Hampton 2005). For example, most of the inspection effort could be directed towards high-risk individuals or consignments, and a portion could be reserved for surveying the rest of the pathway through a random selection process. The endpoint survey for airline passengers is an example of this type of approach.

All profiling strategies need to devote some degree of effort to these types of intelligence or data-gathering activities to be sustainable. Exactly how much effort should go towards targeting intervention activities versus obtaining information to update the risk model will depend on the specific application and the amount of information used for building the risk model. Applications where a small number of variables are used in the risk model could potentially afford to have a smaller number of inspections dedicated to low-risk areas, as there would be fewer dimensions within the data to explore.

Machine learning models also need to be maintained in a similar fashion. With changing patterns of risk over time, on a pathway such as the airline passenger path, it would be expected that the most common items of BRM being brought into the country will also change. This means that the performance of a model that identifies risky items within passenger's bags from X-ray images will likely degrade over time, as it will not have been trained to identify these novel biosecurity risks. When a machine learning model is implemented, a plan for ongoing data collection, labelling, re-training, and review of the models on these new data must be made to prevent this from happening. Having these processes in place will also reduce the risk and allow the identification and correction of problematic or ill-posed predictions in a timely manner (D'Rosario and D'Rosario 2020).

FURTHER BENEFITS OF PROFILING

Profiling based on riskiness is typically introduced by biosecurity agencies to increase the productivity of the inspection process—to increase the amount of non-compliance detected per number of inspections conducted. Profiling can therefore offer a significant improvement in terms of resource allocation (see Chapter 9. Resource Allocation), particularly when coupled with machine learning.

Risk models can also be used to improve the efficiency of other forms of risk-reduction or data-gathering activities, such as education campaigns or stakeholder engagement exercises to reach the individuals most likely to breach biosecurity regulations (Sherring 2021). Not only does profiling allow risk to be minimised given a limited intervention capacity, but it also creates the groundwork to engage with the key contributors of risk, improve their biosecurity practices, and over time, reduces overall risk.

Profiling also has important incentive effects. Without the threat of non-compliance being discovered, agents within the biosecurity system have little incentive to comply with regulations that incur them additional costs (see Chapter 8. Incentives and Chapter 4. Prevent). International travellers may be less willing to dispose of items they have paid for (e.g. fresh fruit, seeds, and cuttings) if there is no penalty or risk of being caught (Sherring 2021). If the threat of non-compliance being discovered is low and the penalties are not sufficiently severe, importers or passengers may be less likely to follow the rules. If non-compliance is discovered, this may result in fines, delays, or other punitive measures, which may cause agents to re-think the costs versus benefit of complying with biosecurity regulations. In this way, intervention activities can be seen as a tool to drive behavioural change within a pathway (see Chapter 8. Incentives). In large pathways, deploying automated tools such as machine learning approaches to screen a greater number of items would likely have a similar effect.

Adaptive inspection schemes target agents on the basis of their compliance history, to provide greater incentives for agents to obey biosecurity rules (see Chapter 4. Prevent and Chapter 5. Screen). In these schemes, agents who fail interventions receive greater scrutiny through more frequent inspections and thus importing becomes relatively more expensive. Given that profiling attempts to *anticipate* which agents are more likely to be non-compliant on the basis of historical patterns of non-compliance, there is potential for adaptive inspection schemes and profiling to be used concurrently.

IN A NUTSHELL

- Profiling uses prior information about a pathway to predict the entities most likely to present a biosecurity risk.
- Profiling is best suited to large pathways (where there are insufficient resources to intervene in every case) and when high-risk entities are clustered within subsets of the pathway.
- The risk models used for profiling need to be transparent and defensible. Risk models are most effective when built using complete and unbiased intervention data.
- For profiling to be sustainable, some intervention effort should be set aside to randomly sample the remainder of the pathway so that changes in risk can be detected and acted upon.
- Machine learning methods have great potential to improve the efficiency and scale at which biosecurity activities can be conducted, especially those involving complex data inputs.

REFERENCES

Altman, D. G., and J. M. Bland. 2014. "Uncertainty and Sampling Error." *BMJ: British Medical Journal* 349: g7064. https://doi.org/10.1136/bmj.g7064.

ANAO. 2012. *Processing and Risk Assessing Incoming International Air Passengers*. Canberra: Australian National Audit Office.

Ashcroft, T. T., D. Nendick, S. M. O'Connor, M. Sarty, D. Gunawardana, and G. Weston. 2008. "Managing the Risk of Invasive Exotic Ants Establishing in New Zealand." In *Surveillance for Biosecurity: Pre-Border to Pest Management*, edited by K. J. Froud, Ian Popay, and S. M. Zydenbos. Hastings, NZ: New Zealand Plant Protection Society.

Barnett, A. 2004. "CAPPS II: The Foundation of Aviation Security?" *Risk Analysis* 24 (4): 909–916. https://doi.org/10.1111/j.0272-4332.2004.00489.x.

BITRE. 2012. Air Passenger Movements Through Capital and Non-Capital City Airports to 2030–31, Report 133. Canberra: Bureau of Infrastructure, Transport and Regional Economics.

BITRE. 2014. Containerised and Non-Containerised Trade Through Australian Ports to 2032–33, Report 138. Canberra: Bureau of Infrastructure, Transport and Regional Economics.

Bloomfield, N. J., S. Wei, B. A. Woodham, P. Wilkinson, and A. P. Robinson. 2021. "Automating the Assessment of Biofouling in Images Using Expert Agreement as a Gold Standard." *Scientific Reports* 11 (1): 2739. https://doi.org/10.1038/s41598-021-81011-2.

Brantingham, P. J. 2018. "The Logic of Data Bias and Its Impact on Place-Based Predictive Policing." *Ohio State Journal of Criminal Law* 15 (2): 473–486.

Casella, G. 1985. "An Introduction to Empirical Bayes Data Analysis." *The American Statistician* 39 (2): 83–87. https://doi.org/10.1080/00031305.1985.10479400.

Cavusoglu, H., Y. Kwark, B. Mai, and S. Raghunathan. 2013. "Passenger Profiling and Screening for Aviation Security in the Presence of Strategic Attackers." *Decision Analysis* 10 (1): 63–81. https://doi.org/10.1287/deca.1120.0258.

Cheke, R. A. 2018. "New Pests for Old as GMOs Bring on Substitute Pests." *Proceedings of the National Academy of Sciences* 115 (33): 8239–8240. https://doi.org/10.1073/pnas.1811261115.

Clarke, S., T. Hollings, N. Liu, G. Hood, and A. Robinson. 2017. "Biosecurity Risk Factors Presented by International Vessels: A Statistical Analysis." *Biological Invasions* 19 (10): 2837–2850. https://doi.org/10.1007/s10530-017-1486-1.

Copas, J. 1999. "Statistical Modelling for Risk Assessment." *Risk Management* 1 (1): 35–49.

Craik, W., D. Palmer, and R. Sheldrake. 2017. *Priorities for Australia's Biosecurity System. An Independent Review of the Capacity of the National Biosecurity System and Its Underpinning Intergovernmental Agreement.* Canberra: Department of Agriculture and Water Resources. [Government Document].

D'Rosario, M., and C. D'Rosario. 2020. "Beyond RoboDebt: The Future of Robotic Process Automation." *International Journal of Strategic Decision Sciences (IJSDS)* 11 (2): 1–24. http://doi.org/10.4018/IJSDS.2020040101.

Dakin, H. A., J. Leal, A. Briggs, P. Clarke, R. R. Holman, and A. Gray. 2020. "Accurately Reflecting Uncertainty When Using Patient-Level Simulation Models to Extrapolate Clinical Trial Data." *Medical Decision Making* 40 (4): 460–473. https://doi.org/10.1177/0272989X20916442.

Davidson, I. C., C. J. Zabin, A. L. Chang, C. W. Brown, M. D. Sytsma, and G. M. Ruiz. 2010. "Recreational Boats as Potential Vectors of Marine Organisms at an Invasion Hotspot." *Aquatic Biology* 11 (2): 179–191. https://doi.org/10.3354/ab00302.

Gelman, A., and J. Hill. 2006. "Logistic Regression." In *Data Analysis Using Regression and Multilevel/Hierarchical Models*, edited by Andrew Gelman and Jennifer Hill, 79–108. Cambridge: Cambridge University Press.

Gilpin, L. H., D. Bau, B. Z. Yuan, A. Bajwa, M. Specter, and L. Kagal. 2018. "Explaining Explanations: An Overview of Interpretability of Machine Learning." 2018 IEEE 5th International Conference on Data Science and Advanced Analytics (DSAA), 1–3 October 2018.

Gómez-Ríos, A., S. Tabik, J. Luengo, A. S. M. Shihavuddin, and F. Herrera. 2019. "Coral Species Identification With Texture or Structure Images Using a Two-Level Classifier Based on Convolutional Neural Networks." *Knowledge-Based Systems* 184: 104891. https://doi.org/10.1016/j.knosys.2019.104891.

Grosholz, E. D. 2018. "New Sources for the Emergence of New Invaders." *Proceedings of the National Academy of Sciences* 115 (10): 2270–2271. https://doi.org/10.1073/pnas.1800257115.

Hampton, P. 2005. Reducing Administrative Burdens. In *London Hampton Report* https://www.regulation.org.uk/library/2005_hampton_report.pdf.

Harrell, F. E. 2015. *Regression Modeling Strategies: With Applications to Linear Models, Logistic and Ordinal Regression, and Survival Analysis.* 2nd ed. Cham: Springer.

Hester, S., A. Rossiter, A. Robinson, J. Sibley, B. Woolcott, C. Aston, and A. Hanea. 2020. CBIS/CSP Sensitivity: Incorporating Pre-Border Information Analysis. In *Centre for Excellence in Biosecurity Risk Analysis.* https://cebra.unimelb.edu.au/__data/assets/pdf_file/0018/3811023/170608-Final-Report-for-web.pdf

Higgins, V., and J. Dibden. 2011. "Biosecurity, Trade Liberalisation, and the (Anti) Politics of Risk Analysis: The Australia-New Zealand Apples Dispute." *Environment and Planning A* 43 (2): 393–409.

Huang, Z., and M. K. Ng. 1999. "A Fuzzy k-Modes Algorithm for Clustering Categorical Data." *IEEE Transactions on Fuzzy Systems* 7 (4): 446–452. https://doi.org/10.1109/91.784206.

IGB. 2019. *Pest and Disease Interceptions and Incursions in Australia.* Canberra: Inspector-General of Biosecurity. Department of Agriculture and Water Resources. https://doi.org/10.1109/CBI.2019.00008.

Lahann, J., M. Scheid, and P. Fettke. 2019. "Utilizing Machine Learning Techniques to Reveal VAT Compliance Violations in Accounting Data." 2019 IEEE 21st Conference on Business Informatics (CBI), 15–17 July 2019.

Lane, S. E., R. Gao, M. Chisholm, and A. P. Robinson. 2017. Statistical Profiling to Predict the Biosecurity Risk Presented by Non-Compliant International Passengers. In *arXiv preprint arXiv:1702.04044*.

Larrazabal, A. J., N. Nieto, V. Peterson, D. H. Milone, and E. Ferrante. 2020. "Gender Imbalance in Medical Imaging Datasets Produces Biased Classifiers for Computer-Aided Diagnosis." *Proceedings of the National Academy of Sciences* 117 (23): 12592–12594. https://doi.org/10.1073/pnas.1919012117.

Lindholm, A., N. Wahlström, F. Lindsten, and T. B. Schön. 2022. *Machine Learning: A First Course for Engineers and Scientists.* Cambridge: Cambridge University Press.

Lum, C., P. Z. Crafton, R. Parsons, D. Beech, T. Smarr, and M. Connors. 2015. "Discretion and Fairness in Airport Security Screening." *Security Journal* 28 (4): 352–373. https://doi.org/10.1057/sj.2012.51.

Magarey, R. D., M. Colunga-Garcia, and D. A. Fieselmann. 2009. "Plant Biosecurity in the United States: Roles, Responsibilities, and Information Needs." *BioScience* 59 (10): 875–884. https://doi.org/10.1525/bio.2009.59.10.9.

Miranda, J. d. R., M. d. C. Alves, E. A. Pozza, and H. Santos Neto. 2020. "Detection of Coffee Berry Necrosis by Digital Image Processing of Landsat 8 Oli Satellite Imagery." *International Journal of Applied Earth Observation and Geoinformation* 85: 101983. https://doi.org/10.1016/j.jag.2019.101983.

Nguyen, G., S. Dlugolinsky, M. Bobák, V. Tran, Á. López García, I. Heredia, P. Malík, and L. Hluchý. 2019. "Machine Learning and Deep Learning Frameworks and Libraries for Large-Scale Data Mining: A Survey." *Artificial Intelligence Review* 52 (1): 77–124. https://doi.org/10.1007/s10462-018-09679-z.

Norouzzadeh, M. S., A. Nguyen, M. Kosmala, A. Swanson, M. S. Palmer, C. Packer, and J. Clune. 2018. "Automatically Identifying, Counting, and Describing Wild Animals in Camera-Trap Images With Deep Learning." *Proceedings of the National Academy of Sciences* 115 (25): E5716–E5725. https://doi.org/10.3390/rs15020417.

Oldroyd, R. A., M. A. Morris, and M. Birkin. 2021. Predicting Food Safety Compliance for Informed Food Outlet Inspections: A Machine Learning Approach. *International Journal of Environmental Research and Public Health* 18, (23): 12635. https://doi.org/10.3390/ijerph182312635

Ristaino, J. B., P. K. Anderson, D. P. Bebber, K. A. Brauman, N. J. Cunniffe, N. V. Fedoroff, C. Finegold, K. A. Garrett, C. A. Gilligan, C. M. Jones, et al. 2021. "The Persistent Threat of Emerging Plant Disease Pandemics to Global Food Security." *Proceedings of the National Academy of Sciences* 118 (23).

Robinson, A., R. Cannon, and G. Sonia. 2012. DAFF Biosecurity Quarantine Operations Risk-Based Approach ACERA 1001 Study H. In *Australian Centre of Excellence for Risk Analysis*.

Rudin, C. 2019. "Stop Explaining Black Box Machine Learning Models for High Stakes Decisions and Use Interpretable Models Instead." *Nature Machine Intelligence* 1 (5): 206–215. https://doi.org/10.1038/s42256-019-0048-x.

Russell, S., and P. Norvig. 2021. *Artificial Intelligence: A Modern Approach, Global Edition.* 4th ed. Harlow, UK: Pearson Education Limited.

Sankhye, S. K. 2020. Machine Learning Methods for Quality Prediction in Manufacturing Inspection. In *Creative Components.* Ames, Iowa: Iowa State University.

Scianni, C., K. Lubarsky, L. Ceballos-Osuna, and T. Bates. 2021. "Yes, We CANZ: Initial Compliance and Lessons Learned from Regulating Vessel Biofouling Management in California and New Zealand." *Management of Biological Invasions* 12 (3): 727–746. https://doi.org/10.3391/mbi.2021.12.3.14.

Sedee, I. 2019. "Creating a Bias in Inspection Data: Exploring the Medium-to Long-Term Effects of Data-Driven Risk-Based Regulation." MSc Engineering and Policy Analysis, Faculty of Technology, Policy and Management, Delft University of Technology.

Sheffield, K. J., J. C. Hunnam, T. N. Cuzner, E. M. Morse-McNabb, S. M. Sloan, J. Nunan, J. Smith, W. Harvey, and H. Lewis. 2018. "Automated Identification of Intensive Animal Production Locations from Aerial Photography." *Australian Veterinary Journal* 96 (9): 323–331. https://doi.org/10.1111/avj.12732.

Sherring, P. 2021. "Declare or Dispose: Keeping Biosecurity Threats Out of New Zealand Using Behaviour Change." In *Broadening Cultural Horizons in Social Marketing: Comparing Case Studies from Asia-Pacific*, edited by Rachel Hay, Lynne Eagle, and Abhishek Bhati, 211–237. Singapore: Springer Singapore.

Stoneham, G., S. M. Hester, J. S.-H. Li, R. Zhou, and A. Chaudhry. 2021. "The Boundary of the Market for Biosecurity Risk." *Risk Analysis* 41 (8): 1447–1462. https://doi.org/10.1111/risa.13620.

Stroup, W. W. 2012. *Generalized Linear Mixed Models: Modern Concepts, Methods and Applications.* Boca Raton: CRC Press.

Sveen, W., M. Dewan, and J. W. Dexheimer. 2022. "The Risk of Coding Racism into Pediatric Sepsis Care: The Necessity of Antiracism in Machine Learning." *The Journal of Pediatrics* 247: 129–132. https://doi.org/10.1016/j.jpeds.2022.04.024.

Turner, R. M., M. J. Plank, E. G. Brockerhoff, S. Pawson, A. Liebhold, and A. James. 2020. "Considering Unseen Arrivals in Predictions of Establishment Risk Based on Border Biosecurity Interceptions." *Ecological Applications* 30 (8): e02194. https://doi.org/10.1002/eap.2194.

Wilson, D., and L. Weber. 2008. "Surveillance, Risk and Preemption on the Australian Border." *Surveillance & Society* 5 (2): 124–141. https://doi.org/10.24908/ss.v5i2.3431.

Zarco-Tejada, P. J., C. Camino, P. S. A. Beck, R. Calderon, A. Hornero, R. Hernández-Clemente, T. Kattenborn, M. Montes-Borrego, L. Susca, M. Morelli, V. Gonzalez-Dugo, P. R. J. North, B. B. Landa, D. Boscia, M. Saponari, and J. A. Navas-Cortes. 2018. "Previsual Symptoms of *Xylella fastidiosa* Infection Revealed in Spectral Plant-Trait Alterations." *Nature Plants* 4 (7): 432–439. https://doi.org/10.1038/s41477-018-0189-7.

Zheng, Y., W. Sheng, X. Sun, and S. Chen. 2017. "Airline Passenger Profiling Based on Fuzzy Deep Machine Learning." *IEEE Transactions on Neural Networks and Learning Systems* 28 (12): 2911–2923. https://doi.org/10.1109/TNNLS.2016.2609437.

14 Map
Creating Maps of Establishment Potential

James S. Camac, Estibaliz Palma, and John B. Baumgartner

ABSTRACT

A strong, efficient, and effective biosecurity system requires the determination of which areas are most susceptible to threat incursion, establishment, and spread. Maps of establishment potential are developed to inform where surveillance resources should be focused, to assess likely spread scenarios, estimate likelihoods of threat absence, create risk maps for prioritising threats and optimally minimise risk. This chapter outlines the three fundamental geographic barriers that influence establishment potential—propagule pressure, abiotic suitability, and biotic suitability. We describe how each barrier is commonly approximated using a range of methods such as species distribution models, pathway analyses, expert elicitation, border screening data, and distributional data of host or habitat. We provide particular emphasis on the various decisions and assumptions that must be made when constructing species distribution models for invasive species. Lastly, we provide a practical and pragmatic approach for integrating these various models and data to create maps of establishment potential. We illustrate this integrated approach using a hypothetical case study on oriental fruit fly (*Bactrocera dorsalis*) entering Australia.

GLOSSARY

Pathway analysis A model that attempts to approximate the propagule pressure or contamination rate of a threat entering a region or point of entry (e.g. ports or airports).

Species distribution model (SDM) A model that approximates a species' niche and maps this niche to the landscape to portray the geographic distribution of suitable conditions.

Establishment likelihood (or establishment potential) A probability that encompasses the three geographic barriers to threat establishment (i.e. the likelihood of arrival or propagule pressure, abiotic suitability, and biotic suitability).

Propagule pressure (also termed introduction rate or contamination rate) A measure of the likelihood or expected number of threat-specific contamination events entering a region.

Abiotic suitability The suitability of the non-living environment (e.g. climate, terrain, or disturbance regime) for a threat.

Biotic suitability The suitability of the living environment (e.g. habitat and interspecies relationships) for a threat. Commonly, this will be based on various habitat requirements (e.g. vegetation type, structure, food source) and sometimes may include interspecific interactions that are facilitative, competitive, or predatory.

Leakage rate The expected number of events (or the likelihood of events) bypassing border screening controls.

DOI: 10.1201/9781003253204-19

Viability rate The likelihood that a contamination event that bypasses border screening is viable for establishment (e.g. based on factors such as pathway survivability or contamination population size).

Model transferability The capacity for a model fit under one set of conditions to be applied to different conditions. In the context of SDMs, a model is said to be transferable if the species-environment relationships estimated under one context remains valid in other contexts.

INTRODUCTION

Threat risk maps are a fundamental decision-support tool in biosecurity, and many governments and industries invest heavily into some form of risk mapping (Elith 2017). Threat risk maps can be used to inform likelihoods of threat absence, where to position finite post-border surveillance resources, trade interventions, and cost-sharing between jurisdictions (see Box 14.1). Threat risk maps can also be used to determine where to prioritise public and stakeholder awareness campaigns for the purposes of early detection or delimitation (i.e. reporting from farmers; see Chapter 6. Detect).

BOX 14.1. USING ESTABLISHMENT LIKELIHOOD MAPS IN BIOSECURITY DECISION MAKING

Maps of establishment likelihoods are fundamental decision-support tools for biosecurity. Specifically, they can be used for:

Informing where to conduct surveillance for early detection

Surveillance for early detection should be prioritised in areas where the likelihood of establishment is highest (see Chapter 6. Detect). Such prioritisation can be done on a threat-by-threat basis using threat-specific establishment likelihood maps. Alternatively, establishment likelihood maps can be combined for multiple threats to systematically identify areas of shared high establishment potential, and thus, inform where to prioritise surveillance for multiple threats (Camac, Baumgartner, Hester, et al. 2021). Maps of establishment likelihood are also useful for assessing the proportion of establishment potential captured by existing or proposed surveillance designs, depending on budget constraints and the spatial arrangement of survey effort (Camac, Baumgartner, Hester, et al. 2021). This is especially important for early detection, which can be expensive to implement and maintain (Camac, Dodd, et al. 2020; see Chapter 6. Detect).

Informing likelihoods of threat absence

To gain and maintain market access, exporting countries may be required to prove that they are free of invasive species and diseases known to affect economic, environmental, and societal values (see Chapter 2. Biosecurity Systems and International Regulations). Freedom of invasive species or disease is commonly assumed if surveillance programs do not detect the threat, especially when they focus on areas of high establishment likelihood or high consequence. A fundamental problem with declaring freedom based solely on the absence of detections is that this approach assumes the surveillance program has perfect sensitivity (i.e. if the invasive species is present, it will always be detected). In practice, it is impossible to be

Map   233

certain a species is absent. Rather, absence can only be inferred in terms of likelihoods (see Chapter 6. Detect). Establishment likelihood maps can be used in combination with surveillance data (i.e. surveillance sensitivity and effort) to infer likelihoods of threat absence that explicitly account for the geographic barriers faced by threats in establishing at a location. Likelihoods of absence can be estimated for locations where surveillance occurs, but also for locations without surveillance to make inferences about absence at regional or national scales (see Chapter 6. Detect).

Initialising spread models

Maps of establishment likelihood can be used to add further realism to spread models for potential exotic threats. Spread models are initialised based on a known incursion point or the location of a known established population (Bradhurst et al. 2021). Initial incursions may be simulated at random across a region of interest, or be specified to occur at a particular location of interest perceived to be at high risk of establishment. Likelihood maps that account for the geographic barriers to establishment (namely, propagule pressure and suitability of biotic and abiotic environments) provide a coherent and transparent method for simulating initial establishment events that seed subsequent spread events. By simulating both the introduction and subsequent spread of a threat, the costs and benefits associated with different pre-border, border, and post-border management strategies can be assessed, ultimately allowing decision-makers to better anticipate and proactively manage future incursions (Bradhurst et al. 2021).

Developing threat risk maps

Establishment likelihood maps are not themselves measures of risk. However, when coupled with spatial estimates of the economic, environmental, and societal/cultural consequences of a threat (see Chapter 3. Anticipate), maps of establishment likelihood can be used to produce risk maps. Risk maps can inform where to conduct surveillance or control measures to minimise risk, for example in areas that exhibit both high likelihoods and consequences of establishment. Risk maps can also be used to prioritise locations for risk mitigation activities, especially those associated with proactive asset protection (see Chapter 7. Prepare, Respond, and Recover).

Depending on how they are built, threat risk maps can describe where an invasive species or disease may arrive, establish and spread, or cause significant economic, environmental, and societal harm. True risk maps describe the product of (1) the likelihood a species will successfully establish in an area and (2) the expected magnitude of harm if establishment occurs in that area. Such maps are particularly useful for identifying locations where both the establishment likelihood and potential consequences of an incursion are high, and thus, where risk mitigation controls should be prioritised.

In practice, however, most so-called risk maps do not incorporate both likelihoods and consequences. Rather, maps tend to focus on estimating likelihoods associated with a particular stage of the invasion process, such as the likelihood of a threat arriving at a port via imported goods (i.e. pathway analysis) or the likelihood an area is environmentally suitable for establishment to occur (Venette et al. 2010).

While many methods exist for constructing likelihood maps, two main approaches are used in biosecurity. The first—pathway analysis—involves estimating the arrival rate of a threat entering a country or arriving at a point in the landscape (e.g. ports). Commonly, pathway analyses are informed by border interception data coupled with movement data (or models) describing vectors of spread (e.g. imported goods, people, wind) from areas of known establishment to locations of interest (Camac, Baumgartner, Garms, et al. 2021; Tingley et al. 2018).

The second commonly used approach is to construct species distribution models (SDMs) that approximate the spatial distribution of suitable environment needed for a threat to establish and persist. SDMs can be parameterised using data on a threat's known physiological tolerances (e.g. temperature tolerances; Kearney et al. 2008), occurrence records from its global or native range (Phillips, Anderson, and Schapire 2006), or a combination of the two (Kriticos et al. 2017). For invasive species, SDMs commonly predict climate suitability as a function of global climatic variables under the assumption that climate is the primary driver limiting a threat's potential geographic distribution.

Despite both pathway analyses and SDMs providing partial (and complementary) information on the barriers to establishment, there have been few attempts to integrate these two model types in biosecurity. Studies that have attempted to link pathway likelihoods with estimates of environmental suitability are often geographically restricted to points of entry (e.g. ports) and do not make inferences about relative likelihoods of establishment beyond these locations (e.g. Tingley et al. 2018).

In this chapter, we outline a general framework for estimating the geographic establishment potential of exotic threats using data types available in most countries. We provide practical guidance on how to (1) distribute pathway risk beyond points of entry using various "rules of thumb" and (2) estimate abiotic and biotic suitability for a threat. We illustrate this mapping framework using a case study of the oriental fruit fly (*Bactrocera dorsalis*) potentially establishing in Australia.

DEVELOPING MAPS OF ESTABLISHMENT LIKELIHOOD

Neither pathway analyses nor SDMs estimate the establishment potential (or establishment likelihood) of a threat. This is because the establishment potential of an exotic threat is governed by three spatial constraints (Figure 14.1), where each must be met for establishment to occur:

1. Can the threat reach the location of interest (i.e. propagule pressure)?
2. Are abiotic conditions suitable (e.g. climate suitability)?
3. Are biotic conditions suitable (e.g. presence of host or food)?

While pathway models inform the likelihood of a species arriving at a location (i.e. propagule pressure), they rarely account for abiotic suitability (e.g. climate) or biotic suitability (e.g. presence of hosts

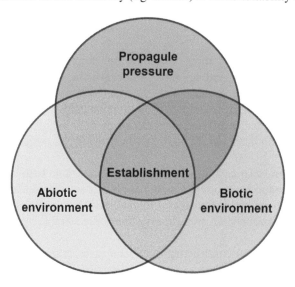

FIGURE 14.1 The three main elements governing the likelihood of establishment of exotic species in the introduced region. (Adapted from Catford, Jansson, and Nilsson 2009).

Map   235

or specific vegetation types). While SDMs can approximate climate suitability and to a lesser degree biotic suitability (if relevant predictors are available), they do not account for propagule pressure.

In the following sections, we provide overviews of each of the three barriers to threat establishment (Figure 14.1) and how they can be estimated using commonly available data and methods. Once estimated, the product (i.e. multiplication) of these three likelihoods give the probability of establishment conditioned on the three barriers.

ESTIMATING PROPAGULE PRESSURE

Most contemporary introductions of exotic threats have been facilitated by the movement of people and goods across continents and international borders (Hulme 2021). To determine the likelihood of an exotic threat arriving at a location (i.e. a pathway analysis), biosecurity practitioners not only require an understanding of how the threat may spread across jurisdictional boundaries, but also an understanding of the main vectors for post-border spread within these regions.

A pathway analysis estimating the expected likelihood of a post-border arrival requires information on: (1) volumes of pathway carriers (e.g. passengers, cargo, or mail) from infected locations to the region of interest; (2) the likelihood a carrier contains a threat viable for establishment; (3) the likelihood contaminated carriers are not detected at the border; and (4) how contaminated carriers disperse post-border.

Pre-border regulation and border screening of people and imported goods entering a country provide data useful for approximating the probability of exotic threat arrival. Data obtained from border screening can be used to fit statistical models that estimate the rate at which a threat arrives at a country's border or points of entry (see Chapter 5. Screen). If information is available regarding the viability of threats (e.g. estimated through transit survival likelihoods or average population size for detected contamination events) and the efficacy of border screening (e.g. by conducting endpoint surveys), statistical models can quantify the expected likelihood that viable introduction events bypass pre-border and border controls.

Estimates of the propagule pressure passing through various points of entry are then paired with information about how the carriers of the threat disperse post-border. For example, if a contaminated shipping container passes border screening at a port, it may be moved a considerable distance before being opened and unpacked, allowing the threat to escape into the wider landscape. Inadequate post-border monitoring of the movement of goods and people, alongside privacy concerns about collecting these types of data, are currently the main barriers to conducting pathway analyses that incorporate post-border movements.

Despite the limited availability of post-border movement data, opportunities exist to use available data to approximate the post-border dispersal of threats. Entry pathways are diverse, but in most cases are related to either human movement or human activities. Depending on the introduction pathway and the availability of national datasets, post-border movement of invasive species and diseases can be approximated in a variety of ways. For example, it can be assumed that international tourists entering a country will concentrate around areas of high tourist accommodation (at least initially) or that returning residents or international mail will distribute throughout a country as a function of human population density or where commerce activity is greatest. Most imported goods (e.g. food, nursery stock, or vehicles) can be assumed to be destined for areas of higher human population density. Post-border distribution of other goods (e.g. farm equipment and fertiliser) can be approximated by coupling land use type with measures of agricultural intensity (e.g. fertiliser usage statistics, farm numbers). There is evidence to support such assumptions, with studies commonly finding land use, road density, and human population density to be strongly correlated with first detections of exotic threats even when correcting for potential survey bias (e.g. Dodd et al. 2016).

ESTIMATING ABIOTIC SUITABILITY

Abiotic barriers limiting a species' potential distribution are diverse in nature and scale and can include climate, disturbance regimes, and terrain attributes. At large geographic scales (e.g. global or continental), the potential distribution of a species is thought to be governed by climate (Araújo and Rozenfeld 2014). The last several decades have seen the development of global databases of climate and biological data as well as a vast range of statistical and mechanistic SDMs. In invasive species management, SDMs are commonly parameterised using global climatic variables (the most readily available global environmental data), and thus, are sometimes referred to as "climate suitability models".[1] These models have become a popular tool in biosecurity for quantifying the potential area of occupancy, which in turn can inform threat prioritization (McGeoch et al. 2016), post-border surveillance design (Camac, Baumgartner, Hester, et al. 2021), and estimates of expected economic impacts (Dodd, Baumgartner, and Kompas 2021; Dodd et al. 2020).

Many biosecurity agencies and associated research institutions actively invest in some form of invasive species climate suitability modelling, whether via initiatives such as the European Union's Pratique risk assessment program, the development of cloud computing systems such as Biosecurity Commons (Biosecurity Commons 2022), the development of proprietary software (e.g. CSIRO's CLIMEX; Kriticos et al. 2016), or governmental platforms such as the United States's NAPPFAST (Magarey et al. 2007) or Australia's CLIMATE/CLIMATCH (ABARES 2020; Climatch 2019).

Despite a vast number of SDM methods (and diverse opinions on how to use them), there remains no clear evidence of a single best approach for predicting invasive species' potential distribution (Barry et al. 2015). Irrespective of the approach chosen to predict climatic suitability for a species, a wealth of subjective decisions and assumptions are made along the model building process, from data sourcing and cleaning to model development and validation. Subjective or arbitrary decisions can lead to a wide variety of potential suitability predictions for a single species, even when the modelling algorithm is held constant (Camac, Baumgartner, Hester, et al. 2021).

In the following sections, we describe the main steps required for biosecurity practitioners to develop and use SDMs (Figure 14.2).

Data Sourcing

Literature Review

When considering whether to build or reproduce an SDM, the first step should be to conduct a short review of the grey literature (national and international governmental reports) and peer-reviewed literature. The literature review should summarise details associated with physiological climate tolerances, existing SDMs, required hosts, and any distributional data reported in the literature. A good place to start a literature review is to examine the Invasive Species Compendium maintained by the Centre for Agriculture and Bioscience International (CABI 2023). The compendium contains data summarising threats' biology, distribution, spread pathways, and associated information sources.

Occurrence Records

Ideally, SDMs should be based on occurrence datasets including both presence and absence localities (i.e. presence-absence methods) or abundance data. Since these are rarely available globally, the majority of SDMs are built using only presence localities (presence-only methods) or by comparing presence locations to a random sample from all available locations (presence-background methods).

In recent years, publicly available occurrence databases have been pivotal in driving the popularity and accessibility of presence-only/presence-background SDM methods. At the global scale, the most comprehensive dataset is the Global Biodiversity Information Facility (GBIF 2023). The data contained within GBIF can be used in different modelling frameworks for either parameterising models or for validating them. Species-specific databases or other expert-curated datasets are another valuable source of occurrence information. These databases are typically compiled and verified by experts and are thus considered highly reliable. In many cases, these data are already integrated into GBIF, however, this is not always the case. As such, it is important that searches are made for such speciality databases.

Map 237

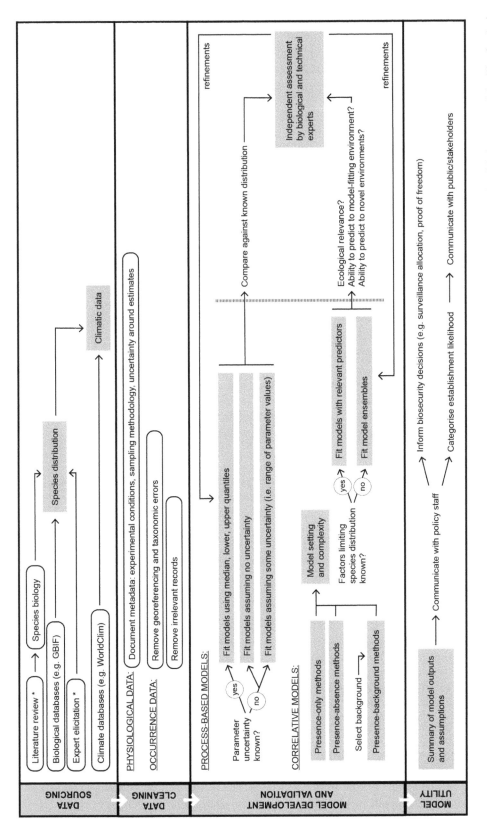

FIGURE 14.2 Steps involved in creating a climatic suitability model, from data sourcing and cleaning to model development and validation, to model utility for decision making. (Adapted from Camac, Baumgartner, et al. 2020). *If the species' biology or distribution is unknown, expert elicitation can be used, or data can be collected from datasets for phylogenetically/morphologically similar species.

Physiological Data

Physiological data that may be incorporated into process-based SDMs are generally derived from controlled laboratory experiments published in the peer-reviewed literature and in academic theses. Several demographic and physiological databases have also been published to simplify parametrisation of process-based models, including:

- *GlobTherm*: A database that includes thermal tolerance metrics for 2,133 species of multicellular algae, plants, fungi, and animals, extracted from published studies (Bennett et al. 2018).
- *AmP*: A database of referenced data on animal energetics and parameter values of models based on Dynamic Energy Budget (DEB) theory for over 1,000 animal species (Marques et al. 2018).
- *COMPADRE*: A database of more than 5,500 population matrix models (and ancillary information) for more than 600 plant species worldwide (Salguero-Gómez et al. 2015).
- *COMADRE*: A database of more than 1,625 population matrix models (and ancillary information) for more than 340 animal species worldwide (Salguero-Gómez et al. 2016).

Climatic Data

Common global climate databases include:

- WorldClim (Fick and Hijmans 2017; WorldClim 2023),
- Climate Research Unit (CRU 2023; Harris et al. 2014),
- CliMond (CliMond 2014; Kriticos et al. 2012), and
- CHELSA (CHELSA 2023; Karger et al. 2017).

Each dataset contains different suites of long-term averaged climatic variables, and estimates long-term climatic averages across the earth using different data sources, interpolation algorithms, and temporal periods. It is therefore important for the user to carefully select the variables (and dataset) that are likely to be most relevant to the threat of interest.

Non-Climatic Abiotic Data

Non-climatic variables (e.g. land use, and disturbance patterns) also play an important role in governing a species' potential distribution. However, such information is generally not available at either the spatial extent or resolution required to parametrise SDMs. Where such information does exist (e.g. land use layers), it is often based on image classification algorithms from global satellite imagery. Consequently, such data are often coarse (either in terms of spatial resolution or categorisation), and thus, may have limited utility in SDM. Over time, the quantity, quality and accessibility of these datasets will undoubtedly improve.

Alternatives for Data-Poor Threats

When there is insufficient biological or distributional data for a given threat, there are three options to estimate relevant physiological tolerances or responses to climatic factors: (1) collate data on a threat with similar phylogeny or similar biological, behavioural, or physiological traits (Morales-Castilla et al. 2017); (2) use a trait-based SDM (e.g. Morin and Lechowicz 2008; Pollock, Morris, and Vesk 2012); or (3) use an expert elicitation framework (e.g. IDEA protocol; Hemming et al. 2018).

Data Cleaning

Occurrence Data

Species occurrence records derived from online databases (e.g. GBIF) are a valuable resource for parametrising and validating SDMs. However, issues with data quality such as errors in georeferencing, taxonomy, temporal periods, irrelevant records (e.g. fossil records, cultivated plant specimens, and zoo animals), and incomplete metadata can introduce significant biases in model fitting and validation (Maldonado et al. 2015). A summary of common errors found in occurrence databases can be found in Table 14.1.

Map 239

TABLE 14.1

Common Errors in Occurrence Databases.

Geo-referencing errors
- Incorrect habitat (marine or terrestrial)
- Coordinates and country mismatches
- Switched or equal latitude and longitude
- Records on country or province centroids
- Records based on rasterised collections
- Records with high spatial uncertainty or strong decimal rounding
- Coordinates entered as (0, 0)

Taxonomic errors
- Misidentifications
- Obsolete or incorrect taxonomy
- Incorrect spelling

Temporal errors
- Incorrect or missing dates
- Records not within climate temporal period

Irrelevant records
- Records from zoos, botanical gardens and museums
- Fossil records
- Migratory records
- Border interceptions
- Records from less reliable sources (e.g. iNaturalist)
- Duplicate records

Standardised tools for cleaning large-scale biodiversity databases include the R package CoordinateCleaner (Zizka et al. 2019), which identifies records from cultivated sources as well as geographic and temporal errors commonly found in biological databases (Table 14.1). Another useful tool is the R package taxize, which helps deal with taxonomic issues such as spelling mistakes, synonyms, or obsolete taxonomy (Boyle et al. 2022; Chamberlain and Szöcs 2013).

Once data have been cleaned, it is critical to remove records that are likely to be transient in nature (i.e. unlikely to be associated with a persistent population). This is because biosecurity applications typically focus on delimiting the area where a threat could establish and spread. CABI (https://www.cabidigitallibrary.org/product/qi) is an example of an expert-verified database that contains the known establishment status of threats at the country or provenance scale. Where appropriate resources are available, all records should be scrutinised by experts in the relevant geographic area to verify whether they are likely to be records of established populations, and if not, they should be removed prior to model development or validation.

Physiological Data

Experimental physiological data suffer from two primary concerns: whether the experimental conditions are transferable to the natural world and whether the individuals used are representative of variability within a species (Briscoe et al. 2019). Both questions are difficult to answer, especially the latter, where intra-specific variability is almost never known. By documenting the experimental approach, the number of individuals used, where they were sourced, and the number of populations sampled, the modeller and consulted experts have the required information to make informed decisions regarding both the representativeness and transferability of experimental findings.

It is also important to record the uncertainty around any experiment-derived physiological responses to allow for the propagation of uncertainty throughout the modelling process. Ultimately, this manifests itself in a range of possible distribution maps, ranging from best to worse case scenarios. It is also important to document physiological processes (e.g. factors governing reproductive maturity or diapause) that lack empirical estimates. Documenting these knowledge gaps highlights areas for future research and provides greater context for the decisions made during model development.

Model Development

There remains no clear evidence of a single, best approach for predicting an invasive species' potential distribution (Barry et al. 2015). To determine which approach performs consistently well requires an understanding of models' ability to predict the potential invasive distribution of a species, which for practical purposes is unknowable (Elith 2017). As such, modellers have mostly evaluated methods based on their predictive performance when predicting to current distribution data (Sequeira et al. 2018, Yates et al. 2018), but these performance metrics do not explicitly examine performance with respect to the true purpose of these models—their ability to predict the potential, currently unoccupied distribution of a species.

Accurate prediction in the region where the model was fitted does not guarantee accurate prediction outside this range (Elith 2017; Fourcade, Besnard, and Secondi 2018). To examine a model's transferability from one region to another, various cross-validation approaches have been used. One commonly used approach is to parametrise a model based on native-range occurrences and use it to predict to the observed invasive region (Yates et al. 2018). This approach is also fraught with problems because the distributional data in the native or invaded range may not be a true indication of a species' ability to persist under environmental conditions not currently occupied (Elith 2017; Elith, Kearney, and Phillips 2010). As with all modelling, while decisions about the most useful method can be informed by existing knowledge, there remains ambiguity in the final choice of model to predict an unobserved event.

Broadly, SDMs can be classified as either process-based models or correlative models. Determining which model to use is ultimately dependent on data availability, tolerance to the assumptions made, and the decision context (Elith 2017; Guillera-Arroita et al. 2015).

Correlative Models

Correlative SDMs exploit the statistical association between spatial environmental data and occurrence records to implicitly capture biological processes limiting the distribution of the species. Irrespective of which correlative approach is used, an "equilibrium assumption" applies—that is, that a species is in equilibrium with their environment (i.e. the species has had the opportunity to establish in all suitable areas within the model-fitting region). A vast range of correlative methods can be used to predict the potential distribution of a species or disease. These methods fall into four categories:

- **Presence-only models** commonly estimate a species' potential distribution in two ways: (1) using occurrence records to delimit the climatic range in which a species may persist (e.g. envelope methods such as BIOCLIM, Booth et al. 2014; and range bagging, Drake 2015); (2) by calculating climatic similarity relative to the climate of known occurrence locations (CLIMATCH; ABARES 2020, CLIMEX; Kriticos et al. 2016). The primary appeal of presence-only models is that they use the most commonly available type of data (presence records) without additional assumptions and decisions associated with other methods (see below). However, presence-only models can be highly sensitive to outliers (e.g. when delimiting climatic ranges) and, like all correlative methods, to sampling biases (i.e. some sites are more likely to be sampled than others; Elith 2017).
- **Presence-background models** derive suitability scores from the frequencies of surveyed occurrences (i.e. positive cases) relative to non-positive cases. The meaning of non-positive cases can vary in important ways (see Box 14.2). For some popular methods, such as Maxent (Phillips et al. 2006), the non-positive case is commonly referred to as the

Map 241

background (i.e. a set of points that randomly sample the landscape or locations available, irrespective of whether a species is present or not).

- **Presence-pseudo-absence models** treat the non-positive cases as implied absences or pseudo-absences (e.g. Genetic Algorithm for Rule-Set Prediction; GARP). These implied absences are either placed (1) anywhere except where presences occur or (2) in geographic or environmental zones considered to be unlikely to contain the species of interest (Elith 2017). The peer-reviewed literature contains several suggestions on how to choose locations for pseudo-absences or define reliable absences for presence-only data (e.g. Le Maitre, Thuiller, and Schonegevel 2008; Lobo, Jiménez-Valverde, and Hortal 2010). Relative to the presence-background approach, this method requires additional subjective decisions about both the position and number of pseudo-absences and is generally perceived as a less rigorous statistical framework (Renner et al. 2015).
- **Presence-absence models** derive suitability scores from relative frequencies of surveyed occurrences and absences in different environmental conditions. Presence-absence models are the ideal correlative model because they can more readily account for imperfect detection and survey effort while also removing the need to arbitrarily select background points (Guillera-Arroita et al. 2015; Lahoz-Monfort, Guillera-Arroita, and Wintle 2014). Unlike other methods, they can be used to estimate true probabilities of presence as opposed to relative probabilities (Guillera-Arroita et al. 2015). If the required data are available, presence-absence models should always be preferred over other correlative approaches. In practice, such data are rarely available at the geographic resolution and extent required to estimate the potential distribution of an invasive species and their use in invasive species management is limited (Elith and Leathwick 2009).

BOX 14.2. CHALLENGES AND ISSUES IN USING CORRELATIVE MODELS FOR INVASIVE SPECIES

Correlative methods that do not utilise data on absences (i.e. presence-only, presence-background, and presence-pseudoabsence models) do not estimate true probabilities of occurrence, as this requires knowing the true prevalence of the modelled organism (Guillera-Arroita et al. 2015). Rather, these correlative models estimate relative probabilities that are commonly assumed to be proportional to the true probabilities. When models are fitted for multiple species, direct comparisons can be made between their general patterns of predictions (e.g. which areas are predicted to have high vs. low suitability) but not between their predicted suitability scores. In other words, a suitability score of 0.9 for one species, may not be the same as a 0.9 for another species.

Presence-background and presence-pseudoabsence models rely on additional subjective decisions about the extent, distribution, and number of background locations or pseudo-absences. The impacts of these decisions are often underappreciated, despite evidence highlighting how different arbitrary modelling decisions and assumptions can lead to substantial differences in model predictions and interpretation (Phillips et al. 2009; Syfert, Smith, and Coomes 2013; Warton and Shepherd 2010).

Ideally, for presence-background models, background points should characterise the range of environments to which the species of interest could have dispersed and should exhibit the same pattern of survey bias as exists in the occurrence records. In practice, defining the extent of potential dispersal is challenging, especially for invasive species with actively expanding ranges (VanDerWal et al. 2009). To reduce the impact of such complications associated with modelling species with expanding invasive ranges, presence-background models often restrict the species' occurrence records to the native range. This decision also simplifies the process of background sampling by limiting the choice of sample points to the extent of the species' native range. Yet, discarding records beyond the species' native range will typically reduce the sample size and may result in potentially unreliable extrapolation when predicting climatic suitability across the species' exotic range (Broennimann and Guisan 2008). For this reason, modellers may be hesitant to discard occurrence records in the invaded range.

When SDMs use records from both the native and the exotic range, different approaches can be used to select background points. The first approach is based on constructing convex hulls around the occurrence records and sampling the background points from within the hulls. The second approach is to sample background points from within a radius from each occurrence record; the size of the radius being defined by the user, usually based on the species' dispersal ability (VanDerWal et al. 2009). The third approach involves sampling background points from a pre-defined set of bioclimatic zones believed to provide suitable environmental conditions for the modelled species.

Occurrence data are often biased towards areas with high population and infrastructure density. To avoid confusion between patterns of survey effort and patterns of climatic suitability, survey bias needs to be accounted for in the SDMs. A simple method to do this is called target-group sampling. This method uses occurrence records from a broad set of species for which sampling bias is likely to be similar to that for the species of interest, due to similar behaviour by those reporting occurrences (Phillips et al. 2009). The sites for all records from all species in the target group then make up the full set of available information on survey effort and can be directly used as background data within the region of interest. A second method extends the target-group approach, using the target-group occurrences to construct an interpolated bias grid that informs the appropriate distribution and intensity of background samples. Alternatively, data describing factors likely to explain geographic survey bias (e.g. distance from roads and cities) can be used directly to calculate such bias grids (Warton, Renner, and Ramp 2013) or can be combined in regression models to estimate them.

While the above methods can efficiently deal with sampling bias, they are not commonly applied in invasive species modelling. This is due partly to lack of technical expertise, and partly to the broad spatial scale of modelling (across countries) that prevents the inclusion of relevant factors at smaller scales (e.g. distance to roads). However, sampling bias can substantially challenge predictions of climatic suitability for invasive species based on presence-only data, and spatial variation in survey effort should be addressed in these models where feasible.

Process-Based Models

Process-based (mechanistic) SDMs are models that explicitly incorporate biological processes that limit distributions. These models range in complexity from simple climatic threshold mapping to complicated biophysical models such as NicheMapR (Kearney and Porter 2017). The commercial software CLIMEX (Kriticos et al. 2016) can also incorporate processes when it is parametrised using experimental data.

The primary advantage of process-based models is that they attempt to model the causal relationships between climate and various vital attributes such as survival, growth, and fecundity. They are therefore assumed to be superior at predicting a species' fundamental distribution (Elith 2017; Kearney and Porter 2009), though limited evidence supports this assumption (Uribe-Rivera et al. 2022). The reliability of process-based models is strongly dependent on whether they account for the fundamental processes delimiting a species' fundamental niche and whether the empirically derived parameters accurately reflect both the responses to climate and intra-specific variability in these responses (Kearney and Porter 2009). In most cases, the climatic factors governing a species' potential distribution are unknown or unknowable, and the resources required to

Map 243

obtain such data (i.e. lab experiments and field validation), coupled with time and in-house technical skills required to build these models, currently make them infeasible for most biosecurity agencies (Briscoe et al. 2019).

Model Settings and Complexity

Correlative methods used to fit SDMs allow the user to fine-tune a variety of settings to achieve the desired model fit and interpretation. Such settings include, among others, the selection and number of background points, the complexity of covariate response curves (i.e. the relationship between the value of an environmental variable, e.g. annual mean temperature, and suitability for the modelled species) and the type of model outputs that can be extracted.

Default settings for SDMs methods are not always appropriate. For example, MAXENT background points are, by default, randomly sampled across the entire extent of the climatic data supplied by the user. Background points should typically be restricted to regions that the species has had an opportunity to reach and occupy (VanDerWal et al. 2009). Thus, when determining appropriate model settings, the species and the problem at hand should always be considered.

Model predictions can be affected by model complexity, which can strongly influence the transferability of the model to conditions outside the range of environmental conditions used to fit the model. Highly complex models (i.e. those with many predictors and interactions) are susceptible to over-fitting to noise and peculiarities of the model-fitting data, especially when the number of occurrence records is low (Breiner et al. 2015). Such models may not generalise well, limiting their ability to accurately predict climatic suitability in regions outside the model-fitting range. This issue can be minimised by optimising a model such that it more accurately predicts data not included in model-fitting.

This can be achieved through the process of cross-validation, whereby the occurrence dataset is split into model-fitting and model-testing subsets, and average performance over the replicate models is evaluated.[2] This process can be repeated for each of a set of competing models that use different predictors and varying complexity, to identify models that predict well and exhibit high transferability. In cases when cross-validation is not feasible (e.g. due to small sample size), opting for simpler models is recommended—with few predictor variables, and few or no interactive terms. Simple models retain greater tractability and are likely to be more easily generalised to new datasets (Merow, Smith, and Silander Jr 2013).

Covariate Uncertainty

Accurately predicting abiotic suitability for species through correlative models requires knowledge of the causal predictors (covariates) that govern the species' niche. However, selection of model predictors remains a significant challenge (Barry et al. 2015). There are no obvious "best" predictive variables, and users tend to rely on variables provided in readily available datasets (e.g. WorldClim). When the choice of predictors is unguided, we recommend that a practitioner ensemble (i.e. model average) a range of simple (2–3 parameter) competing models such that overfitting is less likely and model/covariate uncertainty is encapsulated in the final suitability output. Ensembles of such simple models have shown great promise in predicting distributions of rare (Breiner et al. 2015) and non-native species (Hill et al. 2022).

Model Validation

Irrespective of whether a model is process-based or correlative, the outputs should be validated against independent datasets (i.e. occurrence data not used to parametrise the model). For process-based models, validation is straightforward and involves overlaying localities of known establishments onto maps of predicted suitability and then examining whether patterns broadly match (Kearney et al. 2008; Kriticos et al. 2017). However, validation of correlative SDMs for invasive species faces a variety of difficulties. Validation tools for correlative models range from examination of the model residuals to formal statistical summaries (e.g. area under the receiver operating curve AUC, kappa, explained deviance) that examine the model's ability to accurately predict the fitted, or ideally independent, data (Elith 2017). Unfortunately, the application of these statistical evaluations to invasive species modelling is not particularly appropriate (Elith 2017). This is because the objective of invasive species SDMs is to predict a species' potential (i.e. unobserved) distribution, not just its known range, which is what these statistical measures inform.

To overcome these issues, some authors have fitted the invasive species SDMs using occurrence records from the native range only, and then validated the predictions by either using data in the same region or by using data from the known invaded range. While there is some support for the former validation approach (Barry et al. 2015), using the known invaded region for validation is dependent on: (1) whether limiting factors for the species' distribution in the native and invaded ranges are comparable (i.e. whether the model is indeed transferrable), and (2) whether occurrences in the invaded range are representative of persistence.

Due to the difficulties in evaluating and validating correlative invasive distribution models, we suggest that:

- Models are first assessed for their ecological relevance through: expert knowledge; examining functional response curves; sourcing additional data including physiological information (e.g. temperature tolerances).
- Model predictive ability in the model-fitting region is assessed (e.g. by using AUC) if the method involves estimating response curves (e.g. non-ensemble models).
- Regions of novel environmental space (i.e. environmental conditions not found in the model-fitting region) are examined, especially for suitability maps derived from a correlative model that estimates response curves.
- Model outputs and associated model documentation (i.e. how the model was parametrised, from data sourcing through to validation) be independently assessed by both technical and biological experts. Doing so will ensure key aspects of the species' biology are incorporated and that models are appropriately implemented.

Model Utility for Decision Making

To ensure the appropriate use of a model, fundamental details must be clearly articulated and summarised for biosecurity decision-makers. It is critically important to summarise what data were used, how the model was developed, assumptions that were made, and how one should interpret and can use the output. In essence, the documentation should account for all decisions made in Figure 14.2.

The type of model will have important implications for how it can be used to inform biosecurity decision-making. Correlative models are commonly used to inform a variety of biosecurity decisions. However, while these models may be appropriate in some decision contexts (e.g. spatial prioritisation of surveillance) they may be inappropriate in others. Guillera-Arroita et al. (2015) highlighted that using presence-background models such as Maxent to inform multi-threat prioritisation is often inappropriate because outputs are relative probabilities that are not comparable across species. For a comprehensive guide as to whether relative likelihoods or ranked scores can be used for various biosecurity decision contexts, we recommend readers consult Guillera-Arroita et al. (2015).

It is also common practice to convert maps of continuous predictions (e.g. true and relative probabilities) into a binary (suitable/not suitable) or categorical (high/moderate/low) heat map (Guillera-Arroita et al. 2015). While this is a simple way to communicate to the broader public, we strongly advise against discretising model outputs in biosecurity decision contexts. This is because the decision of where a particular threshold should be is often highly subjective and dependent on both the expected consequences of an invasive species outbreak and the tolerance for incorrectly labelling a suitable location as unsuitable. Discretising model outputs rarely improves decision-making because it degrades the information supplied by the model (Guillera-Arroita et al. 2015).

Map   245

ESTIMATING BIOTIC SUITABILITY

Biotic barriers such as various habitat requirements (e.g. vegetation type, structure, food source) and inter-specific relationships (e.g. facilitative, competitive, or predatory) can have significant impacts on species' persistence, and consequently, their ability to establish in new locations. While such information is sometimes incorporated in SDMs, it is rarely done so for invasive species. This is because: (1) methods best suited to modelling the potential distribution of an invasive species are likely those that estimate the fundamental niche; (2) biotic influences are considered to act on much finer spatial scales than those at which models are commonly applied; (3) biotic predictors are often confounded with climatic model predictors; and (4) biotic factors vary substantially across species and rarely are there relevant standardised global datasets that can be used to approximate such effects (Elith 2017). While data describing global distributions of major biomes and vegetation types are available, their classification systems are typically too broad to meaningfully estimate biotic effects for most species (Brown et al. 2022). Biotic relationships are incredibly complex, and these relationships are mostly unknown in novel environments where new habitats, predators, facilitators, and food sources may play a significant role in species' establishment and long-term persistence.

Despite these challenges in statistically estimating biotic impacts on species' distributions, biosecurity practitioners and governments often use a variety of datasets to inform the availability of suitable habitat or food sources required by potential threats in their regions of interest. These datasets are often derived from direct investment by government departments, and consequently, are country- or region-specific. As these spatial data are not standardised for international use, they are typically developed on finer scales and can include classification systems containing hundreds of classes. Frequently used datasets include highly detailed distributions of land use, vegetation types, known vectors or host material, or estimates of vegetation cover or greenness. For example, the Australian government has invested heavily in the development of a national 50m land use raster containing over 100 land use types that span conservation, agriculture (often identifying the production type), residential/urban, and water land use types (ABARES 2019), with similar datasets available in many other countries.

DEVELOPING A PRAGMATIC ESTABLISHMENT LIKELIHOOD MAP FOR ORIENTAL FRUIT FLY IN AUSTRALIA

Here we provide a pragmatic example of developing an establishment likelihood map for the oriental fruit fly in Australia that explicitly encapsulates the three barriers to establishment (i.e. propagule pressure, abiotic suitability, and biotic suitability).

The oriental fruit fly, *Bactrocera dorsalis*, is a fruit fly native to Asia that has spread and established in over 65 countries, including parts of America, Oceania, and most of sub-Saharan Africa (Figure 14.3). The species poses a significant threat to many horticultural industries worldwide with over 300 different types of fruit and vegetables susceptible (Vargas, Piñero, and Leblanc 2015).

Where oriental fruit fly has yet to establish, countries have implemented networks of pheromone traps with the dual purpose of early detection of incursions and providing evidence to trading partners that the threat is absent (see Chapter 6. Detect). However, the effectiveness of such surveillance networks at detecting new incursions is ultimately dependent on whether traps are positioned in locations where the likelihood of incursion and establishment is greatest, underscoring the relevance and value of maps of establishment likelihood.

ESTIMATING PROPAGULE PRESSURE

The first step in developing an oriental fruit fly establishment likelihood map is to determine the main pathways through which the species may enter the country, and the likelihoods of it (1) evading border screening procedures and (2) remaining viable for an establishment event to occur. Many

FIGURE 14.3 Adult female oriental fruit fly, *Bactrocera dorsalis* (Hendel), laying eggs by inserting her ovipositor in a papaya. (Photograph by Scott Bauer, USDA.)

approaches can be used to estimate these likelihoods, including empirical methods that utilise border interception data, expert elicitation, or a combination of the two.

Here, we use a hypothetical example. Let us assume that oriental fruit fly can enter Australia via four main pathways: international tourists; residents returning from overseas; imported fruit and vegetables; and travellers from the Torres Strait Islands (located between Australia and Papua New Guinea) carrying contaminated fruit or vegetables into Cairns Airport. Let us also assume that estimates exist for annual likelihoods of one or more viable contamination events (i.e. a surviving population size capable of establishing) for each pathway (see Table 14.2).

TABLE 14.2

Hypothetical Likelihoods of Viable Contamination Entering Australia for Four Pathways of Entry of the Oriental Fruit Fly.

Pathway	Probability (Viable contamination | passes border inspection)
International tourists	0.01
Returning residents	0.005
Imported fruit and vegetables	0.001
Passengers from Torres Strait Islands to Cairns	0.05

Note: Here, probability (viable contamination | passes border inspection) refers to the conditional probability that (1) a leakage event occurs and (2) it is viable such that a post-border establishment event could occur.

Map 247

People travelling from the Torres Strait Islands to mainland Australia via Cairns is the pathway with the highest probability (0.05), translating to a 5% chance of a viable contamination event bypassing border screening each year. This pathway has the highest likelihood, due to parts of the Torres Strait regularly experiencing incursions of oriental fruit fly that likely originate from Papua New Guinea, where the species has a persistent established population (Camac, Baumgartner, et al. 2020). The second highest pathway likelihood is associated with international tourists, who may be less aware of Australia's quarantine rules and are therefore more likely to bring in contaminated fruit relative to returning residents (0.01 vs. 0.005, respectively). Imported fruit and vegetables have a low likelihood of a viable contamination (0.001), because they are cold treated prior to export. Assuming pathways are independent, these likelihoods translate to a combined annual probability of approximately 6.5%[3] that one or more viable leakage events occur in a year.

The next step is to determine how pathway contaminations are likely to disperse beyond points of entry into the wider landscape. Ideally, these post-border movements would be informed by empirical data; however, such data are rarely available at the required scale and resolution. It is often necessary to make informed assumptions about how pathway contaminations will disperse through geographic space using pre-existing data and probabilistic functions. Mathematically, this can be depicted as:

$$\Pr\left(Arrival\right)_{i,k} = \Pr\left(Viable\ contamination \mid passes\ border\ inspection\right)_{k} \times f\left(i,k\right) \qquad \text{Eq. (14.1)}$$

Here, $\Pr(Arrival)_{i,k}$ is the expected probability a contaminant from pathway k arrives at location i. It is estimated as a function of the likelihood a contaminant passes border inspection multiplied by $f(i,k)$, a weighting function that describes the proportion of pathway units (e.g. people, containers, etc.) expected to arrive at location i. In the following sections we briefly describe some simple ways for formulating $f(i,k)$, for the four pathways used in this hypothetical example.

Torres Strait Pathway

We assume that most passengers entering Cairns from the Torres Strait Islands are likely to be distributed as a function of population density within the city. We assume that 50% of passengers will remain within 10 km of the airport, which is positioned in close proximity to the city. We account for this by using a negative exponential distance-decay function, where the likelihood of a passenger dispersing to a location decreases with distance from airport, such that 50% of incoming passengers will be distributed within 10 km of the airport. As these individuals are likely to be staying with friends and relatives, we also weight this by the resident human population in each location based on national census data.

The weight (i.e. $f(i,k)$ in Eq. 14.1) for this pathway can be specified as:

$$\Pr\left(Torres\ Strait\ arrival\right)_{i} = e^{\left(-\frac{log(0.5)}{10} \times Distance_{i}\right)} \times Proportion\ of\ population_{i} \qquad \text{Eq. (14.2)}$$

where the 0.5 within the log specifies the proportion of passengers dispersed within a particular distance from the airport, specified as the denominator. When mapped onto a raster grid, this weighted function results in the heat map shown in Figure 14.4.

Returning Residents and Imported Fruit and Vegetables

We assume that Australian citizens returning from overseas and imported fruit and vegetables will both be dispersed as a function of human population density derived from Australian census data. This acknowledges that (1) most returning residents will quickly return to their place of residence

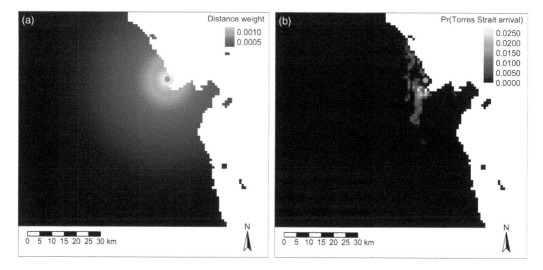

FIGURE 14.4 Weights defining the post-entry distribution of contamination likelihood around Cairns (red point), on the hypothetical Torres Strait Island pathway. Shown are (a) the impact of distance on relative likelihood of oriental fruit fly arrival; and (b) the relative likelihood of arrival after accounting for human population density. Warmer colours signify higher arrival likelihoods/distance weights. Raster data were projected to the Australian Albers (EPSG: 3577) coordinate system and then aggregated from a 1 km to 5 km raster grid to aid visualisation.

irrespective of the distance from airport, and (2) imported goods will go where demand is greatest. As such we specify the weight, $f(i,k)$, for this pathway as:

$$\Pr\left(Returning\ resident\ arrival\right)_i,\ \Pr\left(Imported\ food\ arrival\right)_i = Proportion\ of\ population_i \quad \text{Eq. (14.3)}$$

When mapped onto a raster grid, this weighted function results in the heat map shown in Figure 14.5.

International Tourists

International tourists are not expected to travel significant distances in the days immediately following arrival (i.e. when the risk of a passenger carrying contaminated fruit or other goods is likely greatest). We will assume that 50% of passengers will disperse within 250 km from an international airport, and we assume they will congregate in areas with high densities of tourist accommodation. Here, we will use a similar approach as for the Torres Strait Islands pathway, whereby we assume a negative exponential decline with distance from international airports, weighted by the proportion of tourist rooms in a location (data derived from Camac, Dodd, et al. 2020):

$$\Pr\left(Tourist\ arrival\right)_i = e^{\left(-\frac{log(0.5)}{250} \times Airport\ distance_i\right)} \times Proportion\ of\ tourist\ beds_i \quad \text{Eq. (14.4)}$$

Estimating Likelihoods of Arrival Across Pathways

Once the pathway-specific likelihoods have been distributed across the region of interest, we can then determine the probability of one or more viable contaminations (i.e. propagule pressure) arriving at location i across k pathways of interest:

$$\Pr\left(Arrival\right)_i = 1 - \prod_{k=1}^{n}\left(1 - \Pr\left(Arrival\right)_{i,k}\right) \quad \text{Eq. (14.5)}$$

Where, $\Pr\left(Arrival\right)_{i,k}$ is the likelihood a viable contamination from pathway k arrives at location i (Figure 14.6).

Map   **249**

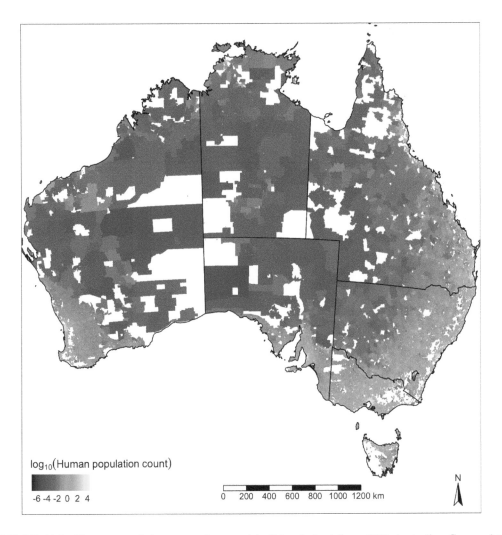

FIGURE 14.5   Human population count ($\log_{10}$ scale). (Map derived from 2016 Australian Census data (Australian Bureau of Statistics 2017)). Warmer colours signify higher human population counts. Raster data were projected to the Australian Albers (EPSG: 3577) coordinate system and aggregated from a 1 km to 5 km raster grid to aid visualisation.

ESTIMATING ABIOTIC SUITABILITY

The potential distribution of the oriental fruit fly has been approximated using a variety of techniques, including CLIMEX (Stephens, Kriticos, and Leriche 2007), Maxent, and GARP (De Meyer et al. 2010). While some broad similarities exist (e.g., Central Africa and northern parts of South America are highly suitable) substantial differences also exist among model predictions. Acknowledging the inherent uncertainty in both model and covariate selection, we opted to approximate the geographic distribution of suitable climate for the oriental fruit fly using a method known as range bagging (Drake 2015). Range bagging estimates the environmental limits of a species' habitat by calculating convex hulls in environmental space, around environmental conditions at known occurrence locations. This process is then repeated using random subsets, of user-defined size, of both occurrence records as well as available environmental covariates (e.g. mean annual rainfall, mean annual temperature, etc.). Suitability is then defined for each raster cell as the proportion of replicates for which

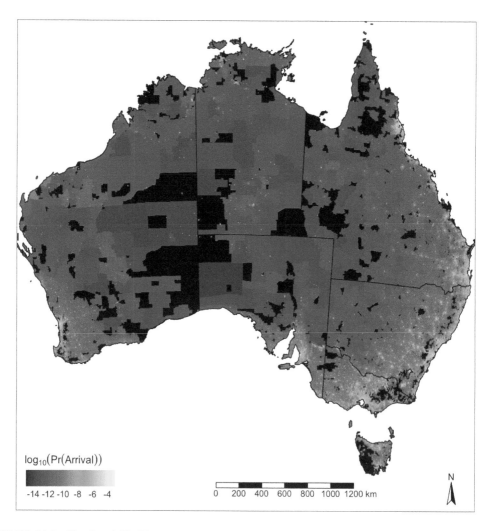

FIGURE 14.6 The ($\log_{10}$) likelihood that one or more viable oriental fruit fly contaminations arrive across the four hypothetical pathways. Warmer colours signify higher arrival likelihoods. Raster data were projected to the Australian Albers (EPSG: 3577) coordinate system and aggregated from a 1 km to a 5 km raster grid to aid visualisation.

the location is labelled as suitable (i.e. falls within the hull). For example, a suitability score of 0.1 would indicate that 10% of the estimated convex hulls deemed that particular location climatically suitable.

The approach has seen recent application to invasion biology and appears promising in the context of biosecurity because no absences or background data are required (Camac, Baumgartner, Garms, et al. 2021, Camac, Baumgartner, Hester, et al. 2021, Camac, Baumgartner, et al. 2020, Camac, Dodd, et al. 2020). This in turn removes several subjective decisions required in the modelling process. The method may also reduce inaccuracies that can arise from projecting to novel environmental conditions (i.e. environmental conditions outside the model-fitting data). This is because, unlike some other methods (e.g. Maxent, GLM, GAM), the method does not attempt to estimate response curves, but rather defines convex hull boundaries in environmental space based on known occurrences. Another major advantage of range bagging is that climatic suitability scores are comparable among species. Finally, range bagging also allows one to explicitly account for uncertainty in covariate selection by specifying low dimensionality (e.g. two dimensions) and allowing the algorithm to randomly select from among a suite of possible covariates—effectively resulting in an ensemble of potentially hundreds or thousands of competing models' parameterisations.

Map   251

Here, we used the range bagging algorithm with dimensionality set to two covariates used in each replicate, the number of bootstrapped models set to 100, and the proportion of occurrence records used per model set at 0.5. We used ensembles of "simple" two-dimensional models to minimise biases associated with model over-fitting and collinearity, and thus, maximise the model's transferability into novel environments (Camac, Baumgartner, Hester, et al. 2021, Camac, Baumgartner, et al. 2020). Ensembles of small models, each with only two variables, have been shown to frequently outperform standard SDM methods (Breiner et al. 2018). We allowed the algorithm to sample from all 19 WorldClim 2 (Fick and Hijmans 2017) bioclimatic parameters (i.e. BIO01 to BIO19) derived from the published 10 arc-minute (approximately 15 km resolution) raster layers.

To parameterise the range bagging models, we obtained global occurrence records for oriental fruit fly from the Global Biodiversity Information Facility (GBIF 2023) and supplemented these data with those collated by Hill, Gallardo, and Terblanche (2017). Acknowledging that such biological databases are susceptible to data quality issues (see Table 14.1), we cleaned these data using routines in the recently published CoordinateCleaner R package (Zizka et al. 2019).

We also removed duplicate records and thinned occurrence records to one point per 15 km (the resolution of the WorldClim version 2 climate data). Following this, we removed all records that occurred in countries lacking known established and persistent populations, based on CABI expert-curated country-level distributional data.

Once all cleaning and cross-referencing was complete, the data were used with the range bagging method to approximate the geographic distribution of suitable average climate (Figures 14.7 and 14.8).

ESTIMATING BIOTIC SUITABILITY

To approximate the geographic distribution of suitable biotic environment within Australia, we used the Australian Land Use and Management Classification raster (ALUM version 8; ABARES 2019) to define the presence and absence of fruit fly host material. Specifically, we created a binary raster whereby any 50 m^2 grid cell containing commodities vulnerable to fruit fly (e.g. tree fruits, vine fruits, olives, citrus, vegetables, herbs, and shrub berries), as well as urban

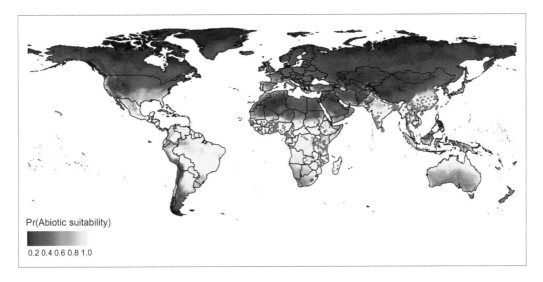

Pr(Abiotic suitability)

0.2 0.4 0.6 0.8 1.0

FIGURE 14.7 Estimated global climate suitability for the oriental fruit fly at the 10 arc-minute (approx. 15 km) scale. Suitability is the proportion of ensembled convex hulls derived from range bagging that identify a location as climatically suitable across bootstrapped combinations of environmental variables. Blue dots signify occurrence records used in range bagging parameterisation. Warmer colours signify higher likelihoods of climatic suitability.

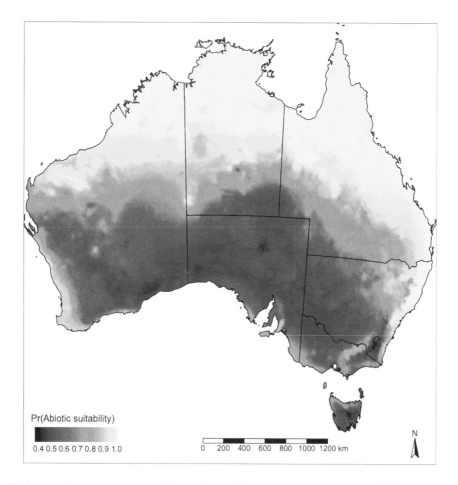

FIGURE 14.8   Estimated climate suitability in Australia for the oriental fruit fly. Suitability is the proportion of ensembled convex hulls derived from range bagging that identify a location as climatically suitable across bootstrapped combinations of environmental variables. Warmer colours signify higher climatic suitability. Raster data were projected to the Australian Albers (EPSG: 3577) coordinate system and resampled to a 5 km raster grid to aid visualisation.

areas, was assumed to contain host material and assigned a value of 1. All grid cells lacking such land uses assumed to contain no fruit fly host material and were classified as 0. This binary raster was converted to a 1 km grid resolution by determining whether any of the aggregated grid cells contained host material.

The Normalised Difference Vegetation Index (NDVI) describes the amount of live green vegetation and therefore can be a good indicator of the amount of suitable habitat or food source. We combined our binary host presence layer with normalised NDVI (scaled to the 0–1 range) by calculating the product of the two layers, thereby scaling host presence according to cell vegetation greenness (Figure 14.9). In effect, this means locations with no host material have a biotic suitability score of zero, and those with host material have a score conditioned on the amount of greenness at that location (our crude measure of host abundance).

DERIVING MAPS OF ESTABLISHMENT POTENTIAL

Now that we have produced three maps that encompass our belief about the probabilities of viable post-border arrivals and the suitability of the biotic and abiotic environments, we can estimate

Map 253

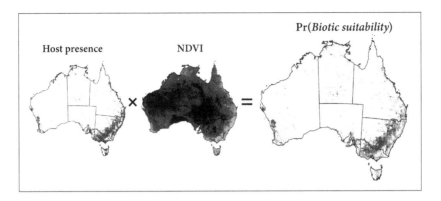

FIGURE 14.9 Biotic suitability was calculated as the product of binary host presence (present vs. absent; blue indicates presence) and NDVI (an indicator of vegetation greenness; warm colours denote higher NDVI).

establishment potential (see Figure 14.1) by taking the product of the three maps, as shown in Figure 14.10.

The result of this product (of the three geographic barriers) is our pragmatic estimate of the likelihood one or more establishment events could occur at a given location in Australia (Figure 14.10). This map can then be used to inform biosecurity decision making.

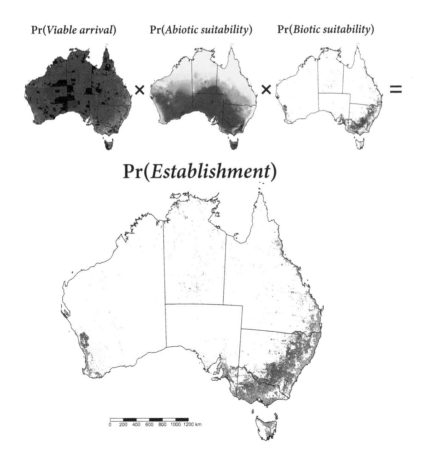

FIGURE 14.10 The product of the three geographic barriers to oriental fruit fly establishment can be used to approximate establishment potential (likelihood of establishment).

IN A NUTSHELL

- Invasive species' establishment potential is a function of three fundamental factors: (1) how likely it is to arrive at a location; (2) whether the climatic environment is suitable; and (3) whether suitable habitat or hosts exist. If there is no chance of arrival at a location, or if condition 2 and/or 3 are unmet, establishment cannot occur.
- Estimating establishment potential is critical for informing post-border surveillance, estimating likelihoods of threat absence, simulating incursion, and spread scenarios, and threat prioritisation.
- Pathway analyses estimate likelihoods of arrival at different points of entry (e.g. ports) and SDMs are often used to predict an invasive species' potential distribution.
- Practical and pragmatic approaches can be used to integrate outputs of SDMs and pathway models by linking them with commonly available spatial datasets and assumptions about post-border movements.
- Despite the limited availability of post-border movement data, opportunities exist to use existing data to approximate the post-border dispersal of threats.

NOTES

1 SDMs are also commonly known as habitat suitability models, bioclimatic envelope models, and ecological niche models, among other names.
2 It is important to note that how well a model predicts the known distribution of a species may not be representative of the model's ability to predict a species' potential (i.e. unobserved) distribution.
3 The conditional probability that one or more viable leakage events occur across pathways is estimated as 1 minus the product of the probability that no viable contaminations occur in each pathway. Mathematically this is written as: $1 - ((1 - 0.01) \times (1 - 0.005) \times (1 - 0.001) \times (1 - 0.05)) = 0.0651$.

REFERENCES

ABARES. 2019. "Catchment scale land use of Australia - Update December 2018." accessed 9 November 2022. https://doi.org/10.25814/5c7728700fd2a.

ABARES. 2020. Climatch v2.0 User Manual, edited by Australian Bureau of Agricultural and Resource Economics and Sciences. Canberra.

Araújo, M. B., and A. Rozenfeld. 2014. "The Geographic Scaling of Biotic Interactions." *Ecography* 37 (5):406–415. https://doi.org/10.1111/j.1600-0587.2013.00643.x.

Australian Bureau of Statistics. 2017. "2074.0 - census of population and housing: Mesh block counts, Australia, 2016." https://www.abs.gov.au/AUSSTATS/abs@.nsf/Lookup/2074.0Main+Features12016.

Barry, S., J. Elith, D. Heersink, P. Caley, M. Kearney, P. Tenant, and A. Arthur. 2015. Final report: CEBRA 1402B Tools and approaches for invasive species distribution modelling for surveillance. Tech. rep., CSIRO: Centre of Excellence for Biosecurity Risk Analysis, The University of Melbourne.

Bennett, J. M., P. Calosi, S. Clusella-Trullas, B. Martínez, J. Sunday, A. C. Algar, M. B. Araújo, B. A. Hawkins, S. Keith, I. Kühn, C. Rahbek, L. Rodríguez, A. Singer, F. Villalobos, M. Ángel Olalla-Tárraga, and I. Morales-Castilla. 2018. "GlobTherm, a Global Database on Thermal Tolerances for Aquatic and Terrestrial Organisms." *Scientific Data* 5 (1):180022. https://doi.org/10.1038/sdata.2018.22.

Biosecurity Commons. 2022. "Biosecurity Commons." https://www.biosecuritycommons.org.au/.

Booth, T. H., H. A. Nix, J. R. Busby, and M. F. Hutchinson. 2014. "Bioclim: The First Species Distribution Modelling Package, Its Early Applications and Relevance to Most Current MaxEnt Studies." *Diversity and Distributions* 20 (1):1–9. https://doi.org/10.1111/ddi.12144.

Boyle, B. L., B. S. Maitner, G. G. C. Barbosa, R. K. Sajja, X. Feng, C. Merow, E. A. Newman, D. S. Park, P. R. Roehrdanz, and B. J. Enquist. 2022. "Geographic Name Resolution Service: A Tool for the Standardization and Indexing of World Political Division Names, With Applications to Species Distribution Modeling." *bioRxiv*:2022.04.25.489424. https://doi.org/10.1101/2022.04.25.489424.

Bradhurst, R., D. Spring, M. Stanaway, J. Milner, and T. Kompas. 2021. "A Generalised and Scalable Framework for Modelling Incursions, Surveillance and Control of Plant and Environmental Pests." *Environmental Modelling & Software* 139:105004. https://doi.org/10.1016/j.envsoft.2021.105004.

Map 255

Breiner, F. T., A. Guisan, A. Bergamini, and M. P. Nobis. 2015. "Overcoming Limitations of Modelling Rare Species by Using Ensembles of Small Models." *Methods in Ecology and Evolution* 6 (10):1210–1218. https://doi.org/10.1111/2041-210X.12403.

Breiner, F. T., M. P. Nobis, A. Bergamini, and A. Guisan. 2018. "Optimizing Ensembles of Small Models for Predicting the Distribution of Species With Few Occurrences." *Methods in Ecology and Evolution* 9 (4):802–808. https://doi.org/10.1111/2041-210X.12957.

Briscoe, N. J., J. Elith, R. Salguero-Gómez, J. J. Lahoz-Monfort, J. S. Camac, K. M. Giljohann, M. H. Holden, B. A. Hradsky, M. R. Kearney, S. M. McMahon, B. L. Phillips, T. J. Regan, J. R. Rhodes, P. A. Vesk, B. A. Wintle, J. D. L. Yen, and G. Guillera-Arroita. 2019. "Forecasting Species Range Dynamics With Process-Explicit Models: Matching Methods to Applications." *Ecology Letters* 22 (11):1940–1956. https://doi.org/10.1111/ele.13348.

Broennimann, O., and A. Guisan. 2008. "Predicting Current and Future Biological Invasions: Both Native and Invaded Ranges Matter." *Biology Letters* 4 (5):585–589.: https://doi.org/10.1098/rsbl.2008.0254.

Brown, C. F., S. P. Brumby, B. Guzder-Williams, T. Birch, S. B. Hyde, J. Mazzariello, W. Czerwinski, V. J. Pasquarella, R. Haertel, S. Ilyushchenko, K. Schwehr, M. Weisse, F. Stolle, C. Hanson, O. Guinan, R. Moore, and A. M. Tait. 2022. "Dynamic World, Near Real-Time Global 10 m Land Use Land Cover Mapping." *Scientific Data* 9 (1):251.https://doi.org/10.1038/s41597-022-01307-4.

CABI. 2023. "CABI compendium invasive species." accessed 2 June 2023. https://www.cabidigitallibrary.org/product/qi

Camac, J. S., J. B. Baumgartner, B. Garms, A. Robinson, and T. Kompas. 2021. Estimating trading partner exposure risk to new pests or diseases. Technical Report for CEBRA project 190606. Centre of Excellence for Biosecurity Risk Analysis, The University of Melbourne.

Camac, J. S., J. B. Baumgartner, S. Hester, R. Subasinghe, and S. Collins. 2021. Using edmaps & Zonation to inform multi-pest early-detection surveillance designs. Technical Report for CEBRA project 20121001: Centre of Excellence for Biosecurity Risk Analysis, The University of Melbourne.

Camac, J. S., J. B. Baumgartner, A. Robinson, and J. Elith. 2020. *Developing Pragmatic Maps of Establishment Likelihood for Plant Pests*. Centre of Excellence for Biosecurity Risk Analysis, The University of Melbourne.

Camac, J. S., A. Dodd, N. Bloomfield, and A. Robinson. 2020. *Sampling to Support Claims of Area Freedom: Technical Report for the Department of Agriculture*. Centre of Excellence for Biosecurity Risk Analysis, The University of Melbourne.

Catford, J. A., R. Jansson, and C. Nilsson. 2009. "Reducing Redundancy in Invasion Ecology by Integrating Hypotheses into a Single Theoretical Framework." *Diversity and Distributions* 15 (1):22–40.https://doi.org/10.1111/j.1472-4642.2008.00521.x.

Chamberlain, S., and E. Szöcs. 2013. "Taxize: Taxonomic Search and Retrieval in R [version 2; Peer Review: 3 Approved]." *F1000Research* 2 (191). https://doi.org/10.12688/f1000research.2-191.v2.

CHELSA. 2023. "Climatologies at high resolution for the earth's land surface areas." accessed 2 June 2023. http://chelsa-climate.org.

Climatch. 2019. "Climatch." accessed 2 June 2023. https://climatch.cp1.agriculture.gov.au/.

CliMond. 2014. "CliMond: global climatologies for bioclimatic modelling." accessed 2 June 2023. https://www.climond.org.

CRU. 2023. "Climatic Research Unit." accessed 2 June 2023. https://www.uea.ac.uk/groups-and-centres/climatic-research-unit.

De Meyer, M., M. P. Robertson, M. W. Mansell, S. Ekesi, K. Tsuruta, W. Mwaiko, J. F. Vayssières, and A. T. Peterson. 2010. "Ecological Niche and Potential Geographic Distribution of the Invasive Fruit Fly *Bactrocera Invadens* (Diptera, Tephritidae)." *Bulletin of Entomological Research* 100 (1):35–48. https://doi.org/10.1017/S0007485309006713.

Dodd, A., J. B. Baumgartner, and T. Kompas. 2021. *Final Report: Stakeholder Perceptions of and Options for Refining the CEBRA 'Value' Model for Use Within DAWE, CEBRA Project 20100401*. Centre of Excellence for Biosecurity Risk Analysis, The University of Melbourne.

Dodd, A. J., M. A. McCarthy, N. Ainsworth, and M. A. Burgman. 2016. "Identifying Hotspots of Alien Plant Naturalisation in Australia: Approaches and Predictions." *Biological Invasions* 18 (3):631–645. https://doi.org/10.1007/s10530-015-1035-8.

Dodd, A., N. Stoeckl, J. B. Baumgartner, and T. Kompas. 2020. *Key Result Summary: Valuing Australia's Biosecurity System, CEBRA Project 170713*. Centre of Excellence for Biosecurity Risk Analysis, The University of Melbourne.

Drake, J. M. 2015. "Range Bagging: A New Method for Ecological Niche Modelling from Presence-Only Data." *Journal of the Royal Society Interface* 12 (107):20150086. https://doi.org/10.1098/rsif.2015.0086.

Elith, J. 2017. "Predicting Distributions of Invasive Species." In *Invasive Species: Risk Assessment and Management*, edited by Andrew P. Robinson, Mark A. Burgman, Mike Nunn and Terry Walshe, 93–129. Cambridge: Cambridge University Press.

Elith, J., M. Kearney, and S. Phillips. 2010. "The Art of Modelling Range-Shifting Species." *Methods in Ecology and Evolution* 1 (4):330–342. https://doi.org/10.1111/j.2041-210X.2010.00036.x.

Elith, J., and J. R. Leathwick. 2009. "Species Distribution Models: Ecological Explanation and Prediction Across Space and Time." *Annual Review of Ecology, Evolution, and Systematics* 40 (1):677–697. https://doi.org/10.1146/annurev.ecolsys.110308.120159.

Fick, S. E., and R. J. Hijmans. 2017. "WorldClim 2: New 1-Km Spatial Resolution Climate Surfaces for Global Land Areas." *International Journal of Climatology* 37 (12):4302–4315. https://doi.org/10.1002/joc.5086.

Fourcade, Y., A. G. Besnard, and J. Secondi. 2018. "Paintings Predict the Distribution of Species, or the Challenge of Selecting Environmental Predictors and Evaluation Statistics." *Global Ecology and Biogeography* 27 (2):245–256. https://doi.org/10.1111/geb.12684.

GBIF. 2023. "Global Biodiversity Information Facility." accessed 2 June 2023. https://www.gbif.org.

Guillera-Arroita, G., J. J. Lahoz-Monfort, J. Elith, A. Gordon, H. Kujala, P. E. Lentini, M. A. McCarthy, R. Tingley, and B. A. Wintle. 2015. "Is My Species Distribution Model Fit for Purpose? Matching Data and Models to Applications." *Global Ecology and Biogeography* 24 (3):276–292. https://doi.org/10.1111/geb.12268.

Harris, I., P. D. Jones, T. J. Osborn, and D. H. Lister. 2014. "Updated High-Resolution Grids of Monthly Climatic Observations – the CRU TS3.10 Dataset." *International Journal of Climatology* 34 (3):623–642. https://doi.org/10.1002/joc.3711.

Hemming, V., M. A. Burgman, A. M. Hanea, M. F. McBride, and B. C. Wintle. 2018. "A Practical Guide to Structured Expert Elicitation Using the IDEA Protocol." *Methods in Ecology and Evolution* 9 (1):169–180. https://doi.org/10.1111/2041-210X.12857.

Hill, M., P. Caley, J. Camac, J. Elith, and S. Barry. 2022. "A novel method accounting for predictor uncertainty and model transferability of invasive species distribution models." *bioRxiv*:2022.03.14.483865. https://doi.org/10.1101/2022.03.14.483865.

Hill, M. P., B. Gallardo, and J. S. Terblanche. 2017. "A Global Assessment of Climatic Niche Shifts and Human Influence in Insect Invasions." *Global Ecology and Biogeography* 26 (6):679–689. https://doi.org/10.1111/geb.12578.

Hulme, P. E. 2021. "Unwelcome Exchange: International Trade as a Direct and Indirect Driver of Biological Invasions Worldwide." *One Earth* 4 (5):666–679. https://doi.org/10.1016/j.oneear.2021.04.015.

Karger, D. N., O. Conrad, J. Böhner, T. Kawohl, H. Kreft, R. W. Soria-Auza, N. E. Zimmermann, H. P. Linder, and M. Kessler. 2017. "Climatologies at High Resolution for the earth's Land Surface Areas." *Scientific Data* 4 (1):170122. https://doi.org/10.1038/sdata.2017.122.

Kearney, M., B. L. Phillips, C. R. Tracy, K. A. Christian, G. Betts, and W. P. Porter. 2008. "Modelling Species Distributions Without Using Species Distributions: the Cane Toad in Australia Under Current and Future Climates." *Ecography* 31 (4):423–434. https://doi.org/10.1111/j.0906-7590.2008.05457.x.

Kearney, M., and W. Porter. 2009. "Mechanistic Niche Modelling: Combining Physiological and Spatial Data to Predict species' Ranges." *Ecology Letters* 12 (4):334–350. https://doi.org/10.1111/j.1461-0248.2008.01277.x.

Kearney, M. R., and W. P. Porter. 2017. "NicheMapR – an R Package for Biophysical Modelling: The Microclimate Model." *Ecography* 40 (5):664–674. https://doi.org/10.1111/ecog.02360.

Kriticos, D. J., J. M. Kean, C. B. Phillips, S. D. Senay, H. Acosta, and T. Haye. 2017. "The Potential Global Distribution of the Brown Marmorated Stink Bug, *Halyomorpha Halys*, a Critical Threat to Plant Biosecurity." *Journal of Pest Science* 90 (4):1033–1043. https://doi.org/10.1007/s10340-017-0869-5.

Kriticos, D. J., G. Maywald, T. Yonow, E. Zurcher, N. Herrmann, and B. Sutherst. 2016. *CLIMEX Version 4: Exploring the Effects of Climate on Plants, Animals and Diseases.* Canberra: CSIRO.

Kriticos, D. J., B. L. Webber, A. Leriche, N. Ota, I. Macadam, J. Bathols, and J. K. Scott. 2012. "CliMond: Global High-Resolution Historical and Future Scenario Climate Surfaces for Bioclimatic Modelling." *Methods in Ecology and Evolution* 3 (1):53–64. https://doi.org/10.1111/j.2041-210X.2011.00134.x.

Lahoz-Monfort, J. J., G. Guillera-Arroita, and B. A. Wintle. 2014. "Imperfect Detection Impacts the Performance of Species Distribution Models." *Global Ecology and Biogeography* 23 (4):504–515. https://doi.org/10.1111/geb.12138.

Le Maitre, D. C., W. Thuiller, and L. Schonegevel. 2008. "Developing an Approach to Defining the Potential Distributions of Invasive Plant Species: A Case Study of Hakea Species in South Africa." *Global Ecology and Biogeography* 17 (5):569–584. https://doi.org/10.1111/j.1466-8238.2008.00407.x.

Map   257

Lobo, J. M., A. Jiménez-Valverde, and J. Hortal. 2010. "The Uncertain Nature of Absences and Their Importance in Species Distribution Modelling." *Ecography* 33 (1):103–114. https://doi.org/10.1111/j.1600-0587.2009.06039.x.

Magarey, R. D., G. A. Fowler, D. M. Borchert, T. B. Sutton, M. Colunga-Garcia, and J. A. Simpson. 2007. "NAPPFAST: An Internet System for the Weather-Based Mapping of Plant Pathogens." *Plant Disease* 91 (4):336–345. https://doi.org/10.1094/PDIS-91-4-0336.

Maldonado, C., C. I. Molina, A. Zizka, C. Persson, C. M. Taylor, J. Albán, E. Chilquillo, N. Rønsted, and A. Antonelli. 2015. "Estimating Species Diversity and Distribution in the Era of Big Data: To What Extent Can We Trust Public Databases?" *Global Ecology and Biogeography* 24 (8):973–984. https://doi.org/10.1111/geb.12326.

Marques, G. M., S. Augustine, K. Lika, L. Pecquerie, T. Domingos, and A. L. M. Kooijman. 2018. "The AmP Project: Comparing Species on the Basis of Dynamic Energy Budget Parameters." *PLOS Computational Biology* 14 (5):e1006100. https://doi.org/10.1371/journal.pcbi.1006100.

McGeoch, M. A., P. Genovesi, P. J. Bellingham, M. J. Costello, C. McGrannachan, and A. Sheppard. 2016. "Prioritizing Species, Pathways, and Sites to Achieve Conservation Targets for Biological Invasion." *Biological Invasions* 18 (2):299–314. https://doi.org/10.1007/s10530-015-1013-1.

Merow, C., M. J. Smith, and J. A. Silander Jr. 2013. "A Practical Guide to MaxEnt for Modeling species' Distributions: What It Does, and Why Inputs and Settings Matter." *Ecography* 36 (10):1058–1069. https://doi.org/10.1111/j.1600-0587.2013.07872.x.

Morales-Castilla, I., T. J. Davies, W. D. Pearse, and P. Peres-Neto. 2017. "Combining Phylogeny and Co-Occurrence to Improve Single Species Distribution Models." *Global Ecology and Biogeography* 26 (6):740–752. https://doi.org/10.1111/geb.12580.

Morin, X., and M. J. Lechowicz. 2008. "Contemporary Perspectives on the Niche That can Improve Models of Species Range Shifts Under Climate Change." *Biology Letters* 4 (5):573–576. https://doi.org/10.1098/rsbl.2008.0181.

Phillips, S. J., R. P. Anderson, and R. E. Schapire. 2006. "Maximum Entropy Modeling of Species Geographic Distributions." *Ecological Modelling* 190 (3):231–259. https://doi.org/10.1016/j.ecolmodel.2005.03.026.

Phillips, S. J., M. Dudík, J. Elith, C. H. Graham, A. Lehmann, J. Leathwick, and S. Ferrier. 2009. "Sample Selection Bias and Presence-Only Distribution Models: Implications for Background and pseudo-Absence Data." *Ecological Applications* 19 (1):181–197. https://doi.org/10.1890/07-2153.1.

Pollock, L. J., W. K. Morris, and P. A. Vesk. 2012. "The Role of Functional Traits in Species Distributions Revealed Through a Hierarchical Model." *Ecography* 35 (8):716–725. https://doi.org/10.1111/j.1600-0587.2011.07085.x.

Renner, I. W., J. Elith, A. Baddeley, W. Fithian, T. Hastie, S. J. Phillips, G. Popovic, and D. I. Warton. 2015. "Point Process Models for Presence-Only Analysis." *Methods in Ecology and Evolution* 6 (4):366–379. https://doi.org/10.1111/2041-210X.12352.

Salguero-Gómez, R., O. R. Jones, C. R. Archer, Y. M. Buckley, J. Che-Castaldo, H. Caswell, D. Hodgson, A. Scheuerlein, D. A. Conde, E. Brinks, H. de Buhr, C. Farack, F. Gottschalk, A. Hartmann, A. Henning, G. Hoppe, G. Römer, J. Runge, T. Ruoff, J. Wille, S. Zeh, R. Davison, D. Vieregg, A. Baudisch, R. Altwegg, F. Colchero, M. Dong, H. de Kroon, J.-D. Lebreton, C. J. E. Metcalf, M. M. Neel, I. M. Parker, T. Takada, T. Valverde, L. A. Vélez-Espino, G. M. Wardle, M. Franco, and J. W. Vaupel. 2015. "The COMPADRE Plant Matrix Database: An Open Online Repository for Plant Demography." *Journal of Ecology* 103 (1):202–218. https://doi.org/10.1111/1365-2745.12334.

Salguero-Gómez, R., O. R. Jones, C. R. Archer, C. Bein, H. de Buhr, C. Farack, F. Gottschalk, A. Hartmann, A. Henning, G. Hoppe, G. Römer, T. Ruoff, V. Sommer, J. Wille, J. Voigt, S. Zeh, D. Vieregg, Y. M. Buckley, J. Che-Castaldo, D. Hodgson, A. Scheuerlein, H. Caswell, and J. W. Vaupel. 2016. "COMADRE: A Global Data Base of Animal Demography." *Journal of Animal Ecology* 85 (2):371–384. https://doi.org/10.1111/1365-2656.12482.

Sequeira, A. M. M., P. J. Bouchet, K. L. Yates, K. Mengersen, and M. J. Caley. 2018. "Transferring Biodiversity Models for Conservation: Opportunities and Challenges." *Methods in Ecology and Evolution* 9 (5):1250–1264. https://doi.org/10.1111/2041-210X.12998.

Stephens, A. E. A., D. J. Kriticos, and A. Leriche. 2007. "The Current and Future Potential Geographical Distribution of the Oriental Fruit Fly, *Bactrocera dorsalis* (Diptera: Tephritidae)." *Bulletin of Entomological Research* 97 (4):369–378. https://doi.org/10.1017/S0007485307005044.

Syfert, M. M., M. J. Smith, and D. A. Coomes. 2013. "The Effects of Sampling Bias and Model Complexity on the Predictive Performance of MaxEnt Species Distribution Models." *PLOS ONE* 8 (2):e55158. https://doi.org/10.1371/journal.pone.0055158.

Tingley, R., P. García-Díaz, C. R. R. Arantes, and P. Cassey. 2018. "Integrating Transport Pressure Data and Species Distribution Models to Estimate Invasion Risk for Alien Stowaways." *Ecography* 41 (4):635–646. https://doi.org/10.1111/ecog.02841.

Uribe-Rivera, D. E., G. Guillera-Arroita, S. M. Windecker, P. Pliscoff, and B. A. Wintle. 2022. "The Predictive Performance of Process-Explicit Range Change Models Remains Largely Untested." *Ecography*:e06048. https://doi.org/10.1111/ecog.06048.

VanDerWal, J., L. P. Shoo, C. Graham, and S. E. Williams. 2009. "Selecting pseudo-Absence Data for Presence-Only Distribution Modeling: How Far Should You Stray from What You Know?" *Ecological Modelling* 220 (4):589–594. https://doi.org/10.1016/j.ecolmodel.2008.11.010.

Vargas, R. I., J. C. Piñero, and L. Leblanc. 2015. "An Overview of Pest Species of *Bactrocera* Fruit Flies (*Diptera: Tephritidae*) and the Integration of Biopesticides With Other Biological Approaches for Their Management With a Focus on the Pacific Region." *Insects* 6 (2):297–318. https://doi.org/10.3390/insects6020297.

Venette, R. C., D. J. Kriticos, R. D. Magarey, F. H. Koch, R. H. A. Baker, S. P. Worner, N. N. Gómez Raboteaux, D. W. McKenney, E. J. Dobesberger, D. Yemshanov, P. J. De Barro, W. D. Hutchison, G. Fowler, T. M. Kalaris, and J. Pedlar. 2010. "Pest Risk Maps for Invasive Alien Species: A Roadmap for Improvement." *BioScience* 60 (5):349–362. https://doi.org/10.1525/bio.2010.60.5.5.

Warton, D. I., I. W. Renner, and D. Ramp. 2013. "Model-Based Control of Observer Bias for the Analysis of Presence-Only Data in Ecology." *PLOS ONE* 8 (11):e79168. https://doi.org/10.1371/journal.pone.0079168.

Warton, D. I., and L. C. Shepherd. 2010. "Poisson Point Process Models Solve the "pseudo-Absence Problem" for Presence-Only Data in Ecology." *The Annals of Applied Statistics* 4 (3):1383–1402. https://doi.org/10.1214/10-AOAS331.

WorldClim. 2023. "WorldClim." accessed 2 June 2023. http://www.worldclim.org.

Yates, K. L., P. J. Bouchet, M. J. Caley, K. Mengersen, C. F. Randin, S. Parnell, A. H. Fielding, A. J. Bamford, S. Ban, A. M. Barbosa, C. F. Dormann, J. Elith, C. B. Embling, G. N. Ervin, R. Fisher, S. Gould, R. F. Graf, E. J. Gregr, P. N. Halpin, R. K. Heikkinen, S. Heinänen, A. R. Jones, P. K. Krishnakumar, V. Lauria, H. Lozano-Montes, L. Mannocci, C. Mellin, M. B. Mesgaran, E. Moreno-Amat, S. Mormede, E. Novaczek, S. Oppel, G. Ortuño Crespo, A. T. Peterson, G. Rapacciuolo, J. J. Roberts, R. E. Ross, K. L. Scales, D. Schoeman, P. Snelgrove, G. Sundblad, W. Thuiller, L. G. Torres, H. Verbruggen, L. Wang, S. Wenger, M. J. Whittingham, Y. Zharikov, D. Zurell, and A. M. M. Sequeira. 2018. "Outstanding Challenges in the Transferability of Ecological Models." *Trends in Ecology & Evolution* 33 (10):790–802. https://doi.org/10.1016/j.tree.2018.08.001.

Zizka, A., D. Silvestro, T. Andermann, J. Azevedo, C. Duarte Ritter, D. Edler, H. Farooq, A. Herdean, M. Ariza, R. Scharn, S. Svantesson, N. Wengström, V. Zizka, and A. Antonelli. 2019. "CoordinateCleaner: Standardized Cleaning of Occurrence Records from Biological Collection Databases." *Methods in Ecology and Evolution* 10 (5):744–751. https://doi.org/10.1111/2041-210X.13152.

Section VI

Conclusion

15 Conclusion

Susan M. Hester, James S. Camac, Edith Arndt,
Sana Bau, Les Kneebone, Evelyn Mannix,
Andrew P. Robinson, and Lucie M. Bland

In an environment of considerable uncertainty, managing a multitude of biosecurity risks across the pre-border, border, and post-border continuum is an increasingly complex challenge. This complexity, in tandem with uncertainty around the number, arrival, and consequences of threats means that biosecurity will continue to be a wicked problem for biosecurity system managers. While the context in which biosecurity systems operate—increasing urbanisation, varied impacts of climate change, increasing international trade and human movement—is largely out of the control of biosecurity regulators, the tools, methodologies, and critical thinking necessary for the system to adapt, certainly are not! In this book we outline a suite of tools and methodologies, available to biosecurity practitioners, as well as critical decision factors and expertise that must be considered when developing a robust, efficient, and resilient biosecurity system. Many tools and methodologies have been available for some time, but often in disciplines not traditionally consulted by biosecurity agencies or well-represented within agencies (e.g. social science, economics, decision science, evaluation science), and as a result their application to biosecurity has yet to become mainstream or routine.

Most biosecurity systems and agencies responsible for mitigating threats are founded on a deep understanding of pest and disease biology and spread. Consequently, biosecurity regulatory staff are often drawn from science disciplines such as veterinary epidemiology, plant pathology, entomology, and ecology. This is particularly the case for agencies that operate in the pre-border and border parts of the biosecurity continuum, and for good reason—information about the underlying biology of the threat is a key part of informing estimates of the likelihood and consequences of entry, establishment, and spread of pests and diseases (namely, import risk analysis). Since the 1995 Agreement on the Application of Sanitary and Phytosanitary Measures, national governments have been required to impose science-based principles on biosecurity risk management—a fundamental change in recent biosecurity practice.

However, a biosecurity system that designs risk assessments based mainly on threat biology, will not guarantee its resilience to tackle future challenges. Signs of biosecurity systems unable to cope with increasing challenges often involve commentary around funding, including about the merit of past and current spending practices, budget deficits, internal competition for funds between agency divisions, and a growing amount of unexpected or high-profile incursions that are costly to manage. While a techno-scientific approach may have been developed with the best intentions and may have served societies well in the past, biosecurity problems and their solutions transcend the scientific subdisciplines traditionally associated with biosecurity. A truly multi-disciplinary, collaborative, and cooperative approach is now required—expertise in statistics, economics, actuarial science, decision science, political science, engineering, molecular biology, organisational theory, artificial intelligence, ecological modelling, performance evaluation, along with the traditional science disciplines all play a role in developing future-ready, efficient, and adaptive biosecurity systems. Collaboration and cooperation between research, industry, and government (and within levels of government) is also required to operate activities across the Biosecurity Continuum.

Economics is key amongst the disciplines that are under-represented in biosecurity agencies, but which could add significant value to existing scientific knowledge and contemporary risk analysis methodologies, and thus the overall operation of biosecurity systems. Perhaps the most obvious use of economics in biosecurity agencies is to guide efficient prioritisation and allocation of funds aimed at reducing biosecurity threats—ensuring investment decisions that provide value for money, rather than investments based mainly on technical feasibility. Biosecurity budgets are often allocated based on subjective criteria, such as the visibility of a threat, historical legacies, or industry pressure. Such decisions rarely account for the value for money of proposed biosecurity activities or policies. When allocation decisions aren't based on risk mitigation activities that offer the greatest return on investment from limited budgets, it is likely that unnecessary or ineffective measures will be implemented. Inefficient budget allocations lead to otherwise avoidable damages to assets, missed opportunities for eradication, and foregone opportunities to invest in other, better value-for-money projects. As such, budget misallocations may contribute to increasing risk and poorer outcomes throughout the biosecurity system as a whole. The basic premise of value for money holds regardless of whether allocation decisions are made to respond to a single threat, across a portfolio of threats, or at different stages of the biosecurity system (namely, pre-border, border, or post-border).

Another less obvious use of the economics discipline is in designing efficient rules and regulations to mitigate risks. Traditionally, risk management in biosecurity agencies largely ignores the human dimension—the way individuals and organisations involved in the supply chain will respond to the incentives that are inherent in *all* rules and regulations to which they are subject. As a result, well-intended interventions—most of which impose a cost burden on supply chain participants— are often counterproductive. Rules may be ignored or circumvented, leading to potentially costly incursions at worst, or at best an ineffective policy that wastes resources. Designing cost-effective and efficient regulations that are robust to "gaming", is the realm of the economics discipline (specifically, market design economics), but the required skillsets are little known and therefore not called upon to design risk management measures. Ideally, rules and regulations would be designed using market design economics, in conjunction with scientific expertise related to the pathway being regulated, and taking advantage of relevant technological advances, including in computing, data analytics, artificial intelligence, genomics, biological engineering, diagnostic capabilities, and epidemiology.

Improving the design and implementation of rules and regulations should introduce welcome efficiencies into the use of biosecurity agency budgets. For example, well-calibrated rules should reduce the burden of expensive post-border responses, freeing up funding for other risk-reduction activities. To understand whether efficiencies are being realised and whether policies are functioning as intended, it is crucially important to systematically and regularly evaluate the outcomes of biosecurity spending. Monitoring, evaluation, and reporting are activities that allow managers to know whether program objectives are being met. These activities also help fulfil public accountability requirements and inform future decision making and resource allocation. Ideally, the approaches to monitoring, evaluation, and reporting should be determined early, during the inception stage of any program. The skillsets best placed to assist with monitoring, evaluation, and reporting activities are found in statistics, economics, evaluation science, and a range of social science disciplines including sociology, political science, and public administration.

Achieving widespread improvements in efficiency requires data that would allow robust statistical analysis of inspection regimes, calculation of the costs of prevention and management activities and their effects on reducing damages for all pests and diseases. This includes data on the likelihood of particular incursions leaking through the border, information on how pests and diseases might spread over time and space in novel environments, feasibility of detection and control, and price and cost data associated with management activities. Many biosecurity agencies around the globe have invested heavily in research associated with these perceived data and knowledge gaps. As a result, capability now exists for maps of establishment potential to be developed to inform where surveillance resources should be focused, to assess likely spread scenarios, estimate likelihoods of threat absence, and to create risk maps for prioritising threats in order to optimally minimise risk.

Capability also exists for agency staff to take advantage of advances in automation using statistical and machine learning methods to improve the efficiency and scale at which biosecurity activities can be conducted. Biosecurity agencies should ensure the existence of an organisational culture that allows findings from research into new methods, often publicly funded, to be incorporated into business as usual practices. Organisational silos can often inhibit the free flow of information from research and knowledge exchange, and therefore reduce the associated benefits from implementing research findings.

Despite advances in spread modelling, statistical methods, and machine learning, that enable more reliable estimations of benefits and costs of risk management scenarios, obtaining the necessary data to compute the most economically efficient set of activities to reduce biosecurity risks posed by a large number of threats across the biosecurity continuum, can be very challenging in practice. For example, in the post-border space, risk management activities must often be designed at early stages, where information about a threat—its likely spread, and impact on environments—is limited or uncertain. The rapidly evolving nature of wicked problems that biosecurity risk mitigation deals with means that data deficiencies are inevitable. Decisions must often be made with incomplete or imperfect information and under time pressure. Invoking expert contributions to the knowledge base in a robust way is gaining momentum. Using repeatable, systematic, and transparent methods for elicitation helps to ensure that data derived from experts are subject to the same standards of rigour as empirical or modelled data. Indeed, recent advances in structured expert elicitation means information gathered in this way can be subjected to the same standards of methodological rigour as field or modelled data.

Despite difficulties posed by data and uncertainty, economic principles can guide even the most well-developed of biosecurity agencies to undertake risk-reducing activities that provide the best value for money. Economic principles can also be applied to calculate the total budget that should be allocated to reducing biosecurity risks, based on what a community deems appropriate in relation to public expenditure on other key areas such as health, education, and defence. As discussed, a range of skills are required in this endeavour. Where biosecurity systems do not appropriately harness the skills, methodology and thinking from a range of disciplines, system resilience will be elusive—a whole of system approach to biosecurity management is required. While this type of approach is likely to emerge organically as biosecurity systems mature, ideas in this book also allow biosecurity managers to take a short cut to future-proofing their systems.

Chapters in this book detail the tools, methodologies, and frameworks for addressing key biosecurity problems that are available for implementation by biosecurity managers: principles from economics to guide efficient allocation of budgets and for designing policies that lead to desired stakeholder behaviour; tools for understanding spatial spread of pests; and to monitor and evaluate the effectiveness of biosecurity actions. The key to building efficient and resilient biosecurity systems is take a whole of system perspective, incorporating the discipline expertise required to solve the multi-dimensional problems posed by the need to reduce global biosecurity risks. If such an approach is not taken, the challenges facing biosecurity agencies will continue to overwhelm available resources.

Index

Note: Page numbers in **bold** and *italics* refer to tables and figures, respectively.

Milton Keynes UK
Ingram Content Group UK Ltd.
UKHW052027141024
449569UK00016B/727

9 781032 181691